Gulliver in the Country of Lilliput: An Interplay of Noncovalent Interactions (Volume II)

Gulliver in the Country of Lilliput: An Interplay of Noncovalent Interactions (Volume II)

Guest Editor

Ilya G. Shenderovich

MDPI

Basel • Beijing • Wuhan • Barcelona • Belgrade • Novi Sad • Cluj • Manchester

Guest Editor
Ilya G. Shenderovich
Faculty of Chemistry and
Pharmacy
University of Regensburg
Regensburg
Germany

Editorial Office
MDPI AG
Grosspeteranlage 5
4052 Basel, Switzerland

This is a reprint of the Special Issue, published open access by the journal *Molecules* (ISSN 1420-3049), freely accessible at: https://www.mdpi.com/journal/molecules/special_issues/4U03D6OMP4.

For citation purposes, cite each article independently as indicated on the article page online and as indicated below:

Lastname, A.A.; Lastname, B.B. Article Title. *Journal Name* **Year**, *Volume Number*, Page Range.

ISBN 978-3-7258-5341-0 (Hbk)
ISBN 978-3-7258-5342-7 (PDF)
https://doi.org/10.3390/books978-3-7258-5342-7

Cover image courtesy of Ilya G. Shenderovich

Contents

About the Editor

Ilya G. Shenderovich

Ilya G. Shenderovich (Dr.Nat.Sci.) received a BSc degree in Physics (1993, Chemical Physics) and an MSc degree in Physics (1995, Physics of Condensed Matter) from the Sankt-Petersburg State University, Russia (Mentor: Dr. G. N. Kuz'min). He became a Candidate of Science in Physics and Mathematics (PhD) in 1999 (Topic: Manifestation of Covalency, Cooperativity, and Symmetry of Strong Hydrogen Bonds in NMR Spectra, Mentor: Prof. Dr. G.S. Denisov) and a Doctor of Science in Physics and Mathematics in 2011 (Topic: Study of Hydrogen Bonds in Amorphous Materials and at Interfaces by NMR). He progressed in his research career under the guidance of Prof. Dr. H.-H. Limbach at the Freie Universität Berlin. He runs the NMR department of the Faculty of Chemistry and Pharmacy at the Universität Regensburg. His main research interests focus on noncovalent interactions in condensed matter. His main research methods are NMR spectroscopy and model DFT calculations.

Preface

Noncovalent interactions form the bridge between the ideal gas abstraction and the real world. In the past, most studies were limited to the analysis of the single strongest interaction in a molecular system under consideration, which was thought responsible for the most important structural properties of the system. The current challenge therefore is to go beyond this limitation. The publications collected in this Reprint highlight the interplay of noncovalent interactions in complex molecular systems, emphasizing the roles of cooperativity and anti-cooperativity, solvation, reaction fields, steric hindrance, intermolecular dynamics, and other subtle yet numerous contributions that collectively define the realized molecular conformation, chemical reactivity, and condensed matter structure.

Ilya G. Shenderovich
Guest Editor

molecules

MDPI

Editorial

Editorial to the Special Issue "Gulliver in the Country of Lilliput: An Interplay of Noncovalent Interactions (Volume II)"

Ilya G. Shenderovich

Institute of Organic Chemistry, University of Regensburg, Universitaetstrasse 31, 93053 Regensburg, Germany; ilya.shenderovich@ur.de

Noncovalent interactions are the bridge between the ideal gas abstraction and the real world. In the past, most studies were limited to the analysis of the single strongest interaction in a molecular system under consideration, which was held responsible for the most important structural properties of the system. The current challenge is to surpass this limitation.

The first edition of the Special Issue, "Gulliver in the Country of Lilliput: An Interplay of Noncovalent Interactions"—which has been published as a standalone book—was successful [1]. As such, the overarching goal of this accompanying Special Issue was to gather insights into the interplay of noncovalent interactions within complex molecular systems, including the roles of cooperativity and anti-cooperativity, solvation, reaction fields, steric hindrance, intermolecular dynamics, and numerous other factors affecting molecular conformation, chemical reactivity, and the structure of condensed matter.

The contributions that comprise this Special Issue can be broadly categorized into three main themes: (i) the identification of specific interactions responsible for the observed or predicted structures of heteromolecular systems [2–4]; (ii) the role of noncovalent interactions in the behavior of molecules under confinement [5–7]; and (iii) the spectral manifestations of such interactions in complex molecular systems [8–11].

The experimental study of Mihrin et al. [2] reports the structures of the phenol dimer, trimer, and monohydrate embedded in inert neon matrices at 4 K. Specifically, large-amplitude hydrogen-bond librational motions were measured and interpreted in terms of the molecular structures of the corresponding complexes using extensive conformational searches and computational optimizations employing both dispersion-corrected DFT and wavefunction ab initio methodologies. Phenol clusters were found to exhibit, in addition to the energy-dominating O–H$\cdots$O cooperative hydrogen bonds, C–H$\cdots\pi$ contacts and competing London dispersion forces between the aromatic rings. In the phenol monohydrate, phenol behaves as the hydrogen-bond donor to water, in contrast to aliphatic alcohols, which act as acceptors [12].

Rosokha et al. [3] investigated the role of halogen bonding in N-chlorosuccinimide (SimCl) reactions with bromide and iodide. X-ray crystallography revealed halogen-exchange products in both cases, exhibiting exceptionally short Br$\cdots$Cl and I$\cdots$Cl contacts. DFT computations showed that SimCl$\cdot$Br$^-$ and SimCl$\cdot$I$^-$ complexes dissociate into succinimide anions and interhalogen molecules (ClBr or ClI) without high-energy Cl$^+$ intermediates, significantly lowering the N–Cl bond dissociation barrier. Consequently, these rearrangements proceed efficiently even at −73 °C. This work highlights that halogen bonds can be strong enough to facilitate chemical reactions [13].

Zhang and Wang [4] used DFT to explore whether 1,2,5-oxadiazole, 2,1,3-benzoxadiazole, and their S-, Se-, and Te-analogues form complexes with fullerene C_{60} via an N→C dative bond [14]. They found that such a bond can form only when assisted by an N–Ch$\cdots C_{60}$ chalcogen bond (Ch = S, Se, Te), driven by σ-holes on the chalcogen. The interaction energies associated with these chalcogen bonds increase by an order of magnitude from S to Te. However, only for 1,2,5-telluradiazole and 2,1,3-benzotelluradiazole does the combined effect of the N→C dative bond and N–Te$\cdots C_{60}$ chalcogen bond overcome the significant repulsive deformation energy of C_{60}, making these structures energetically stable. The structures of the corresponding 1,2,5-oxadiazole and 2,1,3-benzoxadiazole complexes with C_{60} are governed by $\pi\cdots\pi$ stacking interactions.

In their review, Buntkowsky et al. [5] highlight exemplary studies focusing on the intriguing phase behavior exhibited by guest molecules confined within the pores of MCM-41, SBA-15 mesoporous silicas, and their functionalized derivatives. By combining advanced solid-state NMR techniques with molecular dynamics simulations, the authors demonstrate that despite the cramped pore environments, the balance of intermolecular forces—whether for small molecules such as water or octanol, or for surfactants like $C_{10}E_6$ or Triton X-100—frequently drives the formation of ordered structures. Notably, guest diffusion within these mesopores is anisotropic and strongly dependent on the filling level [15]. The authors anticipate that ongoing advances in molecular dynamics will soon enable predictive modeling of these confined structures, which can then be validated through targeted experimental studies.

Gómez-Salazar et al. [6] reported the development of a thiol-functionalized silica adsorbent tailored to remove melanoidin-type compounds from organic wastes produced during ethanol fermentation, alongside a UV–Vis spectroscopic method used to monitor the process. Rapid adsorption kinetics and a high capacity for melanoidin removal arise from the direct interaction of the guest with the thiol functional group, a reaction that is well documented [16], rather than with the pristine silica surface. The adsorbent also exhibits high thermal stability, making it well suited for industrial application and helping to reduce wastewater pollution.

Jabłoński [7] presents a theoretical investigation into the factors that govern the selective encapsulation of small guest molecules within the interior cavity of superphanes. In its simplest form, a superphane consists of two aromatic rings held in a rigid, parallel orientation by six $-(CH_2)_2-$ linkers, creating a barrel-like hollow core. Modified superphanes, however, incorporate composite binding chains that not only adjust cavity size but also provide specific noncovalent interaction sites for guest molecules, enhancing the trapping selectivity [17]. Through energy-decomposition analyses, the results obtained demonstrate that these engineered binding sites exert a greater influence on selectivity than cavity volume alone.

Melandri et al. [8] describe a combined experimental and theoretical study of the rotational spectrum of the 1:1 acrolein–MeOH (and MeOD) complex in the gas phase. Analysis of the measured spectroscopic parameters unambiguously defines the complex's conformation, which agrees with high-level quantum-chemical predictions. A notable feature of methanol spectra is the splitting of rotational lines due to hindered internal rotation of the CH_3 group. In every known 1:1 complex, the barrier to this methyl rotation decreases [18], and in acrolein–MeOH it falls to about 60% of the monomer value. Importantly, this reduction does not correlate with complexation energy but instead arises from coupling with an additional large-amplitude motion. While theory reproduces the qualitative trend, the calculated barrier remains significantly higher than the experimental value. Resolving this quantitative gap will require more sophisticated theoretical treatments in future studies.

Gómez and Cappelli [9] employ aqueous caffeine solutions to illustrate the advantages of a fully polarizable QM/MM approach [19] for (i) predicting the optimal excitation wavelength for resonance Raman spectroscopy and thereby achieving exceptionally low detection limits for caffeine quantification and (ii) elucidating how hydrogen-bond networks shape caffeine's spectral fingerprints. While the strongest solute–solvent interaction is the N···H–O hydrogen bond between water and caffeine's imidazole nitrogen, each of caffeine's two carbonyl groups can simultaneously coordinate up to two water molecules, substantially increasing the stabilization energy of the first hydration shell. It would be highly informative to quantify the entropic penalty associated with this enhanced stabilization and to assess its impact on caffeine's solubility in water.

Shenderovich [10] surveyed experimental data on how noncovalent interactions perturb the isotropic hyperfine coupling constant of the nitroxide radical TEMPO in a solution. Because these interaction-induced shifts exceed the experimental measurement uncertainty by two orders of magnitude, TEMPO serves as an exceptionally sensitive probe for quantifying weak interactions in solution. These data were used to identify a practical DFT functional and basis-set combination capable of reproducing the observed effects and to assess the utility of TEMPO for studying competing hydrogen bonds (e.g., with 2,2′-bipyridinium [20]) and strong halogen-bond donors. The work demonstrates that accurately interpreting the measured coupling constant changes in terms of specific complex populations requires evaluating many candidate structures, whose relative abundances cannot always be inferred from simple heuristics. Notably, like a P=O moiety [21], the N–O group in TEMPO can simultaneously form two hydrogen bonds of comparable strength.

Stoumpos et al. [11] compared the intramolecular vibrations of pyridinium in one-dimensional lead halide perovskites and in pyridinium halide salts by infrared absorption and Raman scattering. Vibrational bands were rigorously assigned through DFT calculations aided by symmetry analysis. The modes that involve N–H motions exhibit pronounced frequency shifts depending on whether the cation resides within the perovskite channels or in the salt lattice, whereas changing the halide (Br^- or I^-) produces only marginal effects. By analyzing these environment-sensitive shifts, one can qualitatively gauge crystal-structure disorder or detect nonequivalent pyridinium sites. Thus, pyridinium cations in hybrid organic–inorganic crystals serve as intrinsic spectroscopic probes, reporting on their local lattice environment through changes in the selected vibrational frequencies. Similarly, quasi-elastic neutron scattering studies of the rotational dynamics of organic cations have proven effective for probing structural features in perovskites [22].

Acknowledgments: I want to sincerely thank everyone who has contributed to this Special Issue. Special thanks go to the entire team of Molecules for their motivation, professional expertise, and support.

Conflicts of Interest: The author declares no conflicts of interest.

References

1. Shenderovich, I.G. (Ed.) *Gulliver in the Country of Lilliput: An Interplay of Noncovalent Interactions*; MDPI: Basel, Switzerland, 2021; Available online: https://www.mdpi.com/books/reprint/3554-gulliver-in-the-country-of-lilliput (accessed on 14 July 2025).
2. Mihrin, D.; Feilberg, K.L.; Larsen, R.W. Self-Association and Microhydration of Phenol: Identification of Large-Amplitude Hydrogen Bond Librational Modes. *Molecules* **2024**, *29*, 3012. [CrossRef]
3. Hardin, M.; Zeller, M.; Rosokha, S.V. Halogen Bonding and Rearrangements in Complexes of *N*-Chlorosuccinimide with Halides. *Molecules* **2025**, *30*, 639. [CrossRef]
4. Zhang, Y.; Wang, W. Chalcogen-Bond-Assisted Formation of the N→C Dative Bonds in the Complexes between Chalcogenadiazoles/Chalcogenatriazoles and Fullerene C_{60}. *Molecules* **2024**, *29*, 2685. [CrossRef] [PubMed]
5. Haro Mares, N.B.; Döller, S.C.; Wissel, T.; Hoffmann, M.; Vogel, M.; Buntkowsky, G. Structures and Dynamics of Complex Guest Molecules in Confinement, Revealed by Solid-State NMR, Molecular Dynamics, and Calorimetry. *Molecules* **2024**, *29*, 1669. [CrossRef] [PubMed]

6. Gonzalez-Valerio, C.; Peregrina-Lucano, A.A.; Manríquez-González, R.; Pérez-Fonseca, A.A.; Robledo-Ortíz, J.R.; Shenderovich, I.G.; Gómez-Salazar, S. Color Compounds Removal from Tequila Vinasses Using Silica Gel Adsorbents Functionalized with Thiol Moieties: Equilibrium and Kinetics Studies. *Molecules* **2024**, *29*, 5910. [CrossRef]
7. Jabłoński, M. Characteristics of Intermolecular Interactions between Encapsulated Molecules and the Lantern-Like Carcerand Superphanes. *Molecules* **2024**, *29*, 601. [CrossRef]
8. Lv, D.; Sundelin, D.; Maris, A.; Evangelisti, L.; Geppert, W.D.; Melandri, S. Intermolecular Interactions between Aldehydes and Alcohols: Conformational Equilibrium and Rotational Spectra of Acrolein-Methanol Complex. *Molecules* **2024**, *29*, 3444. [CrossRef]
9. Gómez, S.; Cappelli, C. The Role of Hydrogen Bonding in the Raman Spectral Signals of Caffeine in Aqueous Solution. *Molecules* **2024**, *29*, 3035. [CrossRef] [PubMed]
10. Shenderovich, I.G. Keep Your TEMPO Up: Nitroxide Radicals as Sensors of Intermolecular Interactions. *Molecules* **2024**, *29*, 5032. [CrossRef]
11. Samsonova, A.Y.; Mikheleva, A.Y.; Bulanin, K.M.; Selivanov, N.I.; Mazur, A.S.; Tolstoy, P.M.; Stoumpos, C.C.; Kapitonov, Y.V. Internal Vibrations of Pyridinium Cation in One-Dimensional Halide Perovskites and the Corresponding Halide Salts. *Molecules* **2024**, *29*, 78. [CrossRef]
12. Andersen, J.; Heimdal, J.; Larsen, R.W. Spectroscopic identification of ethanol-water conformers by large-amplitude hydrogen bond librational modes. *J. Chem. Phys.* **2015**, *143*, 224315. [CrossRef] [PubMed]
13. Miller, D.K.; Loy, C.; Rosokha, S.V. Examining a Transition from Supramolecular Halogen Bonding to Covalent Bonds: Topological Analysis of Electron Densities and Energies in the Complexes of Bromosubstituted Electrophiles. *ACS Omega* **2021**, *6*, 23588–23597. [CrossRef] [PubMed]
14. Nandi, A.; Kozuch, S. History and Future of Dative Bonds. *Chem. Eur. J.* **2020**, *26*, 759–772. [CrossRef]
15. Gedat, E.; Schreiber, A.; Findenegg, G.H.; Shenderovich, I.; Limbach, H.-H.; Buntkowsky, G. Stray Field Gradient NMR Reveals Effects of Hydrogen Bonding on Diffusion Coefficients of Pyridine in Mesoporous Silica. *Magn. Reson. Chem.* **2001**, *39*, S149–S157. [CrossRef]
16. Hofmann, T.; Schieberle, P. Chemical Interactions between Odor-Active Thiols and Melanoidins Involved in the Aroma Staling of Coffee Beverages. *J. Agric. Food Chem.* **2002**, *50*, 319–326. [CrossRef]
17. Li, A.; Xiong, S.; Zhou, W.; Zhai, H.; Liu, Y.; He, Q. Superphane: A new lantern-like receptor for encapsulation of a water dimer. *Chem. Commun.* **2021**, *57*, 4496–4499. [CrossRef]
18. Calabrese, C.; Maris, A.; Vigorito, A.; Mariotti, S.; Fathi, P.; Geppert, W.D.; Melandri, S. Structure, Dynamics, and Accurate Laboratory Rotational Frequencies of the Acrylonitrile--Methanol Complex. *J. Phys. Chem. A* **2020**, *124*, 3601–3608. [CrossRef]
19. Giovannini, T.; Egidi, F.; Cappelli, C. Molecular spectroscopy of aqueous solutions: A theoretical perspective. *Chem. Soc. Rev.* **2020**, *49*, 5664–5677. [CrossRef]
20. Lesnichin, S.B.; Tolstoy, P.M.; Limbach, H.-H.; Shenderovich, I.G. Counteranion-Dependent Mechanisms of Intramolecular Proton Transfer in Aprotic Solution. *Phys. Chem. Chem. Phys.* **2010**, *12*, 10373–10379. [CrossRef]
21. Begimova, G.U.; Tupikina, E.Y.; Yu, V.K.; Denisov, G.S.; Bodensteiner, M.; Shenderovich, I.G. Effect of Hydrogen Bonding to Water on the ^{31}P Chemical Shift Tensor of Phenyl- and Trialkylphosphine Oxides and α-Amino Phosphonates. *J. Phys. Chem. C* **2016**, *120*, 8717–8729. [CrossRef]
22. Lavén, R.; Koza, M.M.; Malavasi, L.; Perrichon, A.; Appel, M.; Karlsson, M. Rotational Dynamics of Organic Cations in Formamidinium Lead Iodide Perovskites. *J. Phys. Chem. Lett.* **2023**, *14*, 2784–2791. [CrossRef] [PubMed]

molecules

MDPI

Article

Halogen Bonding and Rearrangements in Complexes of *N*-Chlorosuccinimide with Halides

Maison Hardin [1], Matthias Zeller [2] and Sergiy V. Rosokha [1,*]

1 Department of Chemistry, Ball State University, Muncie, IN 47306, USA; maison.hardin@bsu.edu
2 Department of Chemistry, Purdue University, West Lafayette, IN 47907, USA; zeller4@purdue.edu
* Correspondence: svrosokha@bsu.edu

Abstract: The role of halogen bonding (HaB) in the reactions of *N*-chlorosuccinimide (SimCl), a versatile reagent in organic synthesis, was investigated through experimental and computational analyses of its interactions with halides. The reactions of SimCl with Br^- or I^- resulted in the crystallization of HaB complexes of chloride with *N*-iodosuccinimide (SimI) or *N*-bromosuccinimide (SimBr). Computational analysis revealed that halogen rearrangements, which occurred even at −73 °C, were facilitated by halogen bonding. The dissociation of SimCl·Y^- (Y = I or Br) complexes into a Sim^- + ClY pair (followed by the rotation and re-binding of the interhalogen molecules) bypassed the formation of the high-energy Sim^- + Cl^+ pair and drastically (about tenfold) reduced the dissociation energy of the N–Cl bond. Furthermore, while the dissociation energy of individual SimCl is higher (and its HaB is weaker) compared to that of SimI or SimBr, the dissociation of the N-Cl bond in SimCl·Y^- requires less energy than in the complexes of SimBr or SimI. The facile cleavage of such bonds in HaB complexes explains the high reactivity of SimCl and its effectiveness as a halogenating agent.

Keywords: halogen bonding; *N*-chlorosuccinimide; halides; X-ray crystallography; DFT calculations

1. Introduction

Halogen bonding, an attraction between electron-deficient halogen atoms and an electron-rich species, has emerged in recent years as a powerful tool in crystal engineering and molecular recognition [1–4]. Numerous studies have also demonstrated its substantial role in the reactions of halogenated molecules and its utility in catalysis [5–8]. In particular, both experimental and computational analyses have shown that strong halogen bonding can facilitate electron and/or halogen transfer between interacting species [9–12]. The latter can be viewed as a subclass of nucleophilic substitution reactions, referred to as halophilic or S_N2X substitutions, in which the nucleophile attacks a halogen substituent instead of the carbon atom in a C–X bond [13].

It should be noted, however, that most studies of halogen bonding have focused on iodo- and bromo-substituted molecules [1,2]. This choice is related to the better polarizability of these heavier halogen atoms, which results in much stronger halogen bonding. Halogen bonding via chlorine in most cases is relatively weak [14–17], and the role of this interaction in the reactions of chloro-substituted molecules remains largely unexplored.

Given this context, it is of interest to evaluate the role of halogen bonding in the reactions of *N*-chlorosuccinimides (SimCl). This molecule is a representative of *N*-haloimides, common halogenating agents in which halogens are bonded to nitrogen atoms [18–20].

Furthermore, *N*-iodosuccinimide (SimI) and *N*-bromosuccinimide (SimBr) are known to be very strong halogen-bond (HaB) donors, forming exceptionally short halogen bonds with a variety of nucleophiles [21–24]. In contrast, despite the presence of a σ-hole on the surface of the chlorine atom in *N*-chlorosuccinimide (SimCl, Figure 1) [25,26], which is comparable in strength (~21 kcal/mol at 0.001 a.u. electron density) to that of the surfaces of iodine and bromine in well-known HaB donors (e.g., bromoform [27]), attempts to obtain HaB complexes of SimCl have been mostly unsuccessful.

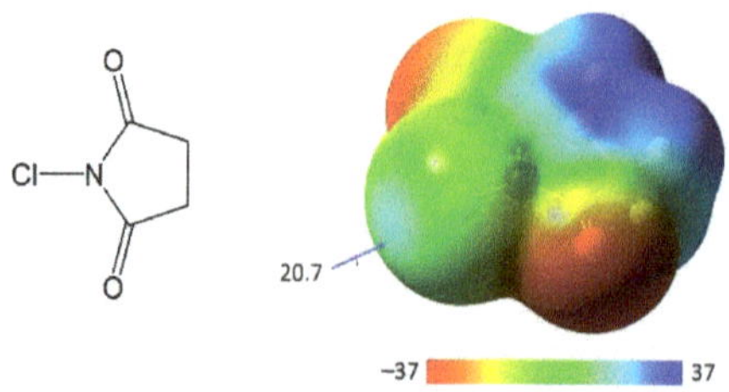

Figure 1. Structure and surface electrostatic potential of SimCl (in kcal/mol at 0.001 a.u. electron density).

For example, Stilinović et al. reported X-ray structures of several HaB complexes of SimI and SimBr with various pyridines but noted that the isolation and characterization of similar complexes with SimCl was hindered by rapid decomposition [25]. Our recent study showed that the interaction of the structurally similar *N*-chlorosaccharin with quinuclidine led to chlorine transfer, yielding HaB complexes between *N*-chloroquinuclidinium cations and chloride instead of a halogen bonded complex of chlorosaccharin [26]. As such, the 2:1 complex of SimCl with chloride remains the only halogen-bonded association of this molecule reported in the Cambridge Crystallographic Data Centre [28]. In comparison, earlier studies on the interaction of SimCl with bromide resulted in the formation of *N*-bromosuccinimide, while reactions of SimI with iodide produced *N*-iodosuccinimide [29,30]. Eberson et al. hypothesized that such halogen exchanges might occur through a halophilic process involving an activated complex with partial bonds between chlorine and bromide or via an intracomplex electron-transfer mechanism [29]. However, the absence of experimental or computational evidence has made it difficult to validate these pathways.

The wide-ranging applications of *N*-chlorosuccinimide as a versatile reagent in organic synthesis [31–33] underscore the importance of clarifying the role, if any, of halogen bonding in its reactions. Moreover, this electrophilic molecule seems ideally suited for exploring the potential effects of halogen bonding in the reactions of chlorinated molecules in general. Therefore, in the current work, we conducted experimental and computational investigations into the interactions of SimCl with the simplest monoatomic nucleophiles—bromide and iodide anions.

2. Results and Discussion

2.1. X-Ray Structural Characterization of the Products of Interaction of SimCl with Bromide and Iodide

The addition of methyltriphenylphosphonium bromide to a solution of SimCl in dichloromethane and the slow evaporation of the resulting mixture at room temperature (see Experimental section) produced colorless crystals **1**. The FT-IR spectrum of these crystals showed a substantial red shift in the C=O vibration frequencies and a blue shift in the C-N stretching frequencies as compared to those of SimCl (Figure S1 and Table S1 in the Supporting Information). X-ray crystallographic analysis revealed that the succinimide moieties in these monoclinic crystals ($P2_1$ space group, see Table S2 in the Supporting Information for crystallographic, data collection, and refinement details) are bonded to bromine atoms, which are, in turn, halogen-bonded to chloride anions (Figure 2). In other words, the interaction of SimCl with bromide resulted in halogen exchange and the formation of 1:1 HaB complexes between the resulting SimBr and chloride.

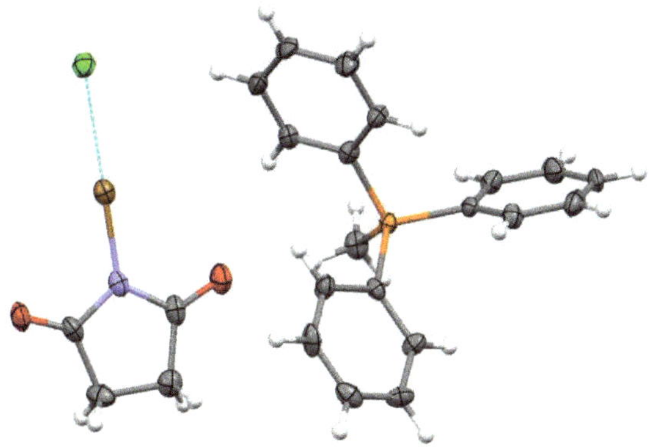

Figure 2. X-ray structure of **1** resulting from the interaction of SimCl with [Me(Ph_3)P]Br. Color code: dark gray—carbon, light gray—hydrogen, blue—nitrogen, red—oxygen, brown—bromine, orange—phosphorous and green—chlorine.

The SimBr·Cl^- complexes are nested between the benzene rings of the bulky counter-ions (see crystal packing in Figure S2 in the Supporting Information). The Br· · · Cl distance of 2.659 Å (Table 1) in these complexes is 26% shorter than the sum of the van der Waals radii of bromine and chlorine atoms [34], indicating a very strong interaction between these species. The N–Br bond length of 1.919 Å in SimBr is elongated compared to the analogous bond in individual SimBr (1.836 Å [35]) and in the previously reported 2:1 complex of SimBr with chloride (~1.88 Å [36]).

Table 1. Geometric characteristics of N-X (intramolecular) and X· · · Cl (intermolecular) bonds in crystals **1–5**.

Crystal	N-X· · ·Cl Contact	$d_{N\text{-}X}$, Å	$d_{X\cdots Cl}$, Å	∠N-X· · ·Cl, deg
1	N-Br· · · Cl	1.919(5)	2.6593(17)	179.4(2)
2	N-I· · · Cl	2.141(3)	2.666(1)	178.99(8)
3	N-I· · · Cl	2.145(7)	2.681(2)	179.1(2)
4	N-Br[Cl]· · · Cl[Br] [a]	1.935(3) [1.76(3)]	2.731(9) [2.75(3)]	177.4(2) [176.9(14)]
5	N-Br[Cl]· · · Cl[Br] [a]	1.926(3) [1.89(5)]	2.687(4) [2.67(7)]	178.74(11) [166(2)]

[a] Cl and Br atoms are disordered between two sites. E.s.d.'s are shown in parenthesis. Minor forms and their characteristics are shown in brackets.

A similar evaporation of a dichloromethane solution containing a mixture of SimCl and iodide (taken as a salt with *n*-tetrapropylammonium as the counterion) produced crystals **2** (see crystal packing in Figure S3 in the Supporting Information). The changes in the IR spectra of these crystals compared to SimCl were similar (e.g., a red shift in the C=O stretching vibrations around 1700 cm^{-1} and a blue shift in the C-N stretching vibrations in 1150–1200 cm^{-1} range) as those in the crystals **1** (Figure S1 in the Supporting Information). These colorless monoclinic plates **2** also comprised the product of halogen exchange, i.e., the SimI·Cl complex in which the succinimide moiety is bonded to iodine atoms, which are, in turn, bonded to chlorine (Figure 3).

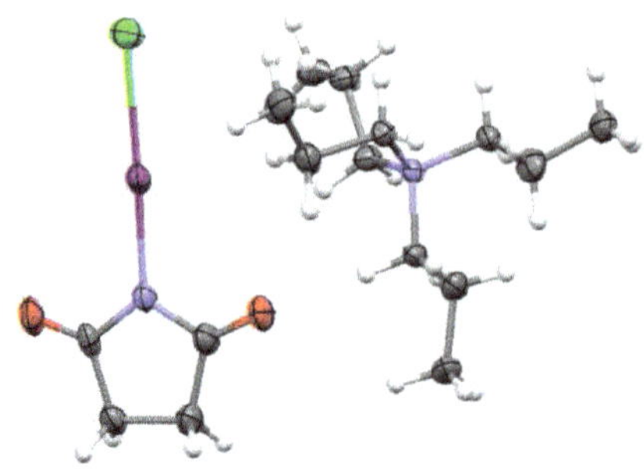

Figure 3. X-ray structure of **2** resulting from the interaction of SimCl with $[Pr_4N]Br$. Color code: dark gray—carbon, light gray—hydrogen, blue—nitrogen, red—oxygen, magenta—iodine and green—chlorine.

The I···Cl distance in the SimI·Cl^- complexes is 29% shorter than the sum of the van der Waals radii of chlorine and iodine and the Br···Cl distance in the complexes of Cl^- with SimBr in crystals **1**. Also, the N–I bond in the SimI·Cl complexes (Table 1) is significantly elongated compared to the bond in individual SimI (2.06 Å [37]). As with the Br···Cl distance, the I···Cl bond in the SimI·Cl complexes is shorter and the N–I bond is longer than that measured in the similar 2:1 complexes [36] and in the associations in which Cl^- is halogen bonded both with SimI and 1,3-bis(2,6-diisopropylphenyl)-2-iodo-imidazolium [38].

To determine whether lowering temperatures (which commonly slow down reactions) can suppress the halogen exchange, we conducted crystallization of the HaB complexes by diffusing hexane into dichloromethane solutions of SimCl with bromide or iodide salts at −70 °C to −75 °C. In the systems containing SimCl together with methyltriphenylphosphonium iodide, such low-temperature crystallizations produced colorless plates (crystals **3**). Similarly to the room-temperature crystallization, these monoclinic crystals (space group $P2_1$) comprised the product of halogen exchange, i.e., a 1:1 HaB complex between SimI and the chloride anion (Figure S4 in the Supporting Information). Although the counter-ions in crystals **2** and **3** were different (Pr_4N^+ vs. $MePh_3P^+$), the geometric characteristics of the SimI·Cl complexes in these crystals were very similar (Table 1).

The analogous low-temperature diffusion of hexane into a solution of SimCl and methyltriphenylphosphonium bromide produced a mixture of colorless prisms (crystals **4**) and plates (crystals **5**). X-ray structural analysis revealed that both types of crystals contained HaB complexes, with chlorine and bromine partially disordered between two sites (Figures S5 and S6 in the Supporting Information). Specifically, one bromine atom was covalently bonded to the nitrogen atom, while the other occupied the terminal halogen position in the complex. The same was applied to chlorine. The product of halogen exchange (where bromine is covalently bonded to nitrogen) represented the predominant form in both crystals **4** and **5**, but the disorder ratios differed between the two. Specifically, the disorder ratio was 0.822(2):0.178(2) in the orthorhombic prisms **4**, whereas the ratio was 0.940(3):0.060(3) in the monoclinic plates **5**. Due to the small fraction of disordered bromine at the terminal positions, the unit cell geometry of the latter was essentially identical to that of the related crystals **1** (see Table S2 in the Supporting Information).

Overall, the results of the low-temperature crystallization demonstrated that addition of iodide or bromide to the solution of SimCl results in halogen exchange even at −70 °C to −75 °C. Under these conditions, the chlorine-to-iodine exchange proceeds much faster than crystallization, so the resulting crystals did not contain the original SimCl. However, the rate of chlorine-to-bromine exchange is apparently comparable to the crystallization kinetics. As a result, the resulting crystals contained minor fractions of the starting SimCl molecules. To clarify the mechanisms underlying these processes and the reasons for the differences observed in the systems with bromide and iodide anions, we conducted a computational analysis of the bonding in these systems.

2.2. Computational Analysis of the Bonding in the Complexes of N-Chlorosuccinimides with Halides

The M06-2X/def2-tzvpp computations (see the Materials and Methods for details) produced HaB complexes between SimBr or SimI molecules and Cl^- anions similar to that measured in the solid state. In particular, the optimized complexes showed Br···Cl and I···Cl separations, which were 23 and 29% shorter, respectively, than the van der Waals separations (Table 2). The N-Br and N-I bonds were elongated as compared to those in individual SimI and SimBr (listed in Table S3 in the Supporting Information). These characteristics indicate strong halogen bonding in these complexes. In comparison, the Cl···I and Cl···Br separations in the optimized SimCl·I^- and SimCl·Br^- complexes were about 10% shorter than the sums of the van der Waals separations, and the N-Cl bond was within 0.03 Å of that of SimCl. Following the distinction in the HaB bond lengths, the bindings energies of the terminal halogens in the SimI·Cl^- and SimBr·Cl^- complexes were substantially larger than in the similar complexes formed by SimCl with various halides.

Table 2. Selected characteristics of the optimized SimX·Y^- complexes.

SimX·Y^-	$d_{N\text{-}X}$, Å	$d_{X\text{-}Y}$, Å	ΔE, kcal/mol [a]	q_Y, [b] e	q_X, [c] e
SimI·Cl^-	2.159	2.659	−15.6	−0.72	0.33
SimBr·Cl^-	1.884	2.793	−7.7	−0.86	0.27
SimCl·I^-	1.685	3.401	−2.3	−0.95	0.22
SimCl·Br^-	1.686	3.179	−2.9	−0.95	0.22
SimCl·Cl^-	1.689	2.986	−3.6	−0.95	0.23

[a] Binding energy of Y^- anion in the HaB complex, ΔE = E(SimX·Y^-) − E(SimX) − E(Y^-), where E(SimX·Y^-), E(SIMX), and E(Y^-) are energies, including zero-point energies, of the optimized complexes, SimX, and halide, respectively. [b] NBO charges on the terminal halogen (Y); [c] NBO charges on the middle halogen (X).

In agreement with the experimental data, the optimized SimI·Cl^- and SimBr·Cl^- complexes showed substantial shifts in the C=O and C-N vibrations as compared to those in the individual SimCl molecule (Figure S1 in the SI). The corresponding shifts in the SimCl·I^- and SimCl·Br^- complexes were much smaller. Such distinctions are related to the differences in the charge transfer from the halides to the succinimide moieties in these complexes (which weaken C=O and strengthen C-N bonds). Specifically, the Natural Bond Orbital (NBO) analysis [39] revealed that the negative charges are predominantly localized on the terminal halogen atoms in all these complexes, while the central halogens exhibit partial positive charges. However, the terminal chlorine atoms in the SimI·Cl^- and SimBr·Cl^- complexes carry noticeably less than a unit negative charge, which is characteristic of halide anions. Accordingly, substantial negative charge also resides on the succinimide moieties (approximately −0.6 e and −0.4 e in SimI·Cl^- and SimBr·Cl^-, respectively). In contrast, the charges on the terminal halogens in the complexes formed by SimCl are within 5% of a unit negative charge, with only about −0.27 e residing on the succinimide moieties.

To further clarify the differences between the X···Cl halogen bonding involving chlorine atoms in the SimCl complexes and those that formed by SimI or SimBr with chloride anions, we performed a QTAIM analysis [40,41] of the optimized SimX·Y^- associations. The electron densities at the bond critical points (BCPs) along the Cl···I and Cl···I bond paths in the SimCl·I^- complexes obtained from this analysis (Table 3) are close to 0.01 a.u.

The energy densities at these points are very small and positive. These characteristics are typical of moderately strong, electrostatically driven supramolecular interactions [41].

Table 3. Electron and energy densities, as well as ELF at the BCPs on the N-X and X$\cdots$Y bond paths in the optimized SimX·Y$^-$ complexes.

	N-X Bond			X$\cdots$Y Bond		
SimX·Y$^-$	**ρ(r)**	**H(r)**	**ELF**	**ρ(r)**	**H(r)**	**ELF**
SimI·Cl$^-$	0.097	−0.038	0.385	0.052	−0.011	0.317
SimBr·Cl$^-$	0.155	−0.100	0.612	0.033	−0.001	0.166
SimCl·I$^-$	0.225	−0.180	0.809	0.013	0.001	0.062
SimCl·Br$^-$	0.225	−0.181	0.807	0.015	0.001	0.062
SimCl·Cl$^-$	0.224	−0.179	0.804	0.018	0.002	0.066

In contrast, the electron densities at the BCPs along the X$\cdots$Cl bond paths in the SimX·Cl$^-$ complexes are significantly higher than 0.01 a.u., and the energy densities at these BCPs are negative. These values suggest a partially covalent nature of the halogen bonding in the SimI·Cl$^-$ and SimBr·Cl$^-$ complexes [41–43]. Consistent with the distinctions in electron and energy densities at the BCPs, the electron localization function (ELF) values along the X$\cdots$Cl bond paths in the SimI·Cl$^-$ and SimBr·Cl$^-$ complexes are much higher than those in the SimCl·Y$^-$ complexes (Table 3).

The increase in the values of ρ(r) and ELF on the X$\cdots$Y bond paths in the SimI·Cl$^-$ and SimBr·Cl$^-$ complexes, compared to the corresponding values in the SimCl complexes, is accompanied by a substantial decrease in these values on the N-X bond paths. This indicates that as the strength of halogen bonding increases, the adjacent covalent bond involving the halogen atom weakens (Note that the strength of the I-Cl bonding in the associations formed by SimI and chloride suggests that it be alternatively considered as a complex between interhalogen ClI and Sim$^-$. Yet, since the data in Table 3 indicate that N-I interaction is somewhat stronger than I-Cl bonding, the SimI·Cl$^-$ seems preferable).

Most importantly, the energies of the optimized SimI·Cl$^-$ and SimBr·Cl$^-$ complexes are lower than those of the corresponding SimCl·I$^-$ and SimCl·Br$^-$ associations with the same composition. Specifically, the halogen rearrangement converting SimCl·I$^-$ into SimI·Cl$^-$ lowers the energy of the system by 15.4 kcal/mol and the free energy by 14.0 kcal/mol. The energy changes resulting from the rearrangement of SimCl·Br$^-$ into SimBr·Cl$^-$ are smaller, with energy decreasing by 2.1 kcal/mol and free energy by 1.5 kcal/mol. The halogen exchange occuring without involvement of the HaB Complexes (i.i., starting from the separate SimCl and I$^-$ or Br$^-$ and producing separate Cl$^-$ and SimI or SimBr) is less termodynamically favorable. Specifically, $\Delta G = -2.1$ kcal/mol in the case of such Cl/I exchange, and the free energy of the separate SimBr and Cl$^-$, is higher by 2.8 kcal/mol than the energy of the intial SimCl and Br$^-$.

These data indicate that halogen bonding resulting in the formation of the SimI·Cl$^-$ and SimBr·Cl$^-$ complexes makes the halogen rearrangement a thermodynamically favored process in both systems. However, besides thermodynamics, the possibility of rearrangement is controlled by the kinetics of these processes. Since the rearrangement involves breaking the N-Cl bonds, the reaction rates are determined by the energies required for their dissociation. It should be noted in this respect that the energy of heterolytic dissociation of SimCl into Sim$^-$ and Cl$^+$ is very large (214 kcal/mol, see Table S3 in the Supporting Information). The homolytic dissociation of this molecule is less energy-intensive, with an energy increase of 79 kcal/mol. Nevertheless, any process involving such dissociations would be too slow to account for the halogen rearrangements observed at room or at low temperatures. Another possible pathway starts with electron transfer from iodide or bromide, followed by the easy dissociation of the $SimCl^{-\cdot}$ anion radical. However, the energy of the pairs resulting from such dissociation (e.g., 88 kcal/mol relative to the starting reactants for $SimCl^{-\cdot}$ + $I^{\cdot}$ pair) are also too high.

Halogen bonding of SimCl with halides, however, dramatically reduces the dissociation energy of the N-Cl bond. For instance, the dissociation of the $SimCl{\cdot}Br^-$ complex into a succinimide anion and a ClBr molecule is accompanied by an energy increase of 24.5 kcal/mol and an even smaller free-energy change (Table 4).

Table 4. Dissociation energy (in kcal/mol) of the N-X bond in the HaB complexes.

	$SimX{\cdot}Y^- \rightarrow Sim^- + XY$		$SimX{\cdot}Y^- \rightarrow Sim^{\cdot} + XY^{\cdot -}$	
$SimX{\cdot}Y^-$	ΔE	ΔG	ΔE	ΔG
$SimI{\cdot}Cl^-$	31.2	21.4	64.5	52.9
$SimBr{\cdot}Cl^-$	26.7	16.6	58.0	46.1
$SimCl{\cdot}I^-$	15.8	7.4	49.1	38.2
$SimCl{\cdot}Br^-$	24.5	15.0	55.9	44.6
$SimCl{\cdot}Cl^-$	26.0	17.7	58.0	47.9

Furthermore, the dissociation of the $SimCl{\cdot}I^-$ complex into a succinimide anion and a ClI molecule results in a free energy increase of only 7.4 kcal/mol. It should be also noted that while the negative charges in the $SimCl{\cdot}I^-$ and $SimCl{\cdot}Br^-$ complexes reside primarily on the terminal halogen atoms (*vide supra*), the homolytic dissociations of these complexes into $Sim^{\cdot}$ radicals and $ClY^{-\cdot}$ anion-radicals require significantly higher energy than the dissociations into succinimide anions and neutral ClY molecules (Table 4).

The analysis of bonding indicates that the halogen rearrangements can be described as dissociation of the $SimCl{\cdot}Y^-$ complexes into succinimide anions and ClY molecules, followed by the rotation and binding of ClY to Sim^- via iodine or bromine which produces more thermodynamically stable complexes. The energy scan (Figure 4, see Materials and Methods for details) along the N-Cl and N-I bonds indicates that the dissociations of the former and formation of the latter occur essentially without barriers.

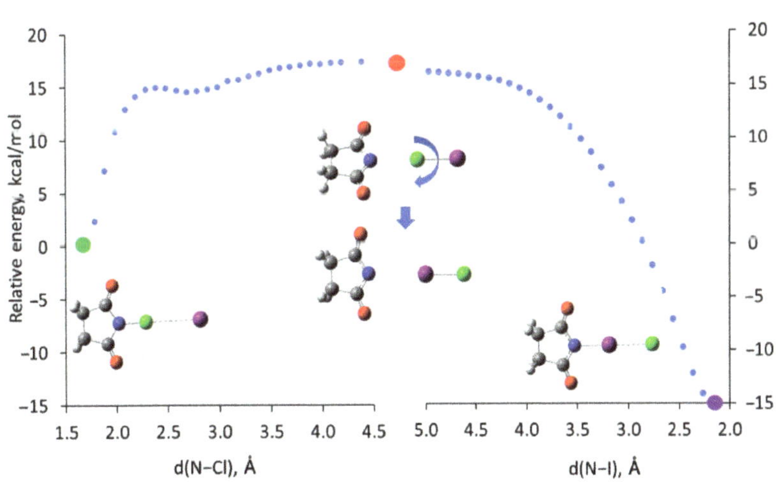

Figure 4. Energy scan along the N-Cl and N-I bonds in the $SimCl{\cdot}I^-$ and $SimCl{\cdot}I^-$ complexes (relative to the energy of $SimCl{\cdot}I^-$). Green and purple circles represent the energies of the optimized $SimCl{\cdot}I^-$ and $SimCl{\cdot}I^-$ complexes, respectively, and the red circle shows the sum of the energies of the individual $SimCl^-$ anion and ClI molecule.

Taking the energies of the pairs of dissociated Sim^- and ClY species as the barriers for the rearrangements allows for estimating the rates of these processes (see the Experimental section for details). In particular, the free energy of the Sim^- + ICl pair relative to the SimCl·I^- complex of 7.4 kcal/mol implies that the rate constants of the rearrangement are on the order of 10^7 s^{-1} at 298 K and 10^4 s^{-1} at 200 K. This corresponds to a half-life of the SimCl·I^- complex of less than a millisecond (even at −73 °C), which is much faster than the rate of crystallization. The estimated rate of the rearrangement of the SimCl·Br^- complex at room temperature, about 10^2 s^{-1} (with $t_1/_2$~0.01 s), is also faster than the crystallization kinetics. However, the estimated rate constant of this process at −73 °C (approximately 10^{-4} s^{-1}) corresponds to a half-life of SimCl·Br^- on the order of a few hours under these conditions. This timescale is comparable to the rates of crystallization and implies (in agreement with the experimental data) that some of the starting complex can co-crystallize with the product of the rearrangement.

The data in Table 4 also show that the dissociation energies (and the corresponding free energies) of the SimCl·I^- and SimCl·Br^- complexes are lower than those of the SimI·Cl^- and SimBr·Cl^- analogs. This is surprising since the heterolytic and homolytic dissociation energies of the individual SimCl molecule are higher than those of SimI and SimBr (Table S3 in the Supporting Information). Moreover, the halogen bonds of SimI or SimBr with chloride are stronger than those of SimCl with bromide or iodide which suggests a more substantial weakening of the N-Br and N-I bonds in the HaB complexes of SimI and SimBr. (This suggestion is consistent with the trend in N-X bond length increase and QTAIM characteristics in Tables 2 and 3). However, the magnitude of the (negative) free energy change in the reaction of the Cl^+ cation (formally released by SimCl) with I^- (or Br^-) anions is much higher than that of the reaction of I^+ (or Br^+) with Cl^-. This difference (approximately 77 kcal/mol for Cl^+/I^- vs. I^+/Cl^- reactions, see Table S4 in the Supporting Information for details) more than compensates for the difference in the N-X bond dissociation energies and explains the counter-intuitive trend in the thermodynamics of the processes with different *N*-halosuccinimides. Furthermore, halogen bonding eliminates the need to form the high-energy Sim^- + X^+ pair resulting from the dissociation of the individual SimCl (since dissociation of the HaB complex proceeds directly to the Sim^- + XY pair without an additional energy barrier), which accounts for the fast kinetics of the rearrangement.

3. Conclusions

The crystallization of SimCl with Br^- and I^- anions results in halogen rearrangements and the formation of HaB complexes of SimI or SimBr with chloride. Computational analysis shows that these rearrangements (which proceed efficiently even at −73 °C) are facilitated by halogen bonding via chlorine atoms. Such SimCl·Y^- complexes dissociate directly into the Sim^- + XY pair (followed by the rotation and re-binding of the interhalogen molecule via heavier atom which leads to the more stable complex). The involvement of HaB complexes makes halogen exchange thermodynamically favorable in the solutions of SimCl with I^- and Br^-. It also eliminates the need for the formation of the high-energy Sim^- + X^+ pair and dramatically (about tenfold) decreases the dissociation energy of the N-Cl bond.

Furthermore, the formation of the HaB complexes and their facile dissociation into succinimide anions and interhalogens represent, in essence, halophilic or S_N2X substitution reactions (in which nucleophiles attack halogen substituents and replace the group originally bonded with this atom). Thus, our results show that even relatively weak halogen bonding via chlorine atoms plays a critical role in these processes.

Notably, the dissociation energy of individual SimCl is higher than that of SimI or SimBr, and the strength of HaB via the Cl atom in SimCl·Y^- complexes is weaker than that in the complexes of SimI and SimBr with chloride. Yet, the dissociation of SimCl·Y^- into Sim^- + ClY pairs requires less energy than that of SimX·Cl^-. This surprising result is attributed to the much more favorable energy changes in the reactions of Cl^+ (formally released by SimCl) with X^- compared to the X^+ + Cl^- reaction. Similar trends are likely observed in the reactions of *N*-halosuccinimides with other nucleophiles. This explains the higher reactivity of SimCl in solutions with various nucleophiles, its efficiency as a halogenating agent, and the difficulties in obtaining halogen-bonded complexes of this halogenated electrophile.

4. Materials and Methods

Commercially available *N*-chlorosuccinimide (SimCl), methyltriphenylphosphonium iodide ($MePh_3PI$), n-tetrapropylammonium iodide (Pr_4NI), methyltriphenylphosphonium bromide ($MePh_3PBr$), hexane, and dichloromethane (all from TCI America), were used without additional purification. Vibrational spectra of were measured with a Perkin-Elmer Frontier FT-IR spectrometer with Universal ATR Sampling Accessory.

Single crystals **1–5** were prepared either by the evaporation of the solutions containing mixtures of reactants at room temperature in dichloromethane or by the diffusion of hexane into dichloromethane solution of the reactants at −73 °C. In particular, to prepare crystals **1**, 27 mg (0.2 mmol) SimCl was dissolved in 5 mL of dichloromethane and 71 mg (0.2 mmol) of $MePh_3PBr$ was dissolved (separately) in 2 mL of dichloromethane, and the solutions were combined. The slow evaporation of the mixture at room temperature resulted in the formation of crystals suitable for the single-crystals X-ray measurements. The X-ray structural analysis showed that they comprised HaB complexes of SimBr and chloride with $MePh_3P^+$ counter-ions. Single crystals **2**, comprising SimI·Cl^- complexes with Pr_4N counter-ions, were prepared in a similar way by the evaporation of a mixture of 27 mg (0.2 mmol) SimCl and 63 mg (0.2 mmol) of Pr_4NI. To prepare single crystals **3**, a dichloromethane solution (5 mL) containing 13 mg (0.1 mmol) of SimCl was cooled down to −78 °C in a Schlenk tube and combined at this temperature with a dichloromethane solution (2 mL) containing 40 mg (0.1 mmol) of $MePh_3PI$. The combined solution was carefully layered with a 1:1 mixture of dichloromethane and hexane and then with hexane. The Schlenk tube was placed in the cool bath and kept at −73 °C. The diffusion of (layered) hexane into the dichloromethane solution of the mixture of SimCl and $MePh_3PI$ produced crystals **3**. A similar diffusion of hexane into a solution containing SimCl and $MePh_3PBr$ at −73 °C resulted in the formation of two types of crystals—(orthorhombic) prisms **4** and (triclinic) plates **5**. Both of them comprised HaB complexes in which chlorine and bromine were disordered between two positions (with somewhat different populations in **4** and **5**). The predominant form in both crystals (with populations of 0.82 and 0.94 in **4** and **5**, respectively) showed bromine atoms bonded to the nitrogen atom of succinimide. The minor fractions of bromine (0.18 and 0.06 in **4** and **5**, respectively) were located at the terminal position of the SimX·Y^- complex. The chlorine showed the opposite distributions between these two sites.

Single crystal structures for **1** and **2** were determined on a Bruker AXS D8 Quest diffractometer with a fixed chi angle, a sealed tube fine focus X-ray tube with Mo Kα radiation (λ = 0.71073 Å), a single crystal curved graphite incident beam monochromator, and a PhotonII area detector. Single crystal structures of **3**, **4**, and **5** were determined on a Bruker AXS D8 Quest 4 circle diffractometer, a microfocus X-ray tube with Cu Kα radiation (λ = 1.54178 Å), and a PhotonIII area detector. Both instruments were equipped with an Oxford Cryosystems low temperature device and examination and data collection of crystals were performed at 150 K. Reflections were indexed and processed, and the files were scaled and corrected for absorption using APEX5 [44]. The space groups were assigned using XPREP within the SHELXTL suite of programs, the structures were solved by dual methods using ShelXT and refined by full-matrix least-squares against F^2 with all reflections using Shelxl2019 using the graphical interface Shelxle [45–48]. Crystals **1–3** were twinned or multi-component, see the SI for details. If not specified otherwise, H atoms attached to carbon and nitrogen atoms were positioned geometrically and constrained to ride on their parent atoms, with C-H bond distances of 1.00, 0.99, and 0.98 Å for aliphatic C-H, CH_2, and CH_3 moieties, respectively. Methyl H atoms were allowed to rotate but not to tip to best fit the experimental electron density. $U_{iso}(H)$ values were set to a multiple of $U_{eq}(C)$ with 1.5 for CH_3, and 1.2 for C-H units, respectively. Crystallographic, data collection and refinement details are listed in Table S1 in the SI.

Geometries of the complexes and their components were optimized without constraints via M06-2X/def2-tzvpp calculations (in dichloromethane, with a polarizable continuum model) using the Gaussian 09 suite of programs [49–52]. Our earlier studies [26,27] showed that such calculations provide good modeling of the HaB complexes, and the results are consistent with the experimental data. The absence of imaginary frequencies confirmed that the optimized structures represent true minima. The binding energies of the halide in the HaB complexes $SimX{\cdot}Y^-$ (i.e., the strength of the $X \cdots Y$ bond) were determined as: $\Delta E = E(SimX{\cdot}Y^-) - E(SimX) - E(Y^-)$, where $E(SimX{\cdot}Y^-)$, $E(SIMX)$, and $E(Y^-)$ are energies, including zero-point energies, of the optimized complexes, SimX and halide, respectively. Since the formation of the complex is accompanied by distortion of the reactants, the binding energy represents a combination of preparation (distortion) energy, E_{prep}, and interaction energy between distorted fragments, ΔE_{int}. The energy scan in Figure 4 was performed using opt = modredundant option in Gaussian 09 starting from the coordinates of the optimized $SimCl{\cdot}I^-$ complex. In these calculations, the N-Cl distance was increased stepwise (starting from 1.685 Å) by 0.1 Å and the complexes were optimized at each N-Cl separation. Initially, the increase in the N-Cl separation was accompanied by the decrease in Cl-I separations from about 3.39 Å to 2.33 Å. During the first several steps, the decrease in Cl-I separations was so significant that the N-I distance also decreased to 4.683 Å (as compared to the N-I distance of 5.076 Å in the initial complex). Once Cl-I bond was fully developed, the N-I distances started to increase. A combination of the N-Cl bond dissociation and the Cl-I bond formation resulted in a small (about 0.3 kcal/mol) bump at steps 7–9. Then, similar calculations were repeated starting from the optimized geometry of the $SimI{\cdot}Cl^-$ complex. In this case, however, the I-Cl bonding in the starting complex was much stronger, so there was no decrease in the N-Cl distances and there was no bump on the energy scan.

Energies, as well as atomic coordinates of the calculated complexes are listed in the ESI. Atomic charges were calculated using the NBO method [38] implemented in the Gaussian 09 suite of programs. QTAIM analyses [39] were performed with Multiwfn [53] using wfn files generated by Gaussian 09. The results were visualized using the molecular graphics program VMD [54]. The rate constants of the halogen rearrangements were estimated via the Eyring (transition-state theory) equation as $k = (kT/h)K^{\neq}$, where k and h are the Boltzmann and Plank constants (transmission coefficient is assumed to be about 1), $K^{\neq} = \exp(-\Delta G^{\neq}/RT)$ [55], and the free energy of the separate $Sim^- + XY$ pairs (which represent the highest-energy point on the reaction pathway, Figure S5 in the Supporting Information) relative to the free energy of $SimX{\cdot}Y^-$ was taken as the activation free (Gibbs) energy, $\Delta G^{\neq}$.

Supplementary Materials: The following supporting information can be downloaded at: https://www.mdpi.com/article/10.3390/molecules30030639/s1, Figure S1. FT-IR spectra of *N*-haloimides and their complexes. Table S1. Selected experimental and calculated vibrational frequencies in individual *N*-haloimides and their complexes; Table S2. Crystallographic, data collection, and refinement details [56,57]. Figures S2–S6. X-ray structures of **1–5**. Table S3. Bond lengths and dissociation energies. Table S4. Calculated free energy changes for $X^+ + Y^-$ reactions. Table S5. Energies, ZPE and (Gibbs) free energies of the optimized complexes and individual SimX molecules. Table S6. Atomic coordinates of the optimized complexes.

Author Contributions: Conceptualization and methodology, S.V.R.; crystallization M.H.; X-ray structural analysis, M.Z., computations, S.V.R.; writing—original draft preparation, S.V.R.; writing—review and editing, M.H., M.Z. and S.V.R.; data curation, visualization, project administration, funding acquisition, S.V.R. All authors have read and agreed to the published version of the manuscript.

Funding: This research was funded by the National Science Foundation, grant number CHE-2404011. Calculations were performed on Ball State University's Beowulf cluster, which is supported by The National Science Foundation (MRI-1726017) and Ball State University. X-ray structural measurements were supported by the National Science Foundation through the Major Research Instrumentation Program under Grant No. CHE-1625543 (funding for the single crystal X-ray diffractometer).

Institutional Review Board Statement: Not applicable.

Informed Consent Statement: Not applicable.

Data Availability Statement: Complete crystallographic data, in CIF format, have been deposited with the Cambridge Crystallographic Data Centre. CCDC 2415447-2415451 contains the supplementary crystallographic data for this paper. These data can be obtained free of charge via www.ccdc.cam.ac.uk/data_request/cif. All computational data are contained within the article or Supplementary Material.

Conflicts of Interest: The authors declare no conflicts of interest.

References

1. Cavallo, G.; Metrangolo, P.; Milani, R.; Pilati, T.; Priimagi, A.; Resnati, G.; Terraneo, G. The halogen bond. *Chem. Rev.* **2016**, *116*, 2478–2601. [CrossRef]
2. Gilday, L.C.; Robinson, S.W.; Barendt, T.A.; Langton, M.J.; Mullaney, B.R.; Beer, P.D. Halogen bonding in supramolecular chemistry. *Chem. Rev.* **2015**, *115*, 7118–7195. [CrossRef]
3. Mukherjee, A.; Tothadi, S.; Desiraju, G.R. Halogen Bonds in Crystal Engineering: Like Hydrogen Bonds yet Different. *Acc. Chem. Res.* **2014**, *47*, 2514–2524. [CrossRef] [PubMed]
4. Metrangolo, P.; Neukirch, H.; Pilati, T.; Resnati, G. Halogen bonding based recognition processes: A world parallel to hydrogen bonding. *Acc. Chem. Res.* **2005**, *38*, 386–395. [CrossRef] [PubMed]
5. Bulfield, D.; Huber, S.M. Halogen Bonding in Organic Synthesis and Organocatalysis. *Chem. Eur. J.* **2016**, *22*, 14434–14450. [CrossRef]
6. Melnyk, N.; Garcia, M.R.; Trujillo, C. Halogen-Bond-Based Organocatalysis Unveiled: Computational Design and Mechanistic Insights. *ACS Catal.* **2023**, *13*, 15505–15515. [CrossRef]
7. Sutar, R.L.; Huber, S.M. Catalysis of organic reactions through halogen bonding. *ACS Catal.* **2019**, *9*, 9622–9639. [CrossRef]
8. Lorpaiboon, W.; Bovonsombat, P. Halogen Bond-Induced Electrophilic Aromatic Halogenations. *Org. Biomol. Chem.* **2021**, *19*, 7518–7534. [CrossRef] [PubMed]
9. Rosokha, S.V.; Vinakos, M.K. Halogen bond-assisted electron transfer reactions of aliphatic bromosubstituted electrophiles. *Phys. Chem. Chem. Phys.* **2014**, *16*, 1809–1813. [CrossRef]
10. Rosokha, S.V.; Traversa, A. From charge transfer to electron transfer in halogen-bonded complexes of electrophilic bromocarbons with halide anions. *Phys. Chem. Chem. Phys.* **2015**, *17*, 4989–4999. [CrossRef] [PubMed]
11. Crugeiras, J.; Ríos, A. Halogen Transfer through Halogen Bonds in Halogen-Bound Ammonia Homodimers. *Phys. Chem. Chem. Phys.* **2016**, *18*, 30961–30971. [CrossRef] [PubMed]
12. Zang, Y. On the role of halogen bond in the halophilic reaction: A theoretical study. *J. Mol. Struct. THEOCHEM* **2010**, *961*, 6–8. [CrossRef]
13. Zefirov, N.S.; Makhon'kov, D.I. X-philic reactions. *Chem. Rev.* **1982**, *82*, 615–624. [CrossRef]
14. Attrell, R.J.; Widdifield, C.M.; Korobkov, I.; Bryce, D.L. Weak Halogen Bonding in Solid Haloanilinium Halides Probed Directly via Chlorine-35, Bromine-81, and Iodine-127 NMR Spectroscopy. *Cryst. Growth Des.* **2012**, *12*, 1641–1653. [CrossRef]

15. Szell, P.M.J.; Bryce, D.L. ^{35}Cl Solid-State NMR and Computational Study of Chlorine Halogen Bond Donors in Single-Component Crystalline Chloronitriles. *J. Phys. Chem. C* **2016**, *120*, 11121–11130. [CrossRef]
16. Widner, D.L.; Knauf, Q.R.; Merucci, M.T.; Fritz, T.R.; Sauer, J.S.; Speetzen, E.D.; Bosch, E.; Bowling, N.P. Intramolecular Halogen Bonding Supported by an Aryldiyne Linker. *J. Org. Chem.* **2014**, *79*, 6269–6278. [CrossRef] [PubMed]
17. Bosch, E.; Ferrence, G.M.; Powell, C.J.; Unruh, D.K.; Krueger, H.R.; Groeneman, R.H. Cooperative Non-Covalent Interactions and Synthetic Feed as Driving Forces to Structural Diversity within Organic Co-Crystals Containing Isosteric Perhalobenzenes. *CrystEngComm* **2022**, *24*, 3841–3845. [CrossRef]
18. Prakash, G.K.S.; Mathew, T.; Hoole, D.; Esteves, P.M.; Wang, Q.; Rasul, G.; Olah, G.A. *N*-Halosuccinimide/BF_3-H_2O, Efficient Electrophilic Halogenating Systems for Aromatics. *J. Am. Chem. Soc.* **2004**, *126*, 15770–15776. [CrossRef] [PubMed]
19. Shi, H.; Zhang, J.; Li, X.; He, J.; Sun, Y.; Wu, J.; Du, Y. Thianthrene/TfOH-Catalyzed Electrophilic Halogenations Using *N*-Halosuccinimides as the Halogen Source. *Chem. Sci.* **2024**, *15*, 13058–13067. [CrossRef] [PubMed]
20. Geng, Y.; Huang, R.; Li, X.; Yu, S. N-Haloimide-enabled halogenation via halogen-bond-assisted C–C activation of alkanols. *Green Chem.* **2023**, *25*, 221–228. [CrossRef]
21. Raatikainen, K.; Rissanen, K. Interaction between amines and *N*-haloimides: A new motif for unprecedentedly short Br··· N and I··· N halogen bonds. *CrystEngComm* **2011**, *13*, 6972–6977. [CrossRef]
22. Raatikainen, K.; Rissanen, K. Breathing molecular crystals: Halogen- and hydrogen-bonded porous molecular crystals with solvent induced adaptation of the nanosized channels. *Chem. Sci.* **2012**, *3*, 1235–1239. [CrossRef]
23. Puttreddy, R.; Jurček, O.; Bhowmik, S.; Mäkelä, T.; Rissanen, K. Very Strong −N–X+··· −O–N+ Halogen Bonds. *Chem. Commun.* **2016**, *52*, 2338–2341. [CrossRef] [PubMed]
24. Makhotkina, O.; Lieffrig, J.; Jeannin, O.; Fourmigué, M.; Aubert, E.; Espinosa, E. Cocrystal or Salt: Solid State-Controlled Iodine Shift in Crystalline Halogen-Bonded Systems. *Cryst. Growth Des.* **2015**, *15*, 3464–3473. [CrossRef]
25. Stilinovic, V.; Horvat, G.; Hrenar, T.; Nemec, V.; Cincic, D. Halogen and Hydrogen Bonding between (*N*-Halogeno)-Succinimides and Pyridine Derivatives in Solution, the Solid State and In Silico. *Chem. Eur. J.* **2017**, *23*, 5244–5257. [CrossRef]
26. Hardin, M.; Zeller, M.; Rosokha, S.V. Complexes of *N*-chloroquinuclidinium with chloride: Strong halogen bonding via chlorine atoms. *CrystEngComm* **2024**, *26*, 6784–6788. [CrossRef]
27. Watson, B.; Grounds, O.; Borley, W.; Rosokha, S.V. Resolving halogen vs hydrogen bonding dichotomy in solutions: Intermolecular complexes of trihalomethanes with halide and pseudohalide anions. *Phys. Chem. Chem. Phys.* **2018**, *20*, 21999–22007. [CrossRef] [PubMed]
28. Ghassemzadeh, M.; Harms, K.; Dehnicke, K.; Magull, J. Mu-2-Chloro Complexes of Succinimide and *N*-Chlorosuccinimide—The Crystal-Structures of PPh_4[Cl(succinimide)$_2$], PPh_4[Cl(*n*-Cl-Succinimide)$_2$] and *N*-Chlorophthalimide. *Z. Naturforsch. B* **1994**, *49*, 506–512. [CrossRef]
29. Finkelstein, M.; Moore, W.; Ross, S.; Eberson, L. Halogen Exchange During Complex-Formation Between an *N*-Halogensuccinimide and a Quaternary Ammonium Halide. *Acta Chem. Scand. B Org. Chem. Biochem.* **1986**, *40*, 402–403. [CrossRef]
30. Vankar, Y.; Kumaravel, G. *N*-Chlorosuccinimide Sodium-Iodide—A Convenient Source of *N*-Iodosuccinimide—Synthesis of Alpha-Iodo Carbonyl-Compounds and Trans-1,2-Iodoacetates. *Tetrahedron Lett.* **1984**, *25*, 233–236. [CrossRef]
31. Einhorn, J.; Einhorn, C.; Ratajczak, F.; Pierre, J.L. Efficient and Highly Selective Oxidation of Primary Alcohols to Aldehydes by *N*-Chlorosuccinimide Mediated by Oxoammonium Salts. *J. Org. Chem.* **1996**, *61*, 7452–7454. [CrossRef] [PubMed]
32. Bovonsombat, P.; Sophanpanichkul, P.; Pandey, A.; Tungsirisurp, S.; Limthavornlit, P.; Chobtumskul, K.; Kuhataparuk, P.; Sathity-atiwat, S.; Teecomegaet, P. Novel Regioselective Aromatic Chlorination via Catalytic Thiourea Activation of *N*-Chlorosuccinimide. *Tetrahedron Lett.* **2015**, *5*, 2193–2196. [CrossRef]
33. Zhang, H.; Wei, Q.; Wei, S.; Qu, J.; Wang, B. Highly Efficient and Practical Thiocyanation of Imidazopyridines Using an *N*-Chlorosuccinimide/NaSCN Combination. *Eur. J. Org. Chem.* **2016**, *2016*, 3373–3379. [CrossRef]
34. Bondi, A. van der Waals volumes and radii. *J. Phys. Chem.* **1964**, *68*, 441–451. [CrossRef]
35. Eraković, M.; Cinčić, D.; Molčanov, K.; Stilinović, V. A Crystallographic Charge Density Study of the Partial Covalent Nature of Strong N···Br Halogen Bonds. *Ang. Chem. Int. Ed.* **2019**, *58*, 15702–15706. [CrossRef] [PubMed]
36. Ghassemzadeh, M.; Harms, K.; Dehnicke, K.; Fenske, D. Mu(2)-Halogeno Complexes of *N*-Bromosuccinimide and *N*-Bromophthalimide—The Crystal-Structures of PPh_4[X(*N*-Bromosuccinimide)$_2$] and PPh_4[X(N-Bromophthalimide)$_2$] with X=Cl and Br. *Z. Naturforsch. B* **1994**, *49*, 593–601. [CrossRef]
37. Padmanabhan, K.; Paul, I.C.; Curtin, D.Y. Structure of *N*-iodosuccinimide. *Acta Cryst.* **1990**, *C46*, 88–92. [CrossRef]
38. Schulz, N.; Sokkar, P.; Engelage, E.; Schindler, S.; Erdelyi, M.; Sanchez-Garcia, E.; Huber, S.M. The Interaction Modes of Haloimidazolium Salts in Solution. *Chem. Eur. J.* **2017**, *24*, 3464–3473. [CrossRef]
39. Weinhold, F.; Landis, C.R. *Discovering Chemistry with Natural Bond Orbitals*; Wiley: Hoboken, NJ, USA, 2012.
40. Bader, R.F.W. A quantum theory of molecular structure and its applications. *Chem. Rev.* **1991**, *91*, 893–928. [CrossRef]

41. Popelier, P.L.A. The QTAIM perspective of chemical bonding. In *The Chemical Bond: Fundamental Aspects of Chemical Bonding*; John Wiley & Sons, Ltd.: Hoboken, NJ, USA, 2014; pp. 271–308.
42. Espinosa, E.; Alkorta, I.; Elguero, J.; Molins, E. From weak to strong interactions: A comprehensive analysis of the topological and energetic properties of the electron density distribution involving X–H···F–Y systems. *J. Chem. Phys.* **2002**, *117*, 5529. [CrossRef]
43. Miller, D.K.; Loy, C.; Rosokha, S.V. Examining a transition from supramolecular halogen bonding to covalent bonds: Topological analysis of electron densities and energies in the complexes of bromosubstituted electrophiles. *ACS Omega* **2021**, *6*, 23588–23597. [CrossRef] [PubMed]
44. Bruker. *Apex5 v2023.9-2, SAINT V8.40B*; Bruker AXS Inc.: Madison, WI, USA, 2023.
45. Bruker AXS Inc. *SHELXTL Suite of Programs, Version 6.14, 2000–2003*; Bruker AXS Inc.: Madison, WI, USA, 2003.
46. Sheldrick, G. Crystal Structure Refinement with SHELXL. *Acta Cryst. C* **2015**, *71*, 3–8. [CrossRef]
47. Sheldrick, G.M. SHELXT—Integrated space-group and crystal-structure determination. *Acta Crystallogr. A* **2015**, *A71*, 3–8. [CrossRef] [PubMed]
48. Hübschle, C.; Sheldrick, G.; Dittrich, B. ShelXle: A Qt Graphical User Interface for SHELXL. *J. Appl. Crystallogr.* **2011**, *44*, 1281. [CrossRef] [PubMed]
49. Frisch, M.J.; Trucks, G.W.; Schlegel, H.B.; Scuseria, G.E.; Robb, M.A.; Cheeseman, J.R.; Scalmani, G.; Barone, V.; Mennucci, B.; Petersson, G.A.; et al. *Gaussian 09, Rev. C.01*; Gaussian, Inc.: Wallingford, CT, USA, 2009.
50. Zhao, Y.; Truhlar, D.G. The M06 suite of density functionals for main group thermochemistry, thermochemical kinetics, noncovalent interactions, excited states, and transition elements: Two new functionals and systematic testing of four M06-class functionals and 12 other functionals. *Theor. Chem. Acc.* **2008**, *120*, 215–241.
51. Tomasi, J.; Mennucci, B.; Cammi, R. Quantum mechanical continuum solvation models. *Chem. Rev.* **2005**, *105*, 2999–3093. [CrossRef] [PubMed]
52. Weigend, F.; Ahlrichs, R. Balanced basis sets of split valence, triple zeta valence and quadruple zeta valence quality for H to Rn: Design and assessment of accuracy. *Phys. Chem. Chem. Phys.* **2005**, *7*, 3297–3305. [CrossRef] [PubMed]
53. Lu, T.; Chen, F. Multiwfn: A Multifunctional Wavefunction Analyzer. *J. Comput. Chem.* **2012**, *33*, 580–592. [CrossRef] [PubMed]
54. Humphrey, W.; Dalke, A.; Schulten, K. VMD—Visual Molecular Dynamics. *J. Mol. Graph.* **1996**, *14*, 33–38. [CrossRef] [PubMed]
55. Atkins, P.; de Paula, J.; Friedman, R. *Physical Chemi: Quanta, Matter and Change*; Oxford University Press: Oxford, UK, 2014; p. 848.
56. Sheldrick, G.M. *CELL_NOW. Version 2008/4*; University of Gottingen: Gottingen, Germany, 2008.
57. Regine, H.-I.; Sheldrick, G.M. Refinement of Twinned Structures with SHELXL97. *Acta Cryst.* **1998**, *B54*, 443–449.

molecules

MDPI

Article

Color Compounds Removal from Tequila Vinasses Using Silica Gel Adsorbents Functionalized with Thiol Moieties: Equilibrium and Kinetics Studies

Carlos Gonzalez-Valerio [1], Alejandro A. Peregrina-Lucano [2], Ricardo Manríquez-González [3], Aida A. Pérez-Fonseca [1], Jorge R. Robledo-Ortíz [3], Ilya G. Shenderovich [4,*] and Sergio Gómez-Salazar [1,*]

[1] Chemical Engineering Department, University of Guadalajara—CUCEI, Blvd. Marcelino García Barragán #1421, esq. Calzada Olímpica, Guadalajara 44430, Jalisco, Mexico; carlos.gonzalez0543@alumnos.udg.mx (C.G.-V.); aida.perez@academicos.udg.mx (A.A.P.-F.)

[2] Pharmacobiology Department, University of Guadalajara—CUCEI, Blvd. Marcelino García Barragán #1421, esq. Calzada Olímpica, Guadalajara 44430, Jalisco, Mexico; aapl69@hotmail.com

[3] Department of Wood, Cellulose, and Paper, University of Guadalajara—CUCEI, Zapopan 45150, Jalisco, Mexico; ricardo.manriquez@academicos.udg.mx (R.M.-G.); jorge.robledo@academicos.udg.mx (J.R.R.-O.)

[4] Faculty of Chemistry and Pharmacy, University of Regensburg, Universitaetstrasse 31, 93053 Regensburg, Germany

* Correspondence: ilya.shenderovich@ur.de (I.G.S.); sergio.gomez@cucei.udg.mx (S.G.-S.)

Abstract: Tequila vinasses are organic wastes generated during ethanol fermentation at elevated temperatures (≥90 °C) and pH ≤ 4.0, making them hazardous to the environment. This paper describes a new, simplified UV–vis spectroscopy-based procedure for monitoring the adsorption of color compounds in tequila vinasses onto silica-based adsorbents, along with an optimized synthesis method to produce the most efficient sol–gel synthesized thiol-functionalized adsorbent. Under optimized conditions, the uptake capacity of this adsorbent reaches 0.8 g g^{-1} in 90 min. Experimental results demonstrate that the adsorbent has a specific affinity for melanoidin-type molecules. The adsorbent demonstrates excellent thermal stability (~316 °C). The results of this work indicate that the adsorbent possesses potential in the treatment of tequila vinasses from wastewater discharges.

Keywords: tequila vinasses; sol–gel; silica gel; adsorbent; adsorption

1. Introduction

Tequila is the most famous Mexican spirituous drink worldwide, produced from the brewed must of an assortment of *Agave tequilana var. Azul*. The production of tequila generates wastes called vinasses. These wastes are generated by the tequila industries from the residual solution of fermented agave sugars after alcohol distillation. It has been reported that about 10–12 L of vinasses are generated by each liter of tequila produced [1]. Tequila vinasses (ca. 80%) are mostly discharged to water bodies and agricultural lands with almost null treatment representing a serious environmental risk [2]. Specifically, tequila vinasses constitute a sort of wastewater possessing high degradation strength and, consequently, a high degree of pollution since they are discharged at elevated temperatures (≥90 °C), pH ≤ 4.0, and very high chemical oxygen demand (COD, 5–150 g L^{-1}) [1–3].

The treatment of tequila vinasses wastes becomes complex due to the contained compound composition and the large amounts generated. The compounds that grant the dark brown color to tequila vinasses are called melanoidins. It is well known that melanoidins are color compounds made up of aromatic heterocyclic structures with ionic moieties that are difficult to degrade by microorganisms; hence, biological treatments, in general, present bad yields, and combined waste treatments must be implemented to improve degradation efficiencies [4]. Amongst the processes used to treat these wastes are physicochemical, coagulation, sand filtration, Fenton oxidation, and membrane filtration [5].

These processes have proved to have high efficiencies for turbidity, and a chemical oxygen demand (COD) removal of up to 70% and 80%. However, they present high operational costs and high demand for reagents, making them less appealing. In contrast, the easy use of functionalized adsorbents makes this process a suitable technology to solve the problem presented by the tequila vinasses discharged into the environment [6]. For instance, there are some reports which indicate that waste vinasses have been used as raw material to obtain activated carbons and applied in dye removal and other components [7,8]; this is attributed to the phenolic and ionic nature present in the carbons. Another example of this methodology was presented by Caqueret et al. [9], who utilized activated carbon to adsorb compounds present in beet vinasses.

Tequila vinasses constitute a complex mixture of water, fermentable sugars, organic acids, esters, alcohols, furans, and phenolic compounds [10]. A current and reliable method of analysis and a quantification of all the components of tequila vinasses is not available yet. As a result, in most tequila vinasses, the unambiguous identification of all the compounds present is far from being a reality. This situation makes it difficult to identify the compounds of tequila vinasses in real samples. The traditional methods of analyzing vinasses use gas chromatography–mass spectrometry (GC-MS) or liquid chromatography (LC) techniques. For example, Lorenzo-Santiago et al. [10] drew upon a set of analytical methods which included ultra-performance liquid chromatography (UPLC) and ultra-performance liquid chromatography mass spectrometry (UPLC-MS) techniques to analyze some compounds (furans and phenolic) and some sugars, respectively, of tequila vinasses from Amatitan, Jalisco, Mexico. However, many tequila vinasses compounds, such as a significant amount of melanoidins, were not analyzed. In another study, Govea-Paz et al. [11] evaluated total sugars, total reducing sugars, and total phenols following the phenol–sulfuric, Miller, and 4-aminoantipyrine standard methods [12,13] when they studied the generation of biofuels using tequila vinasses in a continuous stirred-tank reactor (CSTR) with biomass carriers and an upflow anaerobic sludge blanket reactor (UASB) connected in series. Their results indicated that feeding more concentrated vinasses produced a decrease in hydrogen and methane production, coinciding with a 5 g/L-peak in reducing sugars. These studies indicate the complexity of analytical techniques that must be used to obtain satisfactory results about the tequila vinasses analysis. Based upon these results, in this work, a novel and relatively simple procedure of tequila vinasses analysis is proposed. The procedure involves a full UV-vis scan of tequila vinasses samples whose signals are further related to the drained masses of vinasses to construct the calibration curve of the weight of solids/mL vs. total absorbance. Then, to obtain the corresponding absorbances of all compounds present in the tequila vinasses for each point of the curve, a full UV-vis scan was performed in the range of 350–900 nm, followed by the numerical integration of the whole spectrum using Simpson's 1/3 rule.

To the best of our knowledge, the use of functionalized silica gel adsorbents in treating tequila vinasses wastes has not yet been reported. The novelty of this work includes reporting on the synthesis and application of our newly prepared thiol-functionalized adsorbent in the treatment of tequila vinasses for the first time. Also, the novelty includes the proposal of a new analytical method to assess the tequila vinasses concentrations that is simpler and more robust compared to the conventional ones such as UV-vis or HPLC techniques. So far, no adsorbents of these types have been applied in the treatment of tequila vinasses, nor in the newly proposed analytical method. Using these adsorbents, it is expected to obtain improved removal efficiencies, fast removal kinetics, and substantial improvement in color compounds removal. This work aims to report on the use of a thiol-functionalized silica gel adsorbent to remove tequila vinasses color compounds. A simple analysis procedure of tequila vinasses was proposed, and the adsorption features of real tequila vinasses samples were studied through adsorption equilibrium and kinetic measurements. Various mathematical models were tested with experimental data to elucidate the adsorption mechanism of tequila vinasses on the adsorbent, and the rate-controlling mechanism in the kinetic measurements. Finally, the solid adsorbent was characterized by mass spectrometry, N_2

adsorption, Fourier transform infra-red (FTIR), thermogravimetric analysis (TGA), and solid-state NMR techniques to gain insights concerning the obtained solid features.

2. Results and Discussion

2.1. Synthesis of the MaMPTXX Adsorbents

Several trials were performed to obtain an optimum adsorbent with enhanced characteristics such as minimal water solubility, maximized removal capacity of tequila vinasses compounds, and improved ligand densities. The tests consisted of varying the synthesis parameters, including the molar ratios of the functional precursor (FP) to the cross-linking agent (CLA) and the amount of stabilizing agent (NaCl), as shown in Table 1.

Table 1. Molar ratios used in the synthesis of thiol-functionalized adsorbents for the treatment of tequila vinasses.

Adsorbent	Molar Ratios									TEOS Hydrolysis Time, Min	TEOS/MPTS Co-Condensation Time, Min	Mixture pH	S_{BET}, $m^2\ g^{-1}$	Avg. Pore Diameter D_p, Å	q_{max} [a], $mmol\ g^{-1}$
	TEOS Mixture				MPTS Mixture				TEA						
	TEOS	NaCl	EtOH	H_2O	MPTS	NaCl	EtOH	H_2O							
Ma000	1	0.009	4	4	0	0	0	0	0.09	30	30	3.0	88.0	203.07	–
MaMPT001	1	0.009	4	4	0.25	0	3	1	0.12	30	30	3.0	278.27	56.69	–
MaMPT002	1	0.009	4	4	0.25	0.009	3	1	0.12	30	30	3.0	432.68	47.19	1.01
MaMPT003	1	0.009	4	4	0.5	0	3	1	0.12	30	30	3.0	238.12	39.29	–
MaMPT004	1	0.009	4	4	0.5	0.009	3	1	0.12	30	30	3.0	311.18	34.10	–
MaMPT005	1	0.009	4	4	1	0	3	1	0.12	30	30	3.0	6.44	211.30	–
MaMPT006	1	0.009	4	4	1.5	0	3	1	0.12	30	30	3.0	4.58	104.10	–

[a] Maximum uptake capacity of adsorbent measured by elemental analysis as total sulfur.

2.2. Adsorption Kinetic Results

The amount of tequila vinasses adsorbed on the MaMPT002 adsorbent as a function of time is shown in Figure 1 for the three initial tequila vinasses concentrations studied of 0.0052, 0.0137, and 0.0275 g mL^{-1}. These concentrations were chosen to use an excess of thiol functional groups on MaMPT002 adsorbent and, simultaneously, to ensure adsorbent saturation. The experimental results suggest that it is the initial concentration that defines the adsorption rate, as higher concentrations lead to faster adsorption. From this perspective, the adsorption interaction between tequila vinasses color compounds and the MaMPT002 adsorbent is driven by the propensity of thiol groups to engage in interactions with compounds in the tequila vinasses. In the cases of the two highest concentrations, the equilibrium is attained after ~60 min, whereas in the case of the lowest concentration, the equilibrium is reached after ~140 min. These results reveal that the MaMPT002 adsorbent possesses improved kinetics and makes it appealing for tequila vinasses wastewater treatment applications.

Knowing the adsorption rates of the solid pellets is of paramount importance and integral to the adsorption of tequila vinasses color compounds. Several steps are generally recognized to be involved in the adsorption rate, from the fluid phase to the specific adsorption sites located at the interior of the pellets, and these include the following: (i) interplay with adsorption sites and (ii) diffusion to these sites. Three popular models were tested to ascertain the rate-controlling mechanism of tequila vinasses compounds on the MaMPT002 adsorbent, namely the following: (a) pseudo-first order (PFO), (b) pseudo-second order (PSO), and (c) intraparticle diffusion (IPD) (see Supplementary Materials for model descriptions). The experimental data were tested against each model to extract the corresponding parameters, and the results were used to find the best-suited model to the data. The results of fitting the kinetic models are shown in Figure 1a–d, and the values of the regressed parameters are in Table 2. These parameters demonstrate that the PSO model inadequately simulates the data, as the values of determination coefficients R^2 for the three concentrations are significantly lower than unity, and the maximum adsorption capacities, q_{max}, are not close to the experimental values. The PFO model is observed to match the experimental data better since the values of R^2 are closer to unity and the q_{max} values match the experimental values better. This model indicates that a reversible reaction occurs between tequila vinasses compounds and the thiol surface groups of the adsorbent, and a dynamic equilibrium linking the two phases is set in the adsorption operation. The values of the kinetic parameter of this model (k_1) increase as the tequila vinasses concentration increases due to increased molecular collisions. On the other hand, the PFO and PSO models are not able to predict the mass transfer mechanism caused by diffusion. Because of this, the intraparticle diffusion model (IPD) was tested with kinetic data. A poor fitting between the model and the data can be seen in Figure 1d, which demonstrates that the adsorption of tequila vinasses compounds does not conform to the intraparticle diffusion mechanism. These results demonstrate that the adsorption kinetic data of tequila vinasses more closely follow a PFO model and indicate that the chemisorption is the limiting step of the adsorption rate of tequila vinasses compounds onto the MaMPT002 adsorbent.

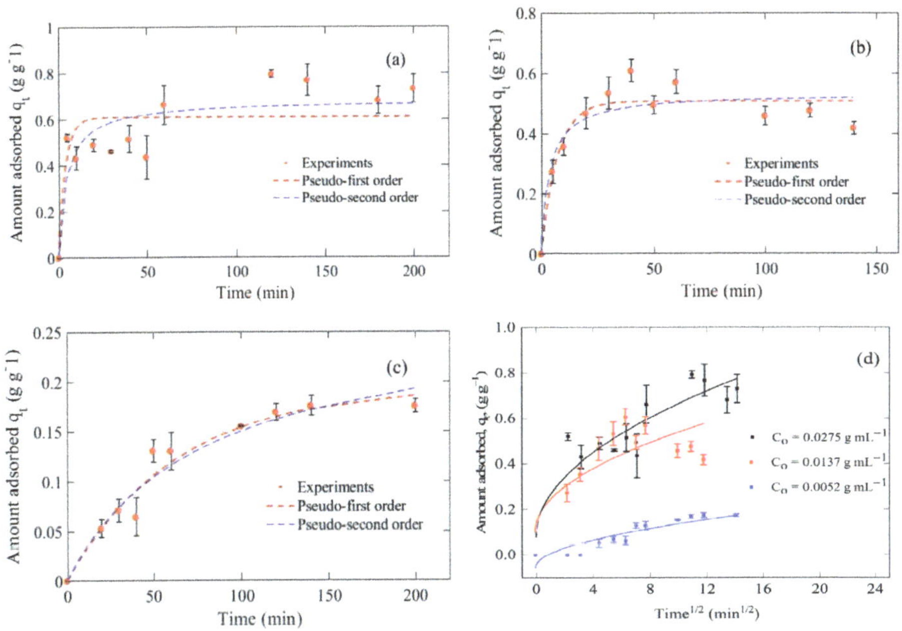

Figure 1. Results of adsorption kinetics of tequila vinasses compounds on the MaMPT002 adsorbent at three different initial conditions and fitting to three kinetic models. (**a**) C_o = 0.0275 g mL^{-1}, (**b**) C_o = 0.0137 g mL^{-1}, (**c**) C_o = 0.0052 g mL^{-1}, (**d**) IPD model; symbols are experimental data, continuous lines are IPD model predictions.

2.3. Adsorption Equilibrium Results

2.3.1. Effect of Tequila Vinasses Solution pH on Adsorption Extent

The capacity of the MaMPT002 adsorbent on the tequila vinasses compounds removal was appraised at three different pH values (3.0, 3.55, and 4.0), distinctive of tequila industry wastewater discharges [14]. Figure 2 shows the tequila vinasses adsorption isotherms on MaMPT002 at 298 K. It is evident from Figure 2a–c that the pH has a slight effect on the tequila vinasses removal capacities as the maximum amounts at all pH levels change only slightly. At pH 3.55, the effective adsorption q_e increases from 0.16 g g^{-1} to 0.801 g g^{-1} as the equilibrium solution concentration C_e changes from 0.0035 g mL^{-1} to 0.026 g mL^{-1}.

Table 2. Results of regressed parameters of the adsorption kinetic models of tequila vinasses compounds on the MaMPT002 adsorbent.

Parameters/Model	Initial Concentration of Tequila Vinasses Compounds, g mL^{-1}		
	0.0275	0.0137	0.0052
$q_{max\ exp}$ (g g^{-1})	0.79	0.47	0.17
		Pseudo-first order	
$q_{max\ calc}$ (g g^{-1})	0.61 ± 0.04	0.51 ± 0.02	0.19 ± 0.02
k_1 (min^{-1})	0.2 ± 0.1	0.14 ± 0.03	0.016 ± 0.003
R^2	0.6474	0.9004	0.9527
		Pseudo-second order	
$q_{e\ calc}$ (g g^{-1})	0.68 ± 0.06	0.53 ± 0.03	0.27 ± 0.04
k_2 (g g^{-1} min^{-1})	0.3 ± 0.2	0.5 ± 0.3	0.05 ± 0.02
R^2	0.7447	0.8587	0.9429
		Intraparticle diffusion	
K_{IPD} (g g^{-1}) × min$^{-1/2}$)	0.18 ± 0.03	0.14 ± 0.03	0.01 ± 0.01
C (g g^{-1})	0.08 ± 0.07	0.10 ± 0.08	−0.06 ± 0.03
R^2	0.8401	0.6476	0.7981

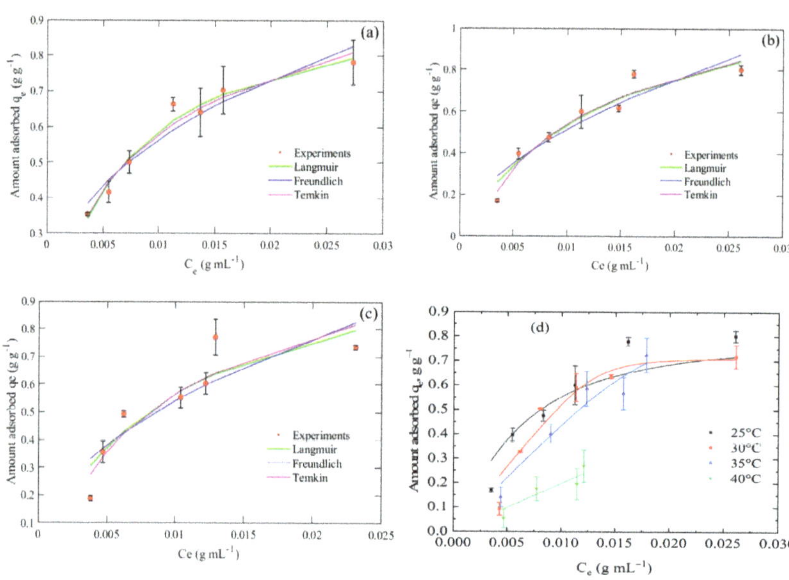

Figure 2. Experimental adsorption isotherms of tequila vinasses on the MaMPT002 adsorbent and simulations using three common isotherm models at three different pH values. (**a**) pH = 3.0, (**b**) pH 03.55, (**c**) pH = 4.0. (**d**) Effect of temperature on tequila vinasses uptake on the MaMPT002 adsorbent at three different temperatures.

An overall picture of the tequila vinasses adsorption mechanism on MaMPT002 has been acquired by the direct application of several adsorption models (see Supplementary Materials). Figure 2a–c show the fitting results of experimental isotherms, and Table 3 depicts the values of the regressed parameters of these models. In all models, the determination coefficient R^2 was used as a criterion to assess the goodness of fit. The Freundlich model is unacceptable because the R^2 values are far from unity and it predicts adsorption capacity values, (q_e), that are not close to the experimental data, especially at the lowest pH value (Figure 2a). In the case of the Langmuir model, despite the highest values of R^2, this model cannot predict satisfactorily the values of q_{max}, also at the lowest pH value (Figure 2a) and is unsatisfactory. In contrast, the Temkin model provides a quite accurate simulation of q_e at all three pH values, with R^2 values close to unity.

Table 3. Fitting results of the adsorption isotherm models to experimental data of tequila vinasses adsorption on the MaMPT002 adsorbent.

Parameters/Model	Initial pH of Tequila Vinasses Solution		
	3.0	3.55	4.0
$q_{max\ exp}$ (g g^{-1})	0.78	0.80	0.77
		Langmuir	
$q_{max\ calc}$ (g g^{-1})	0.99 ± 0.05	1.3 ± 0.2	1.2 ± 0.3
k_1 (min^{-1})	147 ± 20	70 ± 25	95± 46.5
R^2	0.9740	0.9285	0.8430
		Freundlich	
k_F (L^n mg^{-n})	3.2 ± 0.7	6.5 ± 2.9	5.5 ± 3.0
n	2.7 ± 0.3	1.8 ± 0.3	2.0 ± 0.5
R^2	0.9306	0.8808	0.7744
		Temkin	
A_T	1302 ± 297	557 ± 110	667 ± 229
b_t (J mol^{-1})	10948 ± 947	7834 ± 840	8307 ± 1534
R^2	0.9642	0.9461	0.8556

The results of these experiments allowed us to study the interactions between the predominant color compounds present in the tequila vinasses of the melanoidin-type (see Section 2.4), and the thiol functional groups of the MaMPT002 adsorbent. In this regard, it has been reported that the thiol groups can react with melanoidins through a covalent bond via a heterocycle that involves redox reactions through Maillard-derived pyrazinium compounds formed as byproducts of the oxidation of the pyrazinium radical cations [15]. A possible tequila vinasses uptake mechanism by the MaMPT002 adsorbent is shown in Scheme 1. In this mechanism two paths are proposed to occur, (a) in the presence of the nucleophilic thiols of the MaMPT002, melanoidins react to form the melanoidin-thioether bond (undefined position), (b) in a previous step, melanoidins react with water to form pyrazinium compound (such as the 2-hydroxy1,4-dihydropyrazine) which then react with the thiol groups of MaMPT002 to form the corresponding melanoidin-thioether bond via pirazinum ring.

2.3.2. Comparison of Different Effluent Adsorption Performance of the MaMPT002 Adsorbent with Different Adsorbents

For comparison, Table 4 shows the adsorption capacities reported in the literature of the selected adsorbents applied in removing several types of vinasses based on the maximum amount of vinasse removed for each adsorbent (q_{max}) and their specific surface areas. It is noticed that the two activated carbons (Picachem 150 and Picachem 120 PN) performed better than the bagasse fly ash but worse than both the peanut shell and the MaMPT002 adsorbent. Despite their higher specific surface areas, low vinasse compounds removal performance is also notorious in these carbons. These values suggest that the surface area is not a determining factor in the quantity of vinasse removed and does not confer high uptake efficiency.

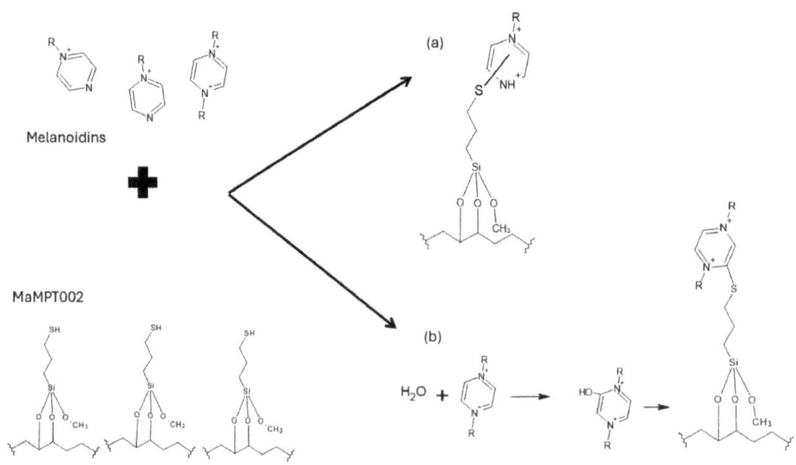

Scheme 1. Proposed adsorption mechanism scenarios of color compounds (e.g., melanoidins) from tequila vinasses on the MaMPT002 adsorbent: (a) undefined thioether bond position in pirazinium, and (b) by the thioether bond in 2 ring position [15].

Alternatively, the MaMPT002 adsorbent possesses a high specific surface area (432.68 $m^2\ g^{-1}$), characteristic of mesoporous silica gel adsorbents synthesized by the sol–gel method. Moreover, the MaMPT002 adsorbent shows a similar vinasse removal capacity but with a simpler synthetic procedure than a peanut shell adsorbent. Finally, the thiol moiety in the MaMPT002 adsorbent confers a versatile alternative in the removal of other complex systems, including several heavy metal ions such as mercury [16,17], and lead [18]. The use of the sol–gel method for the synthesis of silica-based adsorbents allows obtaining materials with a predetermined large pore diameter and a significant internal surface area. Even pure silicon materials of this type significantly affect the mobility of adsorbed molecules due to various non-covalent interactions and hydrogen bonding [19]. Although the energy of such non-covalent interactions is low and they are unstable under ambient conditions, they still determine the average conformation of the partners [20]. Moreover, the adsorption capacity of such materials can be selectively and adaptively significantly increased by the chemical functionalization of their internal surfaces [21–23].

Table 4. Comparison of vinasse uptake using adsorbent MaMPT002 with different materials.

Adsorbent	Effluent Removed	S_{BET} ($m^2\ g^{-1}$)	q_{max} ($g\ g^{-1}$)	Reference
Picachem 150	Sugar beet vinasse	671.0	0.041	[9]
Picachem 120 PN	Sugar beet vinasse	634	0.1	[9]
Peanut shell	Sugarcane vinasse	40.71	0.562, 0.796	[24]
Bagasse fly ash	Sugarcane vinasse	–	0.017	[25]
MaMPT002	Tequila vinasse	432.68	0.801	This work

2.3.3. Effect of Tequila Vinasses Temperature on Adsorption Extent

The effect of tequila vinasses temperature on the adsorption extent of the MaMPT002 adsorbent was investigated at three temperatures (25, 30, and 40 °C) and a pH of 3.55 using optimal conditions. It is apparent from Figure 2d that the removal capacity at 25–35 °C undergoes minor changes; however, it decreased as the temperature increased to 40 °C. The decrease in removal capacity demonstrates that the nature of the adsorption process was exothermic and chemical [26].

The thermodynamic variables standard free energy change (ΔG^o), standard enthalpy change (ΔH^0), and standard entropy change (ΔS^o) for the adsorption of tequila vinasses compounds on the MaMPT002 adsorbent, were determined by using the Van't Hoff equation in the non-linear form as follows:

$$K_c = e^{[\frac{\Delta S^o}{R} - (\frac{\Delta H^o}{R})\frac{1}{T}]} \quad (1)$$

By the third principle of thermodynamics, ΔG^o can be calculated by the following:

$$\Delta G^o = \Delta H^o - T\,\Delta S^o \quad (2)$$

where R is the universal gas constant (8.314 J K^{-1} mol^{-1}), T is the absolute temperature, and K_c is the dimensionless thermodynamic equilibrium constant. The dimensionless form of K_c was obtained from the following definition [27]:

$$K_c = \frac{1000 \times K_t \times \text{molecular weight of tequila vinasses} \times \text{tequila vinasses concentration}}{\gamma} \quad (3)$$

where γ = tequila vinasses activity coefficient = 1 for a diluted solution; K_t represents the best isotherm model fitted constant (e.g., K_t, the Toth equilibrium constant), determined for each dataset corresponding to each temperature. Equation (1) was used to obtain the thermodynamic parameters ΔH^o and ΔS^o by non-linear regression of the data using the Levenberg–Marquardt algorithm. All calculations were conducted using the Origin V9.8 2021® software.

Figure 3 shows the results of non-linear fitting where the values of the Toth model constants, K_t, were used for the calculations since this model was the one that best predicted the data for all temperatures. Table 5 shows the values of the thermodynamic variables obtained from Equations (1) and (2). It is evident from this figure that as the temperature increases, the thermodynamic equilibrium constant (K_c) decreases, thus confirming that the adsorption process of tequila vinasses compounds is exothermic [26,27]. This is also confirmed by the negative sign of ΔH^o, which indicates an exothermic process. The magnitude of −89.18 kJ mol^{-1} suggests that a chemical adsorption process occurs between the tequila vinasses compounds and the surface of the MaMPT002 adsorbent. The negative value of ΔG^o is characteristic of a spontaneous and favorable adsorption process. A decrease in randomness at the liquid–solid interface takes place since ΔS^o shows a negative value. This reveals that the adsorption of tequila vinasses compounds occurs from a disordered phase to an ordered phase on the adsorbent surface. Consequently, the adsorption process of tequila vinasses compounds generates a more orderly rearrangement on MaMPT002 and the loss of the degree of freedom for tequila vinasses compound molecules. The slight entropy loss depicted in Table 5 suggests that the molecules of tequila vinasses compounds were not entirely adsorbed, but they were confined in the liquid contained in the pore spaces in an analogous manner when gas molecules, such as CO_2, are adsorbed on coals [28].

Table 5. Thermodynamic parameters obtained by the non-linear regression of Equation (4).

T (K)	K_c (mL g^{-1})	K_c	ΔS° (kJ/mol)	ΔH° (kJ/mol)	ΔG° (kJ/mol)
–	–	–	−0.266	−89.18	–
298	121.64	55.48	–	–	−10.02
303	75.27	34.33	–	–	−8.69
308	51.94	23.69	–	–	−7.36
313	0.35	0.16	–	–	−6.04

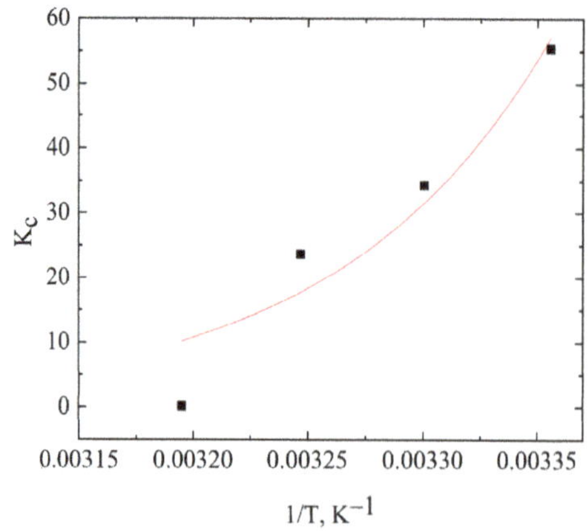

Figure 3. K_c as a function of the inverse of temperature and the non-linear fitting to Equation (4).

2.4. Approximate Selectivity of the MaMPT002 Adsorbent for Different Tequila Vinasse Compounds

Chromatograms in Figure 4 indicate a significant intensity decrease in a peak at a time of ~6.4 min after adsorption (indicated by the circle in this figure). This is indicative of the adsorption of compounds present in the tequila vinasses. The mass spectra of Figure 4 indicate a significant decrease in the intensity of most of the signals after adsorption (Figure 4c), except for the signal at ~680 m/z. This is because the mass spectrometer does not use a standard, but instead it uses the most abundant molecule or fragment as a reference, and the intensity scale is organized around the most abundant fragment with respect to the rest of the fragments. This suggests that the adsorbent shows little selectivity for these molecules. It is worth noting that the complexity of the sample and the large variety of possible molecules corresponding to these masses limit the presentation of one proposal only based on their fragmentations and reported structures. The mass spectrum of Figure 4b shows that the peak of the chromatogram at ~6.4 min corresponds to molecules of 412.31 m/z. The removal of these molecules is further confirmed by a signal decrease in the mass spectrum after adsorption (Figure 4c). On the other hand, to gain insights about the identity of the molecules being adsorbed, the fragmentation pattern corresponding to the molecules with 412.31 m/z was used (Table 6). The fragments with a greater probability of coinciding with this mass are shown in Figure 5. According to the results of the software Optimizer, the mass of 412.31 coincides with a type of molecule called melanoidin. This proposal is based on reports by Hofman and Schieberle [15], which included melanoidin structures capable of interacting with thiol groups. Melanoidins are compounds generated by the Maillard reaction which grants a dark brown color to tequila vinasses. Melanoidins show high chelation capacities toward different metal ions and organic compounds. This poses an environmental problem since these properties make them hazardous when discharged to water bodies and fertile lands because they can chelate metallic ions needed for the flora and fauna of the ecosystem.

Table 6. Fragmentation pattern result of the molecular ion with 412.31 m/z from the Optimizer software of the adsorption process of tequila vinasses on the MaMPT002 adsorbent.

The Final Product Ion, m/z	Abundance
227.1	92341
165.1	13044
121.1	27677
57.2	18350

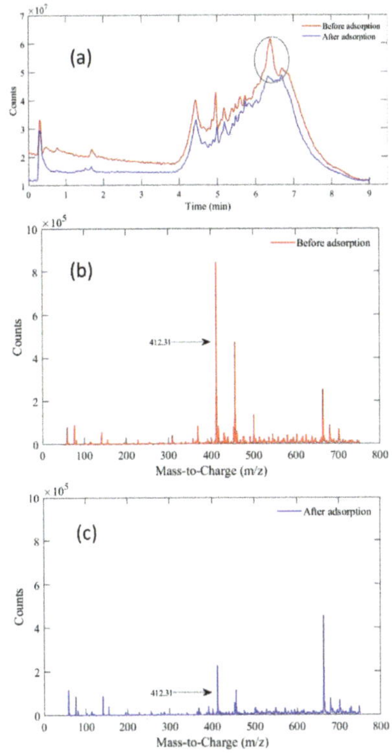

Figure 4. Selectivity results of tequila vinasses adsorption on the MaMPTS002 adsorbent. (**a**) HPLC chromatograms before and after adsorption; (**b**) mass spectrum before adsorption; (**c**) mass spectrum after adsorption.

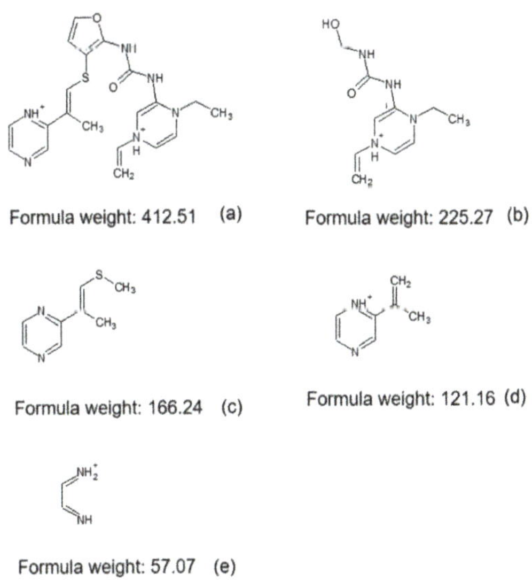

Figure 5. Fragmentation proposal of product ion with 412.31 m/z (based on the closest structure reported of 412.51 m/z) obtained from the Optimizer software of the adsorption of tequila vinasses on the MaMPT002 adsorbent.

2.5. Physicochemical and Spectroscopic Characterization Results

2.5.1. Textural Properties

Figure 6 shows the N_2 adsorption isotherms of the synthesized adsorbents. The isotherms of samples MaMPT001-MaMPT004 (Figure 6a) are of the type IVa, conforming to the IUPAC classification [29], and typical of mesoporous materials. In this case, capillary condensation is accompanied by a hysteresis loop that resembles the H2b type more according to this classification, which occurs when the pore width exceeds a given critical size. This type of hysteresis is characteristic of pores with ink-bottle geometries and is often presented by mesocellular silica foams and a few mesoporous ordered silicas obtained after thermal treatment. Some authors have reported that the hysteresis loops start appearing at pore widths of 4 nm [29,30]. This is confirmed in Figure 6b, where the average pore sizes are ~4.2 nm. Samples MaMPT000, MaMPT005, and MaMPT006 exhibit type II isotherms (Figure 6c) and are typical of the physisorption processes on nonporous or macroporous adsorbents. As can be seen from Table 1, these materials presented the lowest specific surface areas. Regarding the pore size distributions, in the case of samples MaMPT001-MaMPT004 (Figure 6b), they depict narrow distributions with a monomodal shape in the mesopore zone of ~2.0–5.0 nm due to the presence of the hysteresis loop observed in the isotherm; a broad part of the distribution is also observed in the region of macropores for these samples. These results imply that adsorbents possessing structures with uniform pores were formed. The sharp slope of the desorption branch of the isotherm proves the uniformity of these pore structures. The opposite situation is obtained in the samples MaMPT000, MaMPT005, and MaMPT006 (Figure 6d), where broad distributions are depicted in the range of pore sizes of ~5.0–145 nm, thus covering a significant amount of the macropores region. A small amount of mesopores is observed in the case of samples MaMPT005 and MaMPT006 represented by a narrow part of the distribution in the zone of ~2.6–4.0 nm.

Different parameters were investigated during the synthesis of adsorbents that can significantly impact the generation of specific surface area, and in the tequila vinasses compounds uptake capacity. These included the following: (1) TEOS:MPTS molar ratios, (2) amount of NaCl, and (3) presence/absence of functionalizing compound (see Table 1). One sample was prepared as described in Section 2.2 but without MPTS. The purpose of synthesizing this sample was to take it as a reference point to compare the effect of the presence of the MPTS molecule on the specific surface area and the tequila vinasse compounds uptake capacities of the adsorbent. A positive effect of obtaining an increased specific surface area can be observed in this table. Also, it is observed that the TEOS:MPTS molar ratio plays an important role in forming specific surface area. In this sense, high TEOS:MPTS molar ratios produce high specific areas, and low TEOS:MPTS molar ratios produce small specific areas. This can be due to the existence of more and greater functional group oligomers, which generate larger cavities around ligand functional groups, which, in turn, contribute to a higher specific surface area (Figure 6e). Another parameter studied was the presence of the stabilizing agent NaCl. The adsorbents MaMPT001 (278.27 $m^2\ g^{-1}$) and MaMPT002 (432.68 $m^2\ g^{-1}$) differ by the presence of NaCl and present a difference in specific surface areas of 154.56 $m^2\ g^{-1}$. Murakata et al. [31] report that the presence of salts significantly affects the formation of the surface area. Some authors attribute this phenomenon to the fact that when a solid surface becomes in contact with an aqueous phase, the surface becomes electrically charged due to the ionization or dissociation of functional groups present on the surface, by the adsorption of ionic groups on the uncharged surface, or the electrochemical polarization. In the case of silanol groups under the presence of water, these groups can be dissociated to form $Si - O^-$ in alkaline media or $Si - OH_2^-$ in acidic media. The thickness of the electrical double layer is affected by the electrolytes present and can alter the pore size distribution.

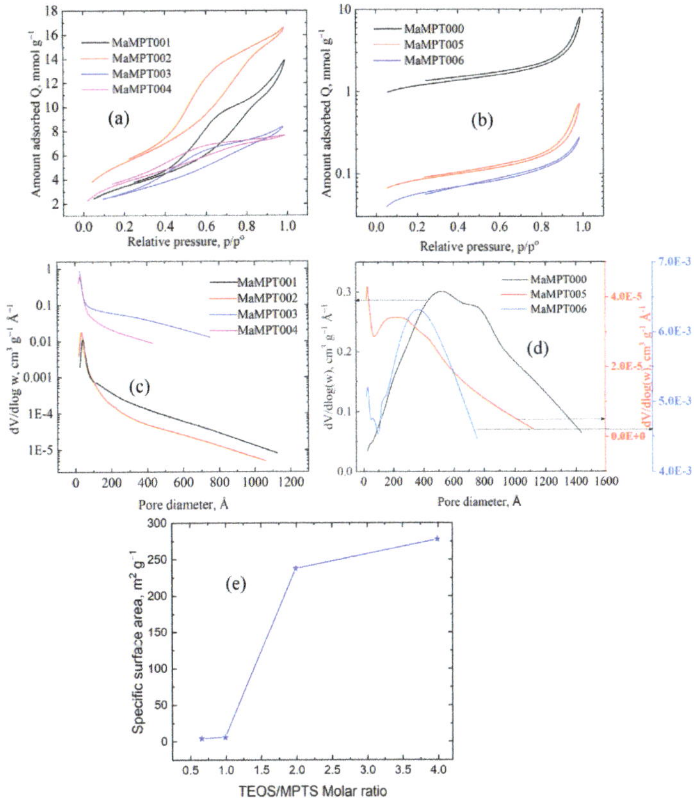

Figure 6. Nitrogen adsorption isotherms and pore size distributions of synthesized adsorbents. (**a**) Isotherms of samples MaMPT001 through MaMPT004; (**b**) isotherms of samples MaMPT000, MaMPT005, and MaMPT006; (**c**) pore size distributions of samples MaMPT001 through MaMPT004; (**d**) pore size distributions of samples MaMPT000, MaMPT005, and MaMPT00, (**e**) TEOS:MPTS molar ratio effect on specific surface area.

2.5.2. TGA Results

Both thermal stability and thermal changes taking place inside of the MaMPT002 adsorbent were monitored by TGA before and after tequila vinasses removal. According to Almaghrabi et al. [32], the silica matrices present a thermal stability of up to 600 °C. Below 200 °C, a decrease in weight % is observed and associated with the loss of physically adsorbed water (Figure 7a). Above 250 °C, the observed weight % loss is associated with the decomposition of the aliphatic chains of MPTS [33]. A steep slope begins at ~316 °C attributed to the depletion of thiol components [34], constituting about 9.5% of the total mass (corresponding to 3.23% of total sulfur). This result indicates that MaMPT002 adsorbent is thermally stable up to 316 °C. On the other hand, after the tequila vinasses compounds uptake, Figure 7b depicts almost identical weight % loss patterns as before the adsorption of tequila vinasses. In this case, only minor changes are observed in the amount of weight lost and are associated with the adsorbed tequila vinasses. By comparing both graphs, it can be concluded that the thermal stability of the MaMPT002 adsorbent is not affected by the presence of tequila vinasses.

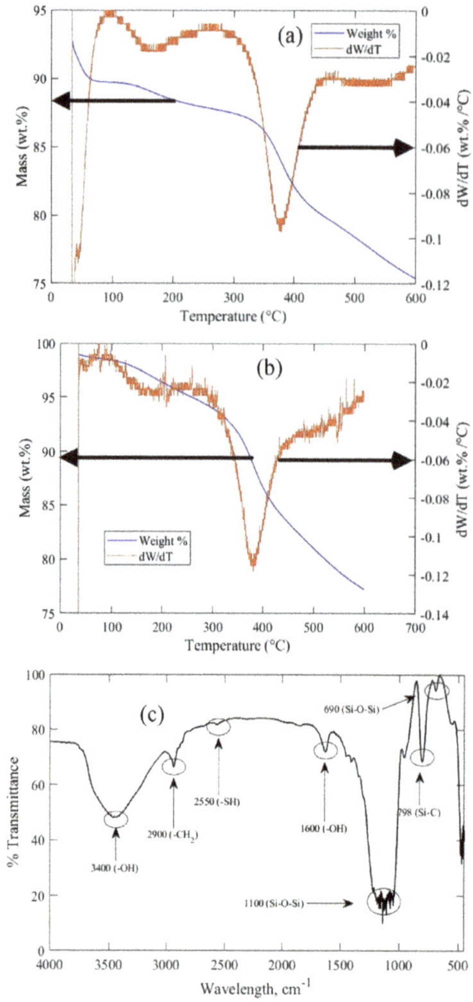

Figure 7. (**a**) TGA results on sample MaMPT002 before tequila vinasses adsorption. (**b**) TGA results on sample MaMPT002 after tequila vinasses adsorption. (**c**) FTIR spectrum of the MaMPT002 adsorbent.

2.5.3. FTIR Study Results

Figure 7c depicts the results of the FTIR analysis of the MaMPT002 adsorbent. Two bands appear at 3400 cm^{-1} and 1600 cm^{-1} and correspond to the stretching and bending of the -OH groups of silanol groups and adsorbed H_2O, respectively. The signal at 1100 cm^{-1} is due to the stretching vibration of siloxane groups Si-O-Si present in the silica matrix. The bands at 2900, 1400, and 690 cm^{-1} correspond to asymmetric stretching, scissoring movement, and bending of the methylene group CH_2, respectively, of the MPTS molecule. A signal appears at 798 cm^{-1} corresponding to the Si-C bonds. The functional thiol group (S-H) presents a weak stretching vibration signal at 2550 cm^{-1}, and finally, the signal observed at 798 cm^{-1} corresponds to the Si-C bonds.

2.5.4. Scanning Electron Microscopy (SEM)

The structural and morphological changes arising from contacting a tequila vinasses solution with the MaMPT002 adsorbent were investigated by the SEM technique. The results indicate that the surface of the adsorbent comprises a clean and even collection of particles (Figure 8a) before the contact with tequila vinasses. A significant change in the surface is observed after tequila vinasses adsorption, where the adsorbent presents a wrinkled and rugged clustered surface with some flat areas (Figure 8b).

Figure 8. SEM results of (**a**) tequila vinasses-free MaMPT002 adsorbent, and (**b**) MaMPT002 adsorbent after tequila vinasses adsorption.

2.5.5. NMR Results

The propyl thiol-grafted moieties, and the silicon species attached and composing the silica matrix of the MaMPT002 adsorbent, were measured by ^{13}C and ^{29}Si solid–state NMR spectroscopy, respectively, as shown in Figure 9. In Figure 9a the zone around 12 ppm depicts a broadened signal of two overlapping peaks associated with the carbon C-3 which presents different chemical environments to those of the ligand. Carbons C-1 and C-2 of the MPTS chain are identified by a signal at 28 ppm, and the one corresponding to carbon C-4 of the unreacted and non-hydrolyzed methoxyl groups and present on the matrix silica surface, appears at 45 ppm [18]. On the other hand, in the case of the ^{29}Si spectrum (Figure 9b), two main regions of the spectrum are clearly identified, one at −111 ppm and the other one at −63 ppm which correspond, in general, to the Q^m and T^m species. Particularly, the signals shown in the zone of −111 ppm correspond to the Q^4 species, and the one at −101 ppm is attributed to a Q^3 species [18]. This last signal confirms the presence of free silanol groups, SiOH. In addition, the absence of signals at −92 ppm demonstrates the absence of geminal silanols (Si $(OH)_2$) in the adsorbent. Additionally, the signals located at −65 ppm correspond to the T^3 species, whereas the signal at −57 ppm is assigned to the T^2 species. These results confirm the presence of the ligand groups MPTS to the silica matrix either by two or three covalent bonds, as shown in the schemes of Figure 8a. Furthermore, these results demonstrate that the surface density of the thiol functional groups is higher than that of the residual Q^3 species, Figure 8b.

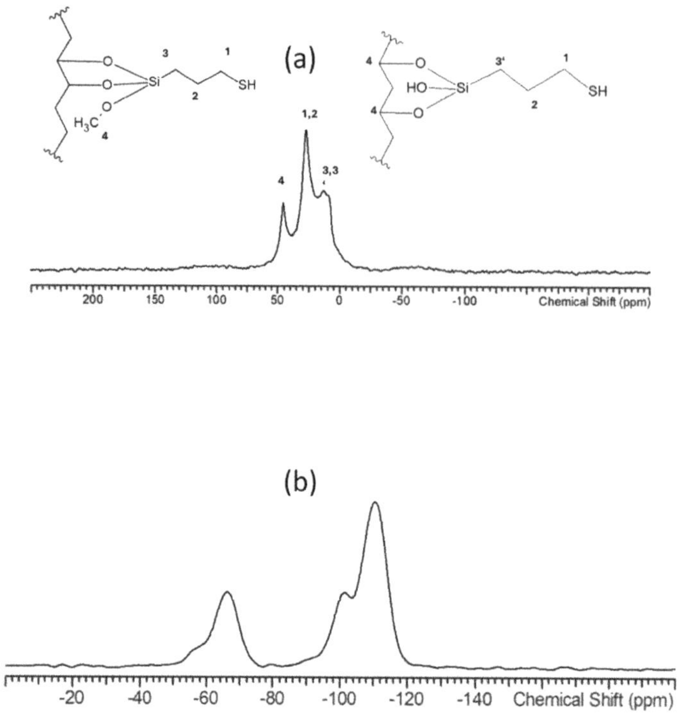

Figure 9. (**a**) ^{13}C CP-MAS NMR spectrum of MaMPT002. (**b**) ^{29}Si MAS NMR spectrum of MaMPT002.

3. Materials and Methods

3.1. Materials

(3-Mercaptopropyl) trimethoxysilane (MPTS) $\geq$ 99% was used as a functional precursor (FP), and tetraethyl orthosilicate (TEOS) $\geq$ 99% as a cross-linking agent (CLA); these reagents, sodium chloride, and triethylamine (TEA) $\geq$99.5%, were all purchased from Sigma Aldrich (Toluca, Mexico). Hydrochloric acid $\geq$ 70% was purchased from Fermont (Monterrey, Mexico); tequila vinasses were collected from a cooling reservoir in a tequila facility at Amatitan, Jalisco, Mexico.

3.2. Synthesis of the Optimum Adsorbent

A set of adsorbents was prepared according to a method described by Lee et al. [35], using the sol–gel modified process. In a typical synthesis, the FP and CLA were hydrolyzed and partially homocondensed by mixing each separately with deionized H_2O, NaCl, and EtOH. All reactions were performed at a pH of 3 adjusted with 3.0 M of HCl dropwise (used as a catalyst). The FP and CLA solutions were stirred for 30 min at room temperature. Afterward, the FP and CLA solutions were mixed and stirred for another 30 min to allow the co-condensation of the silanes. TEA was added to the mixture to induce gelation of the system. The obtained gel was aged for 24 h at room temperature and dried in an oven at 60 °C for 24 h. After that, the obtained material was subjected to acetone/water washings at 60 °C. A drying process was repeated at 60 °C for 24 h. Finally, the adsorbents were ground and sieved to a particle size of 125–180 μm to minimize the external mass transfer resistances, as reported elsewhere [36,37]. Different molar ratios of FP:CLA were studied during the synthesis to select the optimum hybrid material from the set of adsorbents, which had the largest specific surface area and the highest tequila vinasses removal capacity. The optimum adsorbent was used for full characterization. The adsorbents were identified as MaMPTXX where XX corresponds to the test number conducted at the different synthesis conditions. In a preliminary experiment, the seven adsorbents synthesized were assessed for the maximum tequila vinasses adsorption from a 0.033 mg mL^{-1} tequila vinasses stock solution. The results of this experiment indicated that the highest uptake of 0.80 g g^{-1} was obtained from the MaMPPT002 adsorbent, which can be attributed to its high specific surface area (Table 1). This adsorbent was used for the subsequent characterization.

3.3. *Analysis Procedure of Tequila Vinasses Color Solution*

The colored solutions have been studied using several analyses based on the Lambert–Beer law. It is well known that because tequila vinasses are a complex mixture of chemical compounds, they present complicated UV-vis absorption profiles due to the set of different molar extinction coefficients, thus displaying several peaks within a wide range of wavelengths. This presents a problem in assessing the accurate concentration of tequila vinasses for a given component. To overcome this problem, in this work, a procedure is proposed whereby the integration of the areas under the curve of UV-vis signals (using a Thermo scientific Genesys 150 UV–visible spectrophotometer, Waltham, Massachusetts, USA) of a full scan spectrum of wavelengths between 350 nm and 900 nm, were related to the drained mass of the solution using different vinasse concentrations. A calibration curve of the weight of the solids/mL vs. total absorbance was constructed by first placing an accurately measured fixed volume of tequila vinasses (2 mL $\pm$ 0.01 mL) in two sets of 4 ceramic crucibles, each at different vinasse:water volume ratios (2:0; 1.5:0.5; 1:1; 0.5:1.5; 0:2 mL of tequila vinasses mL^{-1} of water). The first set of crucibles was dried until constant weight in a vacuum oven at 50 °C. After that, the total dry weight was registered. This weight was then divided by the original volume of tequila vinasses used so that the amount of the solids of vinasses per mL was obtained and expressed in g of solids mL^{-1} of vinasses. For each crucible of the second set, a full UV-vis scan was performed in the range of 350–900 nm. The whole spectrum was numerically integrated using Simpson's 1/3 rule (see Table S1 of the Supplementary Materials) to obtain the corresponding absorbance of all compounds present in the tequila vinasses for each point of the calibration curve. In the case of all adsorption experiments (kinetics, and isotherms fixed bed), the previously diluted samples were analyzed by performing the full scan in the UV-vis to find the corresponding total absorbance and referred to the calibration curve to determine the concentration.

3.4. *Kinetics of Vinasses Compounds Adsorption Experiments*

Prior to the experiments, the solution of tequila vinasses was filtered and centrifuged to remove the suspended solids and other components. Three different tequila vinasses concentrations (0.0052, 0.0137, and 0.0275 g mL^{-1}) were used to study the effect of the initial solution concentration on the adsorption rate. In total, 20 mg of the MaMPT002 adsorbent and 2 mL of the vinasses solution at a pH of 3.55 were contacted in Eppendorf tubes at varying times (5–200 min) and stirred at 25 °C in a temperature-controlled bath. After adsorption, the final suspensions were filtered, and the collected solutions (final solutions) were measured for tequila vinasses concentrations. Three replicate experiments were conducted for each curve. The amount of tequila vinasses compounds adsorbed was calculated by Equation (4):

$$q = \frac{V\left(C_i - C_f\right)}{m} \tag{4}$$

where q (g g^{-1}) is the amount of tequila vinasses compounds adsorbed per mass unit of adsorbent at a given time t, C_i and C_f (g L^{-1}) are the initial and final solution of tequila vinasses concentrations, respectively; V (L) is the solution volume, and m (g) is the mass of adsorbent.

3.5. *Vinasses Compounds Adsorption Isotherm Experiments*

The adsorption equilibrium of tequila vinasses on the MaMPT002 adsorbent was investigated using the batch mode to evaluate the maximum uptake capacity of adsorbent at three different pH values (3.0, 3.55, and 4.0) common of the vinasse discharges from tequila industries [17]. Tequila vinasses solutions at varying initial concentrations (0.031–0.097 g mL^{-1}) and fixed initial pH were placed in contact in Eppendorf tubes with 20 mg of adsorbent and stirred for 90 min in the temperature-controlled bath at 25 °C. The suspensions were filtered, and the initial and final tequila vinasses solution concentrations were measured. The amount of vinasses compounds adsorbed were calculated by Equation (1). The values reported were the averages of three replicate experiments.

3.6. *Selectivity of MaMPT002 for Tequila Vinasses Compounds*

Tequila vinasses are a complex mix of more than two hundred diverse types of compounds (e.g., monosaccharides, VFA, alcohols, minerals, melanoidins, among other compounds) [38]. Therefore, due to the difficulty of identifying and quantifying each compound that is adsorbed on the surface of the adsorbent, an approximate qualitative measurement was implemented in this work. To this aim, a tequila vinasses solution (10 mL, initial solution) was contacted with the MaMPT002 adsorbent (5 g) for 24 h at 25 °C to assess the selectivity of the adsorbent for the different tequila vinasses compounds. The resulting suspension was filtered, and the collected solution was called the final solution. Then, the initial and final solutions were diluted 1/100 and analyzed by HPLC-MS/MS. By observing the differences in the presence/absence of the characteristic signals of the spectra and chromatograms before and after tequila vinasses adsorption, one can gain preliminary insights about the type of compound(s) being removed and qualitatively assess the affinity/selectivity of the adsorbent for a group of compounds interacting with the solid. An Agilent Technologies 1200® instrument coupled to a 6430B Mass Spectrometer was used. A C18 Zorbax Eclipse XDB column (50 mm × 2.1 mm × 3.5 μm) was used. The mobile phase was 0.1% v/v of formic acid in water and acetonitrile gradient, starting with a ratio of 90% water and 10% acetonitrile, which after 5 min ended up in a ratio of 10% water and 90% acetonitrile, at a rate flow of 0.5 mL min^{-1}. The fragmentation of the molecule of interest was identified with the aid of software Optimizer V9.09 from Agilent Technologies Company (Santa Clara, CA, USA).

3.7. *Physicochemical and Spectroscopic Characterization of the Adsorbent*

3.7.1. Textural Properties of Adsorbent

Textural properties of the MaMPTXX adsorbents were measured by N_2 adsorption–desorption experiments at 77 K using a N_2 adsorption equipment ASAP 2020 KMP Micromeritics (Norcross, GA, USA). Preceding the measurements, about 0.2 g of the samples were conditioned at 423 K and a vacuum of 10 μmHg for 12 h to detach all possible species adsorbed on the surface of the adsorbent, such as water and CO_2 and all that could interfere with the measurement. The specific surface area was determined by the Brunauer–Emmett–Teller (BET) method using adsorption data in the relative pressure range of $0 < P/P_o < 0.3$, the total pore volume was determined by adsorption at $P/P_o = 0.995$, and the pore size distribution was obtained by the Barrett–Joyner–Halenda (BJH) method.

3.7.2. Thermogravimetric Analysis (TGA)

The thermal stability of the MaMPT002 adsorbent was studied by TGA. Dried samples were analyzed using a Discovery model analyzer from TA Instruments (Amarillo, TX, USA) using dried samples. About 10–15 mg of the sample was placed in a platinum pan; a temperature scanning range of 25–600 °C was used; the heating rate was 10 °C min^{-1} under a nitrogen atmosphere at 60 mL min^{-1}.

3.7.3. Fourier Transform Infrared Measurements (FTIR)

The thiol functional groups on the surface of the adsorbent were investigated by FTIR using the KBr method. A KBr salt and an amount of adsorbent were dried at 333 °K, mixed, and compressed to obtain a KBr pressed disc. FTIR spectra were obtained with a FTIR Spectrum One Perkin Elmer instrument equipped with an IR source and a KBr splitter. The spectra of the samples were obtained using 64 scans from 4000 to 500 cm^{-1} and a resolution of 4 cm^{-1}.

3.7.4. Scanning Electron Microscopy (SEM) Study

Changes in the textural properties and morphology of the MaMPT002 adsorbent before and after tequila vinasses adsorption were studied using the SEM technique. A JEOL JSM6610LV microscope (Akishima, Tokyo, Japan) was employed and operated at 15 kV at 14.92–15.02 mm using a gold coating for 20 s.

3.7.5. Solid–State NMR Spectroscopy

^{13}C CP-MAS NMR was used to observe the carbon presence of the MPTS ligand groups on the adsorbent; ^{29}Si MAS NMR was conducted to obtain information about the structure of the silicon matrix, and the grafting degree of the adsorbent through the Q and T species. Preceding all experimental measurements, samples were desiccated during 48 h at 333 K. Solid–state NMR assessments were conducted at ambient temperature using an Infinityplus spectrometer apparatus (Agilent) maintained under a magnetic field of 7 T and equipped with a changing temperature Chemagnetics–Varian 6 mm pencil cross-polarization (CP) magic angle spinning (MAS) probe. $^{13}C\ \{^1H\}$ CP MAS NMR spectra were acquired using a cross-polarization contact time of 3 ms, a 90° pulse with a length of 5.0 μs, a relaxation delay of 5 s, with MAS at 7 kHz. $^{29}Si\{^1H\}$ CP MAS NMR spectra were acquired using a cross-polarization contact time of 5 ms, a 90° pulse with a length of 5.0 μs, a relaxation delay of 1 s, under MAS at 7 kHz. In general, the advantage of the cross-polarization method in significantly increasing the signal-to-noise ratio is partially offset by the fact that the signal intensity is no longer proportional to the concentration of the corresponding structural units in the sample. This means that the ratio between the Q^3 and Q^4 units cannot be measured using $^{29}Si\{^1H\}$ CP NMR. Such a measurement would require a very long measurement time [39,40]. However, if the cross-polarization contact time is long enough, the ratio between Q^2, Q^3, T1, and T^2 units can be measured quite accurately [41,42].

4. Conclusions

A propyl thiol-functionalized silica gel adsorbent was synthesized and applied in the removal of tequila vinasses color compounds from tequila industries. Due to the difficulty of unambiguously evaluating the amounts and types of tequila vinasses components adsorbed, a simple analysis procedure was proposed and applied to analyze tequila vinasses concentrations. The removal of tequila vinasses achieved a maximum uptake of 0.801 g g^{-1} at a pH of 3.55. The adsorption of tequila vinasses demonstrated rapid kinetics (~90 min). The kinetic data matched the PFO model. The adsorption equilibrium isotherm data were explained by the Temkin model. The MaMPT002 adsorbent revealed that vinasses uptake capacities were enhanced/comparable to adsorbents possessing diverse complexing functional groups found in other published works. The MaMPT002 adsorbent showed preferential reactivity toward melanoidin-type compounds, as observed in the HPLC-MS/MS study. A plausible approach to the identity of the removed melanoidin-type compound with 412.31 m/z was proposed. The textural properties of the MaMPT002 adsorbent include enhanced specific surface areas and pore sizes suitable for the easy diffusion of tequila vinasses molecules to the adsorption sites. TGA results indicated that this adsorbent is thermally stable up to 316 °C and the presence of tequila vinasses did not affect its thermal stability. FTIR and solid–state NMR results confirmed the chemical functionalization and the matrix solid structure of the MaMPT002 adsorbent. The results of this work suggest that such adsorbents can be effective in treating tequila vinasses compounds from industrial wastewater. The present work represents the first integral approximation to the development of selective functionalized material to adsorb melanoidin-type compounds from the vinasses of the tequila industry as well as the evaluation of its efficiency as adsorbent in packed-bed systems.

Supplementary Materials: The following supporting information can be downloaded at: https://www.mdpi.com/article/10.3390/molecules29245910/s1, Supplementary S1. Calculation example of the area under the curve of the UV-vis scanning spectra of tequila vinasses using Simpson's 1/3 rule. Supplementary S2. Description of kinetics, isotherms adsorption models. Supplementary S2.1. Adsorption kinetic models [43–45]. Supplementary S2.2. Adsorption isotherm models [46].

Author Contributions: C.G.-V.: sample synthesis, carried out the BET, FTIR, TGA, NMR experiments, data processing, and conceived the original idea and writing. A.A.P.-L.: supervised the project. R.M.-G.: supervised the project, NMR data analysis. A.A.P.-F.: supervised the project, tequila vinasses measurements. J.R.R.-O.: supervised the project, tequila vinasses measurements, and data analysis. I.G.S.: supervised the project and manuscript editing. S.G.-S.: conceived the original idea, supervised the project, and performed writing. All authors have read and agreed to the published version of the manuscript.

Funding: This work was supported by the University of Guadalajara through the program PROSNI-2022.

Institutional Review Board Statement: Not applicable.

Informed Consent Statement: Not applicable.

Data Availability Statement: Data are contained within the article and Supplementary Materials.

Acknowledgments: Carlos Gonzalez-Valerio is grateful for the scholarship No. 779825 from Mexico's National Council of Humanities, Science, and Technology, CONAHCyT.

Conflicts of Interest: The authors declare no competing financial interests.

References

1. García-Depraect, O.; Diaz-Cruces, V.F.; León-Becerril, E. Upgrading of anaerobic digestion of tequila vinasse by using an innovative two-stage system with dominant lactate-type fermentation in acidogenesis. *Fuel* **2020**, *280*, 118606. [CrossRef]
2. Toledo-Cervantes, A.; Guevara-Santos, N.; Arreola-Vargas, J.; Snell-Castro, R.; Méndez-Acosta, H.O. Performance and microbial dynamics in packed-bed reactors during the long-term two-stage anaerobic treatment of tequila vinasses. *Biochem. Eng. J.* **2018**, *138*, 12–20. [CrossRef]
3. Arreola-Vargas, J.; Snell-Castro, R.; Rojo-Liera, N.M.; González-Álvarez, V.; Méndez-Acosta, H.O. Effect of the organic loading rate on the performance and microbial populations during the anaerobic treatment of tequila vinasses in a pilot-scale packed bed reactor. *J. Chem. Technol. Biotechnol.* **2018**, *93*, 591–599. [CrossRef]
4. Arellano-García, L.; Velázquez-Fernández, J.B.; Macías-Muro, M.; Marino-Marmolejo, E.N. Continuous hydrogen production and microbial community profile in the dark fermentation of tequila vinasse: Response to increasing loading rates and immobilization of biomass. *Biochem. Eng. J.* **2021**, *172*, 108049. [CrossRef]
5. Carvajal-Zarrabal, O.; Nolasco-Hipólito, C.; Barradas-Dermitz, D.M.; Hayward-Jones, P.M.; Aguilar-Uscanga, M.G.; Bujang, K. Treatment of vinasse from tequila production using polyglutamic acid. *J. Environ. Manag.* **2012**, *95*, S66–S70. [CrossRef] [PubMed]
6. Mahmoudabadi, Z.S.; Rashidi, A.; Maklavany, D.M. Optimizing treatment of alcohol vinasse using a combination of advanced oxidation with porous α-Fe_2O_3 nanoparticles and coagulation-flocculation. *Ecotoxicol. Environ. Saf.* **2022**, *234*, 113354. [CrossRef] [PubMed]
7. Li, L.; Wu, M.; Song, C.; Liu, L.; Gong, W.; Ding, Y.; Yao, J. Efficient removal of cationic dyes via activated carbon with ultrahigh specific surface derived from vinasse wastes. *Bioresour. Technol.* **2021**, *322*, 124540. [CrossRef]
8. Wei, S.; Tan, Z.; Liu, Z.; Zuo, H.; Xia, Y.; Zhang, Y. Removal of methyl orange dye by high surface area biomass activated carbon prepared from bamboo fibers. *Ind. Crops Prod.* **2024**, *218*, 118991. [CrossRef]
9. Caqueret, V.; Bostyn, S.; Cagnon, B.; Fauduet, H. Purification of sugar beet vinasse—Adsorption of polyphenolic and dark colored compounds on different commercial activated carbons. *Bioresour. Technol.* **2008**, *99*, 5814–5821. [CrossRef]
10. Lorenzo-Santiago, M.A.; Rodríguez-Campos, J.; Rendón-Villalobos, R.; García-Hernández, E.; Vallejo-Cardona, A.A.; Contreras-Ramos, S.M. Thermal treatment to obtain 5-hydroxymethyl furfural (5-HMF), furfural and phenolic compounds from vinasse waste from agave. *Molecules* **2023**, *28*, 1063. [CrossRef]
11. Govea-Paz, Y.; Arellano-García, L.; Aguilar-Juárez, O. Simultaneous hydrogen and methane production by tequila vinasses dark fermentation in series with anaerobic digestion. *Bioresour. Technol. Rep.* **2024**, *25*, 101779. [CrossRef]
12. American Public Health Association. *Standard Methods: For the Examination of Water and Wastewater*, 17th ed.; American Public Health Association: Washington, DC, USA, 1989.
13. Dubois, M.; Gilles, K.A.; Hamilton, J.K.; Rebers, P.A.; Smith, F. Colorimetric method for determination of sugars and related substances. *Anal. Chem.* **1956**, *28*, 350–356. [CrossRef]
14. López-López, A.; Davila-Vazquez, G.; León-Becerril, E.; Villegas-García, E.; Gallardo-Valdez, J. Tequila vinasses: Generation and full-scale treatment processes. *Rev. Environ. Sci. Bio/Technol.* **2010**, *9*, 109–116. [CrossRef]

15. Hofmann, T.; Schieberle, P. Chemical Interactions between Odor-Active Thiols and Melanoidins Involved in the Aroma Staling of Coffee Beverages. *J. Agric. Food Chem.* **2002**, *50*, 319–326. [CrossRef] [PubMed]
16. Nam, K.H.; Gomez-Salazar, S.; Tavlarides, L.L. Mercury (II) Adsorption from Wastewaters Using a Thiol Functional Adsorbent. *Ind. Eng. Chem. Res.* **2003**, *42*, 1955–1964. [CrossRef]
17. Nam, K.H.; Tavlarides, L.L. Mercury Removal from Acidic Waste Solutions Using a Thiol Functional Organo-Ceramic Adsorbent. *Solvent Extr. Ion Exch.* **2003**, *21*, 899–913. [CrossRef]
18. Quirarte-Escalante, C.A.; Soto, V.; De La Cruz, W.; Porras, G.R.; Manríquez, R.; Gomez-Salazar, S. Synthesis of hybrid adsorbents combining sol–gel processing and molecular imprinting applied to lead removal from aqueous streams. *Chem. Mater.* **2009**, *21*, 1439–1450. [CrossRef]
19. Gedat, E.; Schreiber, A.; Findenegg, G.H.; Shenderovich, I.; Limbach, H.-H.; Buntkowsky, G. Stray Field Gradient NMR Reveals Effects of Hydrogen Bonding on Diffusion Coefficients of Pyridine in Mesoporous Silica. *Magn. Reson. Chem.* **2001**, *39*, S149–S157. [CrossRef]
20. Lesnichin, S.B.; Tolstoy, P.M.; Limbach, H.-H.; Shenderovich, I.G. Counteranion-Dependent Mechanisms of Intramolecular Proton Transfer in Aprotic Solution. *Phys. Chem. Chem. Phys.* **2010**, *12*, 10373–10379. [CrossRef]
21. Verma, P.; Kuwahara, Y.; Mori, K.; Raja, R.; Yamashita, H. Functionalized mesoporous SBA-15 silica: Recent trends and catalytic applications. *Nanoscale* **2020**, *12*, 11333–11363. [CrossRef] [PubMed]
22. Xu, S.; Ning, S.; Wang, Y.; Wang, X.; Dong, H.; Chen, L.; Yin, X.; Fujita, T.; Wei, Y. Precise separation and efficient enrichment of palladium from wastewater by amino-functionalized silica adsorbent. *J. Clean. Prod.* **2023**, *396*, 136479. [CrossRef]
23. Lofgreen, J.E.; Ozin, G.A. Controlling morphology and porosity to improve performance of molecularly imprinted sol–gel silica. *Chem. Soc. Rev.* **2014**, *43*, 911–933. [CrossRef] [PubMed]
24. Lima, E.C.; Hosseini-Bandegharaei, A.; Moreno-Piraján, J.C.; Anastopoulos, I. A critical review of the estimation of the thermodynamic parameters on adsorption equilibria. Wrong use of equilibrium constant in the Vant Hoof equation for calculation of thermodynamic parameters of adsorption. *J. Mol. Liq.* **2019**, *273*, 425–434. [CrossRef]
25. Chingono, K.E.; Sanganyado, E.; Berea, E.; Yalala, B. Adsorption of sugarcane vinasse effluent on bagasse fly ash: A parametric and kinetic study. *J. Environ. Manag.* **2018**, *224*, 182–190. [CrossRef] [PubMed]
26. Lima, E.C.; Gomes, A.A.; Tran, H.N. Comparison of the nonlinear and linear forms of the van't Hoff equation for calculation of adsorption thermodynamic parameters (ΔS° and ΔH°). *J. Mol. Liq.* **2020**, *311*, 113315. [CrossRef]
27. de Aquino Lima, F.; dos Santos Júnior, A.C.; Sarrouh, B.; Zanetti Lofrano, R.C. Application of sugarcane bagasse and peanut shell in natura as bioadsorbents for vinasse treatment. *Clean Technol. Environ. Policy* **2024**, *26*, 401–416. [CrossRef]
28. Du, X.; Cheng, Y.; Liu, Z.; Yin, H.; Wu, T.; Huo, L.; Shu, C. CO_2 and CH_4 adsorption on different rank coals: A thermodynamics study of surface potential, Gibbs free energy change and entropy loss. *Fuel* **2021**, *283*, 118886. [CrossRef]
29. Thommes, M.; Kaneko, K.; Neimark, A.V.; Olivier, J.P.; Rodriguez-Reinoso, F.; Rouquerol, J.; Sing, K.S.W. Physisorption of gases, with special reference to the evaluation of surface area and pore size distribution (IUPAC Technical Report). *Pure Appl. Chem.* **2015**, *87*, 1051–1069. [CrossRef]
30. Monson, P.A. Understanding adsorption/desorption hysteresis for fluids in mesoporous materials using simple molecular models and classical density functional theory. *Microporous Mesoporous Mater.* **2012**, *160*, 47–66. [CrossRef]
31. Sato, S.; Murakata, T.; Suzuki, T.; Ohgawara, T. Control of pore size distribution of silica gel through sol-gel process using water soluble polymers as additives. *J. Mater. Sci.* **1990**, *25*, 4880–4885. [CrossRef]
32. Almaghrabi, M.; Alqurshi, A.; Jadhav, S.A.; Mazzacuva, F.; Cilibrizzi, A.; Raimi-Abraham, B.; Royall, P.G. Evaluating thermogravimetric analysis for the measurement of drug loading in mesoporous silica nanoparticles (MSNs). *Thermochim. Acta* **2023**, *730*, 179616. [CrossRef]
33. Wu, J.; Ling, L.; Xie, J.; Ma, G.; Wang, B. Surface modification of nanosilica with 3-mercaptopropyl trimethoxysilane: Experimental and theoretical study on the surface interaction. *Chem. Phys. Lett.* **2014**, *591*, 227–232. [CrossRef]
34. Díaz, I.; Márquez-Alvarez, C.; Mohino, F.; Pérez-Pariente, J.; Sastre, E. A novel synthesis route of well ordered, sulfur bearing MCM-41 catalysts involving mixtures of neutral and cationic surfactants. *Microporous Mesoporous Mater.* **2001**, *44*, 295–302. [CrossRef]
35. Lee, J.S.; Gomez-Salazar, S.; Tavlarides, L.L. Synthesis of thiol functionalized organo–ceramic adsorbent by sol–gel technology. *React. Funct. Polym.* **2001**, *49*, 159–172. [CrossRef]
36. Gomez-Salazar, S.; Lee, J.S.; Heydweiller, J.C.; Tavlarides, L.L. Analysis of Cadmium Adsorption on Novel Organo-Ceramic Adsorbents with a Thiol Functionality. *Ind. Eng. Chem. Res.* **2003**, *42*, 3403–3412. [CrossRef]
37. Gomez-Gonzalez, S.E.; Carbajal-Arizaga, G.G.; Manriquez-Gonzalez, R.; De la Cruz-Hernandez, W.; Gomez-Salazar, S. Trivalent chromium removal from aqueous solutions by a sol–gel synthesized silica adsorbent functionalized with sulphonic acid groups. *Mater. Res. Bull.* **2014**, *59*, 394–404. [CrossRef]
38. Franco-León, J.J.; Arriola-Guevara, E.; Suárez-Hernández, L.A.; Toriz, G.; Guatemala-Morales, G.; Corona-González, R.I. Influence of supplemented nutrients in tequila vinasses for hydrogen and polyhydroxybutyrate production by photofermentation with Rhodopseudomonas pseudopalustris. *Bioresour. Technol.* **2021**, *329*, 124865. [CrossRef] [PubMed]
39. Shenderovich, I.G.; Limbach, H.-H. Solid State NMR for Nonexperts: An Overview of Simple but General Practical Methods. *Solids* **2021**, *2*, 139–154. [CrossRef]

40. Akcakayiran, D.; Mauder, D.; Hess, C.; Sievers, T.K.; Kurth, D.G.; Shenderovich, I.; Limbach, H.-H.; Findenegg, G.H. Carboxylic Acid-Doped SBA-15 Silica as a Host for Metallo-supramolecular Coordination Polymers. *J. Phys. Chem. B* **2008**, *112*, 14637–14647. [CrossRef]
41. Shenderovich, I.G.; Mauder, D.; Akcakayiran, D.; Buntkowsky, G.; Limbach, H.-H.; Findenegg, G.H. NMR Provides Checklist of Generic Properties for Atomic-Scale Models of Periodic Mesoporous Silicas. *J. Phys. Chem. B* **2007**, *111*, 12088–12096. [CrossRef]
42. Gutierrez-Ortega, J.A.; Gomez-Salazar, S.; Shenderovich, I.G.; Manríquez-Gonzalez, R. Efficiency and lead uptake mechanism of a phosphonate functionalized mesoporous silica through P/Pb association ratio. *Mater. Chem. Phys.* **2020**, *239*, 122037. [CrossRef]
43. Chen, H.; Wang, A. Adsorption characteristics of Cu(II) from aqueous solution onto poly(acrylamide)/attapulgite composite. *J. Hazard. Mater.* **2009**, *165*, 223–231. [CrossRef] [PubMed]
44. Ho, Y.S.; McKay, G. A Comparison of Chemisorption Kinetic Models Applied to Pollutant Removal on Various Sorbents. *Process Saf. Environ. Prot.* **1998**, *76*, 332–340. [CrossRef]
45. Weber, W.J.; Morris, J.C. Kinetics of Adsorption on Carbon from Solution. *J. Sanit. Eng. Div.* **1963**, *89*, 31–59. [CrossRef]
46. Foo, K.Y.; Hameed, B.H. Insights into the modeling of adsorption isotherm systems. *Chem. Eng. J.* **2010**, *156*, 2–10. [CrossRef]

molecules

MDPI

Article

Keep Your TEMPO Up: Nitroxide Radicals as Sensors of Intermolecular Interactions

Ilya G. Shenderovich

Institute of Organic Chemistry, University of Regensburg, Universitaetstrasse 31, 93053 Regensburg, Germany; ilya.shenderovich@ur.de

Abstract: This study examines experimental data on the influence of the surrounding medium and non-covalent interactions on the isotropic hyperfine coupling constant, $A_{iso}(^{14}N)$, of the stable nitroxide radical 2,2,6,6-Tetramethylpiperidin-1-yl)oxyl (TEMPO) in solution. The data were used to identify a density functional theory functional/basis set combination that accurately reproduces the experimental $A_{iso}(^{14}N)$ values. The variations in $A_{iso}(^{14}N)$ due to external factors are two orders of magnitude greater than the accuracy of its experimental measurements, making $A_{iso}(^{14}N)$ a highly sensitive experimental probe for quantifying these effects. Additionally, it was found that the proton-accepting ability of the N-O$^{\bullet}$ moiety in TEMPO resembles that of the P=O moiety, enabling the simultaneous formation of two equally strong hydrogen bonds.

Keywords: TEMPO; EPR; hyperfine coupling constant; hydrogen bonding; non-covalent interactions

1. Introduction

The interaction of molecules with nitroxide radicals induces several characteristic spectral changes. This interaction affects the chemical shifts and relaxation times of the nuclei, as well as the appearance of the unpaired electron spin density on these nuclei. The most important actual application of stable radicals is the technique of dynamic nuclear polarization (DNP) [1]. This technique has many applications for molecular systems of varying complexity [2–6].

Paramagnetic shifts induced by nitroxide radicals in surrounding molecules depend on their interaction [7]. The most commonly studied nucleus for such analysis has been the ^{13}C nucleus [8–11]. The method of ^{13}C NMR of TEMPO-functionalized polymers has been used to study the thermoreversible phase transition of aqueous solutions of these polymers [12]. Effects have also been observed on the ^{1}H, ^{19}F, and ^{15}N nuclei [13]. The method of ^{1}H NMR was used to study the pressure effect on intermolecular interactions with nitroxide radicals [14].

Intermolecular interactions with nitroxide radicals can be characterized by studying their effect on the relaxation times of their diamagnetic partners [15–23]. The most definitive confirmation of complex formation between diamagnetic molecules and nitroxide radicals is the appearance of unpaired electron spin density on the nuclei of these molecules [24,25]. Nitroxide radicals act as proton acceptors, with hydrogen bonding being the most prominent mechanism for the formation of such complexes [26,27]. Complexes with alcohols [28–34], anilines [35,36], amines [24], acids [37,38], water [39], and phenols [40,41] are among the best studied.

On the other hand, intermolecular interactions cause two characteristic spectral changes in the nitroxide radicals themselves. These interactions manifest in changes in the isotropic hyperfine coupling constant $A_{iso}(^{14}N)$ [42,43] and variations in the relative linewidths of the components of the radical EPR signal. This difference in linewidth depends on the rate of rotational diffusion of the radical and can be used to study the physical properties of viscous liquids [44], ionic liquids [45,46], polymers [47], micelles [48,49], and inclusion complexes [50].

All the above parameters can be used either individually or in combination when utilizing radicals as probes to study complex molecular systems. However, in this publication, only the effect of molecular complexation on $A_{iso}(^{14}N)$ will be considered and discussed. This parameter is easy to measure, and, apparently, its changes are detectable even in cases of weak and transient interactions [51]. $A_{iso}(^{14}N)$ can be used to analyze steric effects on the accessibility of active sites [52–56] in a wide variety of weak proton donors and local micropolarity [57,58]. Such an analysis may provide a better understanding of the factors affecting the reactivity of these centers, whether through intramolecular interactions [59,60] or external influences [61,62]. Other studies [63,64] suggest that $A_{iso}(^{14}N)$ can also serve as a sensor for halogen bond properties [65,66].

The objectives of this study are as follows: (i) to identify a density functional theory (DFT) functional/basis set pair that accurately reproduces the experimental values of $A_{iso}(^{14}N)$ in aprotic solvents, which will then be used to interpret $A_{iso}(^{14}N)$ observed experimentally (ii) in water and alcohols, and (iii) in toluene solutions of phenols; (iv) to theoretically estimate the maximum possible value of $A_{iso}(^{14}N)$; and (v) to theoretically estimate the effect of halogen bonds on $A_{iso}(^{14}N)$. The radical (2,2,6,6-Tetramethylpiperidin-1-yl)oxyl (TEMPO) will be used as the model nitroxide radical. Figure 1 shows the molecular structure of this stable radical and its EPR spectrum in toluene. Wherever possible, calculated $A_{iso}(^{14}N)$ values are compared with the corresponding experimental values available in the literature.

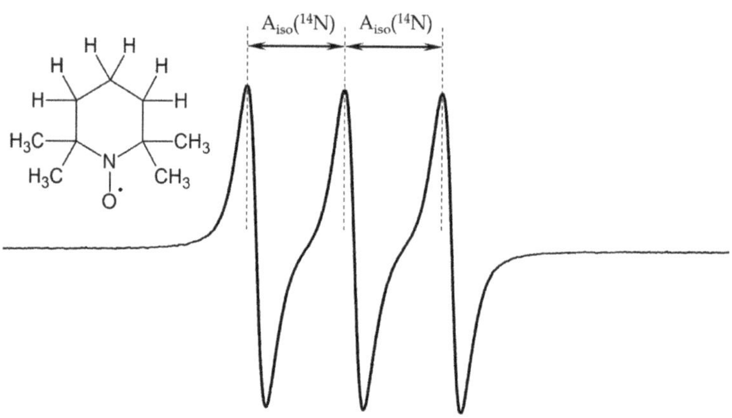

Figure 1. EPR spectrum of 5×10^{-4} M (2,2,6,6-Tetramethylpiperidin-1-yl)oxyl (TEMPO) in toluene at 300 K. $A_{iso}(^{14}N) = 1.55$ mT, g = 2.0023.

2. Results and Discussion

2.1. *The Best Functional/Basis Set Pair for Calculating $A_{iso}(^{14}N)$*

Different DFT functional/basis set pairs have been tested in the past for calculations related to radicals. [67–69]. Here, several DFT functionals and basis sets were evaluated to calculate $A_{iso}(^{14}N)$ for TEMPO in toluene. Both the geometry of TEMPO and its $A_{iso}(^{14}N)$ were calculated using the polarizable continuum model approximation (PCM) with SCRF=(Solvent=toluene). The observed experimental value of $A_{iso}(^{14}N)$ decreases with increasing radical concentration [35,70]. Therefore, reliable calculated values of $A_{iso}(^{14}N)$ should equal or exceed the experimental value. Table 1 reports the calculated values of $A_{iso}(^{14}N)$ for selected DFT functional/basis set pairs.

The uPBE1PBE/LanL2DZ, uHSEH1PBE/LanL2DZ, and uHSEH1PBE/LanL2DZ pairs give results that are very close to the experimental value of 1.551 mT. However, in all cases, the calculated values are slightly lower than the experimental value. A possible reason for this deviation is that the effect of the reaction field created by the polarization of the surrounding dielectric cannot be correctly modeled within the PCM approach [58,71].

This can be addressed using a fictitious external electric field [71–73]. However, since the required field strength is solvent-dependent and cannot be estimated a priori, further calculations will be performed using the PCM approximation. If this explanation is correct, then the deviation between the theoretical and experimental values should increase with increasing solvent polarity.

Table 1. $A_{iso}(^{14}N)$ for TEMPO in toluene calculated for selected DFT functional/basis set pairs.

DFT Functional	Basis Set	$A_{iso}(^{14}N)$, mT
—	—	1.551 ± 0.002 [1]
uPBE1PBE	LanL2DZ	1.5147
uPBE1PBE	LanL2DZdp	1.4802
uPBE1PBE	EPR-III	1.0214
uPBE1PBE	6-31G-J	1.0222
uPBE1PBE	def2tzvp	0.9477
uPBE1PBE	aug-cc-pVqZ	0.8147
uHSEH1PBE	LanL2DZ	1.5154
uB3LYP	LanL2DZ	1.4629
uTPSSh	LanL2DZ	1.4233
uwB97XD	LanL2DZ	1.4559
uM06	LanL2DZ	1.1013

[1] Experimental value [52].

2.2. Aprotic Solvents

Table 2 reports the available experimental values of $A_{iso}(^{14}N)$ for TEMPO in aprotic solvents. For some of these solvents, $A_{iso}(^{14}N)$ was calculated in the uPBE1PBE/LanL2D approximation. Both the geometry of TEMPO and its $A_{iso}(^{14}N)$ were calculated in the PCM approximation using the solvent under study. In most cases, the calculated values are lower than the experimental ones. However, the deviations are not significant. The obtained data do not enable one to confirm or refute the assumption above that the deviation between the theoretical and experimental values increases with increasing solvent polarity. Nevertheless, the calculations correctly reproduce the trend for $A_{iso}(^{14}N)$ to increase with increasing solvent polarity. Notably, these changes are substantial enough to track even small variations in polarity. The experimental measurement error is on the order of 0.001 mT, and the absolute value changes by 0.1 mT when transitioning from hexane to (methanesulfinyl)methane (DMSO).

Table 2. $A_{iso}(^{14}N)$ for TEMPO in aprotic solvents.

Solvent	uPBE1PBE/LanL2DZ, mT	Experiment, mT
Hexane ($\varepsilon \approx 1.88$)	1.4996	1.520 [70] 1.50 [74]
Cyclohexane ($\varepsilon \approx 2.02$)	1.5042	1.540 [70] 1.51 [74]
CCl_4 ($\varepsilon \approx 2.23$)	1.5107	1.563 [75] 1.550 [70] 1.568 [35]
Benzene ($\varepsilon \approx 2.27$)	—	1.554 [76]
Toluene ($\varepsilon \approx 2.37$)	1.5147	1.551 [52]
Oxolane (THF, $\varepsilon \approx 7.43$)	1.5671	1.557 [77]
Dichloromethane ($\varepsilon \approx 8.93$)	—	1.587 [78]
DMSO ($\varepsilon \approx 46.83$)	1.5959	—
Ionic Liquid	—	1.604 [46]
Liquid CO_2	—	1.549 [79]
Supercritical CO_2	—	1.545 [79]
Mineral oil	—	1.539 [46]

2.3. Protic Solvents

Table 3 reports the available experimental values of $A_{iso}(^{14}N)$ for TEMPO in protic solvents. For some of these solvents, $A_{iso}(^{14}N)$ in 1:1 complexes of TEMPO and a solvent molecule was calculated in the uPBE1PBE/LanL2D approximation. Both the geometry of the complexes and $A_{iso}(^{14}N)$ were calculated in the PCM approximation using the solvent under study. The structures of some of these complexes are shown in Figure 2.

Table 3. $A_{iso}(^{14}N)$ in 1:1 complexes of TEMPO and selected solvent molecules.

Hydrogen Bond Donor	uPBE1PBE/LanL2DZ, mT	Experiment, mT
2-methylpropan-2-ol [1]	1.7108	1.60 [74]
Pentan-1-ol	—	1.61 [74]
3-methylbutan-1-ol	—	1.61 [74]
2-methylbutan-2-ol	—	1.58 [74]
Butan-2-ol	1.7256	1.60 [74]
Butan-1-ol	—	1.61 [74]
Propan-2-ol	—	1.61 [74]
Propan-1-ol	—	1.62 [74]
Ethanol	1.7337	1.632 [77] 1.62 [74]
Methanol [1]	1.7371	1.63 [74]
Water [1]	1.7274	1.769 [44] 1.732 [75]

[1] $A_{iso}(^{14}N)$ for TEMPO, calculated using the PCM approximation at ε = 78.36 (water), ε = 32.61 (methanol), and ε = 12.47 (2-methylpropan-2-ol), are 1.5984 mT, 1.5933 mT, and 1.5802 mT, respectively.

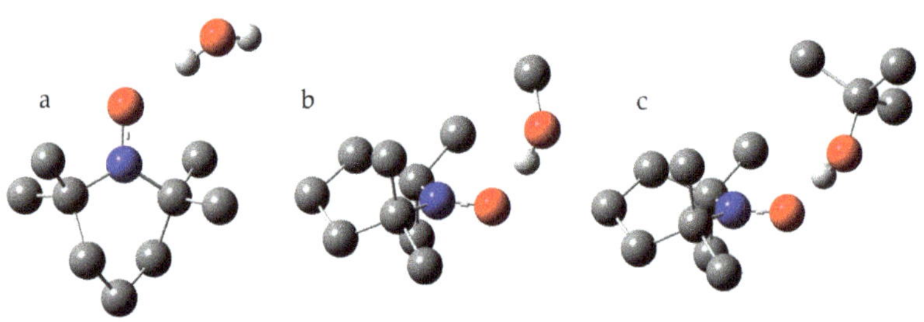

Figure 2. The model structures of the selected 1:1 complexes of TEMPO with (**a**) water, (**b**) methanol, and (**c**) 2-methylpropan-2-ol. Oxygen atoms are represented in red, nitrogen atoms in blue, carbon atoms in grey, and exchangeable protons in white.

Figure 2a shows the structure of the complex with a water molecule. The calculated $A_{iso}(^{14}N)$ value is lower than the experimental $A_{iso}(^{14}N)$ value in water. There are several possible reasons for this. First, the effect of the reaction field created by the polarization of the surrounding dielectric cannot be correctly modeled within the PCM approach [71–73]. Second, the proton-donating ability of a water molecule in water is greater than that of a single water molecule [80]. Given these limitations, the calculation result can be considered satisfactory and unequivocally indicates that the oxygen atom of TEMPO is involved in the hydrogen bond network of water. Note that the calculated $A_{iso}(^{14}N)$ value in the PCM SCRF=(solvent=water) approximation is only 1.5984 mT. Therefore, the hydrogen bond leads to an increase in $A_{iso}(^{14}N)$ by more than 0.12 mT.

Figure 2b,c shows the structures of two 1:1 complexes of TEMPO with alcohol molecules. For all studied alcohols, the calculated $A_{iso}(^{14}N)$ values in such complexes are similar to those with water but significantly exceed the available experimental values of $A_{iso}(^{14}N)$ in these alcohols, as seen in Table 3. On the other hand, these experimental values significantly exceed the calculated $A_{iso}(^{14}N)$ values in the PCM SCRF=(solvent=alcohol) approximation. The hydrogen bonds in these 1:1 complexes lead to an increase in $A_{iso}(^{14}N)$ by more than 0.13 mT. The observed difference between the calculated and experimental values indicates that in alcohols, TEMPO may or may not be hydrogen-bonded. Greatly simplifying the real situation and assuming that there is an equilibrium between TEMPO molecules that do not participate in any way in hydrogen bonding and TEMPO molecules forming the 1:1 complexes of optimal geometry with the alcohol molecules, one can write the following: $A_{iso}(^{14}N)^{exp} = x_H \cdot A_{iso}(^{14}N)^H + x_f \cdot A_{iso}(^{14}N)^f$, where $A_{iso}(^{14}N)^{exp}$, $A_{iso}(^{14}N)^H$, and $A_{iso}(^{14}N)^f$ stand for the observed experimental value, the calculated value in the 1:1 complex, and the calculated value for TEMPO in the PCM approximation, respectively. x_H and x_f are the mole fractions of the hydrogen-bonded and free TEMPO molecules, respectively, and $x_H + x_f = 1$. Thus, $x_H = (A_{iso}(^{14}N)^{exp} - A_{iso}(^{14}N)^f)/(A_{iso}(^{14}N)^H - A_{iso}(^{14}N)^f)$. For TEMPO in methanol, $x_H = (1.63 - 1.5933)/(1.7371 - 1.5933) \approx 0.25$. For TEMPO in 2-methylpropan-2-ol, $x_H \approx 0.15$. Of course, these mole fractions must largely depend on the proton-donating ability of the alcohol molecule and the temperature.

2.4. Interaction of TEMPO with Phenols in Toluene

Table 4 presents the available experimental values of $A_{iso}(^{14}N)$ for TEMPO in solutions of phenols in toluene. For some of these phenols, $A_{iso}(^{14}N)$ in 1:1 complexes of TEMPO and a phenol molecule was calculated in the uPBE1PBE/LanL2D approximation. Both the geometry of the complexes and $A_{iso}(^{14}N)$ were calculated in the PCM approximation in toluene. The structures of these complexes are shown in Figure 3.

Table 4. $A_{iso}(^{14}N)$ for TEMPO in 1:1 complexes with phenols in toluene. [1] The experimental values were obtained for TEMPO/phenol concentrations of 5×10^{-4} M/0.01 M and 5×10^{-4} M/0.05 M.

Phenol	**uPBE1PBE/LanL2DZ, mT**	**Experiment [52], mT**	
		0.01 M	**0.05 M**
Phenol	1.7415	1.565	1.607
4-nitrophenol	1.7852	1.595	1.633
2,5-dinitrophenol	1.8023	1.556	1.574
2,4,6-trinitrophenol	1.6242	1.552	1.557
2,6-di-*tert*-butylphenol	—	1.552	1.554

[1] The experimental and calculated $A_{iso}(^{14}N)$ values for TEMPO in toluene are 1.551 mT [52] and 1.5147 mT, respectively.

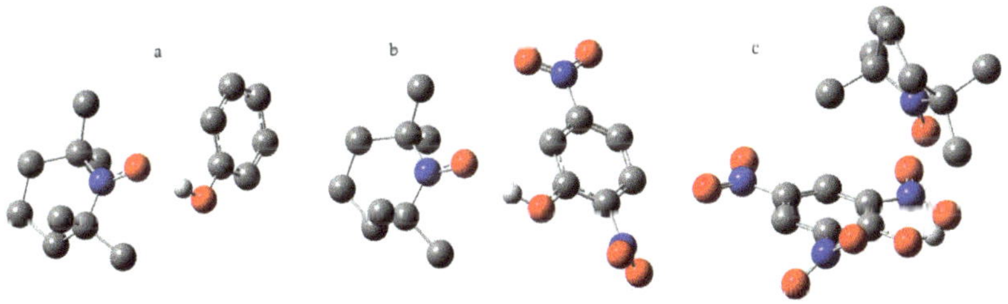

Figure 3. The model structures of the 1:1 complexes of TEMPO with (**a**) phenol, (**b**) 2,5-dinitrophenol, and (**c**) 2,4,6-trinitrophenol. Oxygen atoms are represented in red, nitrogen atoms in blue, carbon atoms in grey, and exchangeable protons in white.

Phenol and 4-nitrophenol form with TEMPO hydrogen-bonded complexes, as seen in Figure 3a. The mole fractions of these complexes in the studied solution can be estimated using $x_H = (A_{iso}(^{14}N)^{exp} - A_{iso}(^{14}N)^{Tol})/(A_{iso}(^{14}N)^H - A_{iso}(^{14}N)^{Tol})$, where $A_{iso}(^{14}N)^{Tol}$ stands for the experimental value in toluene, 1.551 mT [52]. x_H depends on the phenol concentration. For concentrations of 0.01 M and 0.05 M, x_H is approximately 0.07 and 0.29 for phenol and 0.19 and 0.35 for 4-nitrophenol, respectively.

The hydrogen bond with 2,5-dinitrophenol should be even stronger, as seen in Table 4 and Figure 3b. However, the estimated values of x_H are only about 0.02 and 0.09 for concentrations 0.01 M and 0.05 M, respectively. This indicates that the presence of a nitro group in the *ortho* position greatly hinders the formation of the complex. If both *ortho* positions are occupied by nitro groups, 2,4,6-trinitrophenol, the hydroxyl group prefers to form an intramolecular hydrogen bond with one of them, as seen in Figure 3c. The calculated value of $A_{iso}(^{14}N)$ in such a complex is low, as seen in Table 4. The absence of hydrogen bonding between 2,4,6-trinitrophenol and TEMPO in toluene is also confirmed by the low experimental value of $A_{iso}(^{14}N)$. The same is true for 2,6-di-*tert*-butylphenol, as seen in Table 4. Therefore, *tert*-butyl groups at the *ortho* positions block the proton-donating ability of phenol just as they block the proton-accepting ability of pyridine [81].

Note that with a 5-fold increase in phenol concentration, the increase in x_H values depends on the type of phenol. For phenol, x_H increases 4-fold, for 4-nitrophenol it doubles, and for 2,5-dinitrophenol, it increases 5-fold. One can speculate that these changes reflect the instantaneous concentration of the TEMPO/phenol hydrogen-bonded complexes of different geometry in solutions at phenol concentrations of 0.01 M. Phenol is a weak proton donor, resulting in short lifetimes for its hydrogen-bonded complexes at room temperature. Consequently, most TEMPO molecules remain free when the phenol concentration is low. 2,5-dinitrophenol is a stronger proton donor, but the lifetime of its hydrogen-bonded complexes is limited by steric interference from its ortho substituent. Therefore, most TEMPO molecules remain free in this case, as well when phenol concentration is low. For both phenols, the energy gain from hydrogen bonding is offset by entropy losses. When most TEMPO molecules in solution remain free, a 5-fold increase in phenol concentration leads to a 5-fold increase in the concentration of short-living hydrogen-bonded complexes. In contrast, 4-nitrophenol is a strong proton donor with no steric hindrances to its hydrogen bonding, so the instantaneous concentration of TEMPO/4-nitrophenol hydrogen-bonded complexes is already significant at a phenol concentration of 0.01 M. In this case, an increase in the phenol concentration causes only a limited increase in the instantaneous concentration of the hydrogen-bonded complexes.

2.5. Interaction of TEMPO with Strong Proton Donors

What is the largest possible value for $A_{iso}(^{14}N)$ of TEMPO? The largest experimentally measured values are around 2.18−2.19 mT [76,78], attributed to TEMPO-H$^+$. Strong acids can cause chemical degradation of the nitroxide paramagnetic center [82–85], so the actual existence of the complexes discussed below cannot be confirmed here. The current goal is to estimate the Aiso(^{14}N) values that can be expected for complexes between TEMPO and strong proton donors.

The calculated $A_{iso}(^{14}N)$ value for the TEMPO-H$^+$ cation is 2.42 mT, as seen in Table 5, with both its geometry and $A_{iso}(^{14}N)$ value calculated using the PCM approximation in water. The orientation of proton donors relative to the N-O$^\bullet$ moiety of TEMPO, as shown in Figures 2 and 3, suggests that its proton-accepting ability may be similar to the P=O moiety. This allows for the simultaneous formation of two hydrogen bonds of equal energy [86,87]. This feature of the P=O moiety significantly influences the structure of its hydrogen-bonded adducts [88–92]. The calculated $A_{iso}(^{14}N)$ for the TEMPO-2H^{2+} cation is 2.45 mT, as seen in Table 5. Both its geometry and $A_{iso}(^{14}N)$ value were calculated using the PCM approximation in water.

Table 5. The calculated $A_{iso}(^{14}N)$ values of TEMPO in mT in complexes with selected proton donors.

Hydrogen-Bonded Complex	uPBE1PBE/LanL2DZ	O...H Distance, Å
TEMPO-H$^+$ in water	2.4242 [1]	0.97
TEMPO-2H^{2+} in water	2.4540 [1]	0.99
TEMPO-HF in water	1.8508 [1]	1.53
TEMPO-2HF in water	2.0509 [1]	1.59
TEMPO-[H-pyridine]$^+$ in toluene	1.8620 [2]	1.61
TEMPO-[H-BiPy]$^+$ in toluene	1.7468 [2]	1.87
TEMPO-[H-Phen]$^+$ in toluene	1.8145 [2]	1.71

[1] The calculated $A_{iso}(^{14}N)$ value of TEMPO in water is 1.5984 mT. [2] The calculated $A_{iso}(^{14}N)$ value of TEMPO in toluene is 1.5147 mT.

To analyze the difference between TEMPO adducts with one and two hydrogen bond donors, its complexes with hydrogen fluoride (HF) were examined. Figure 4a,b shows their structures. Both the geometry of these adducts and their $A_{iso}(^{14}N)$ value were calculated using the PCM approximation in water. Unlike the TEMPO cations discussed earlier, there is a significant difference in $A_{iso}(^{14}N)$ for these adducts, as seen in Table 5. The formation of the second hydrogen bond in the TEMPO...2HF adduct leads to a lengthening of the O...H distances. Is the formation of two hydrogen bonds energetically favorable? The hydrogen bond energy in these adducts can be defined as the difference between (i) the sum of the electronic energies of a noninteracting TEMPO molecule and one or two noninteracting HF molecules, and (ii) the electronic energy of the TEMPO...HF or TEMPO...2HF adducts, all calculated using the PCM approximation in water. For the TEMPO...HF adduct, the hydrogen bond energy is 49 kJ/mol, while for the TEMPO...2HF adduct, it is 87 kJ/mol $\approx 2 \times 44$ kJ/mol. Therefore, similar to the P=O moiety, the N-O$^{\bullet}$ moiety in TEMPO can form two equally strong hydrogen bonds simultaneously.

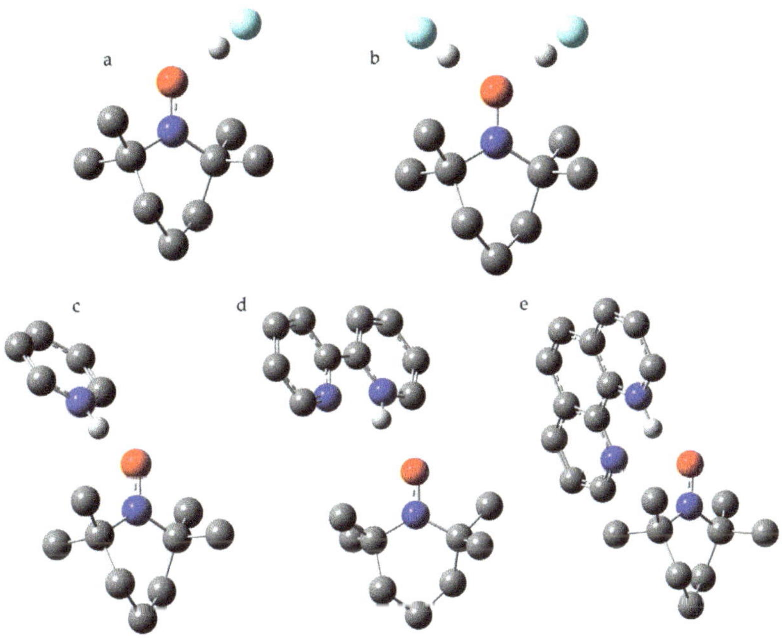

Figure 4. The model structures of the complexes of TEMPO with strong proton donors. (**a**) Hydrogen fluoride. (**b**) Two molecules of hydrogen fluoride. (**c**) Pyridine-H$^+$ cation. (**d**) 2,2′-bipyridine-H$^+$ cation. (**e**) 1,10 phenanthroline H$^+$ cation. Oxygen atoms are represented in red, nitrogen atoms in blue, carbon atoms in grey, fluorine atoms in cyan, and exchangeable protons in white.

This has important implications: in water and other protic solvents, adducts with two hydrogen bonds per TEMPO molecule are likely to exist. $A_{iso}(^{14}N)$ in these adducts is larger than in 1:1 adducts. This may be another reason why the experimental $A_{iso}(^{14}N)$ in water is greater than the calculated value for the 1:1 complex, as seen in Table 3. It also implies that the actual concentration of hydrogen-bonded TEMPO forms in alcohols is lower than what is estimated from the calculated $A_{iso}(^{14}N)$ for 1:1 complexes.

Figure 4c shows the structure of the 1:1 complex of TEMPO with the pyridine-H$^+$ cation. Both the geometry and $A_{iso}(^{14}N)$ value were calculated using the PCM approximation in toluene. This cation is a stronger proton donor than alcohols and phenols. As a result, the hydrogen bond in this complex leads to an increase in the $A_{iso}(^{14}N)$ value by more than 0.3 mT, as seen in Table 5. The presence of such complexes in solution should be easy to detect. The O...H distance in this complex is similar to the O...H distances in pyridine complexes with benzoic acid dimers [93]. However, the formation of the 1:2 adducts discussed earlier cannot be ruled out. In experimental studies of such systems, a balance must be struck between competing factors: the solvent must be sufficiently inert so as not to compete with the cation for interaction with TEMPO, while still maintaining sufficient solubility of the cation.

The *cis*-conformation of the 2,2′-bipyridine-H^+ ([H-BiPy]$^+$) cation is stabilized by a weak intramolecular hydrogen bond [94]. This bond plays a critical role in facilitating a reversible intramolecular proton transfer [95]. Figure 4d,e shows the structures of the 1:1 complexes of TEMPO with [H-BiPy]$^+$ and [H-Phen]$^+$. Both the geometry and $A_{iso}(^{14}N)$ value were calculated using the PCM approximation in toluene. The significant differences in $A_{iso}(^{14}N)$ and the O...H distances in these complexes are likely attributed to the effect of TEMPO on their intramolecular bonds, as seen in Table 5. In [H-Phen]$^+$, this effect appears to be more pronounced compared to [H-BiPy]$^+$. Nevertheless, the changes in $A_{iso}(^{14}N)$ should be readily detectable experimentally for both complexes. It would be instructive to experimentally investigate whether a weak interaction with TEMPO can influence the energy balance in [H-BiPy]$^+$ and alter its conformation from *cis* to *trans*.

2.6. Interaction of TEMPO with a Strong Halogen Bond Donor

Figure 5 shows the structure of the 1:1 complex of TEMPO with iodo 2,2,2-trifluoroacetate. Both the geometry and $A_{iso}(^{14}N)$ value were calculated using the PCM approximation in hexane. The choice of such a strong halogen bond donor [96] and a low polarity solvent was made deliberately to evaluate the maximum changes in $A_{iso}(^{14}N)$ expected for halogen bonding. The calculated $A_{iso}(^{14}N)$ value of TEMPO in hexane, equal to 1.4996 mT, increases to 1.8635 mT in this complex. The presence of such complexes in solution should be easily detected, even if their mole fraction is low.

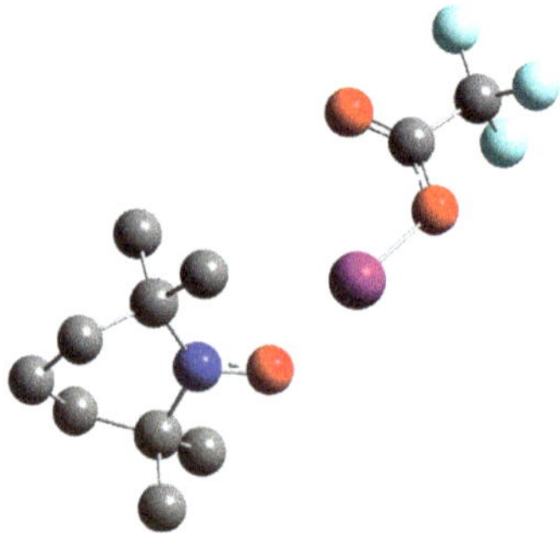

Figure 5. The model structure of the 1:1 complexes of TEMPO with iodo 2,2,2-trifluoroacetate in hexane. The calculated $A_{iso}(^{14}N)$ value in this complex is 1.8635 mT. The calculated $A_{iso}(^{14}N)$ value of TEMPO in hexane is 1.4996 mT. Oxygen atoms are represented in red, the nitrogen atom in blue, carbon atoms in grey, and the iodine atom in purple.

3. Materials and Methods

The Gaussian 09.D.01 program package was used for geometry optimizations and EPR calculations [97], employing the ωB97XD/def2tzvp and uPBE1PBE/LanL2DZ DFT functional/basis set pairs, respectively [98–101]. The optimized geometries and the results of selected EPR calculations are reported in the Supplementary Materials. All calculations were performed using the PCM approximation. Although the results of such calculations are not highly sensitive to the specific value of the dielectric constant, applying this correction is important to account for the effects of the surrounding medium [102]. This study does not address whether other solvent models, such as the Solvation Model based on Density (SMD), the Conductor-like Polarizable Continuum Model (CPCM), or the Conductor-like Screening Model (COSMO), provide better approximations for medium effects. There is no reason to expect that TEMPO is a particularly convenient model system for answering this question. It is quite possible that when using one of these models, the uPBE1PBE/LanL2DZ DFT functional/basis set pair will no longer be the most suitable for EPR calculations. However, a more suitable pair can be easily found for the solvent model in question using the approach outlined in Table 1.

4. Conclusions

The isotropic hyperfine coupling constant $A_{iso}(^{14}N)$ of the stable nitroxide radical TEMPO can be measured experimentally with an accuracy of 0.001 mT. The variation in this constant due solely to the polarity of the solvent is about 0.1 mT. The variations due to halogen and hydrogen bonding are greater than 0.3 mT and 0.5 mT, respectively. Therefore, $A_{iso}(^{14}N)$ serves as a highly sensitive experimental probe for detecting the effects of the surrounding medium and non-covalent interactions.

$A_{iso}(^{14}N)$ can be calculated for model systems of required complexity with reasonable accuracy using the uPBE1PBE/LanL2DZ approximation, which requires only moderate computational effort. If higher accuracy is desired, more sophisticated methods can be employed [103].

However, the most challenging aspect of such calculations is not the accuracy of the computational method but accounting for the effects of the surrounding medium and the presence of a variety of non-covalently bound complexes in solution, with diverse compositions and geometries. Each of these complexes exhibits a different $A_{iso}(^{14}N)$ value. The structure and linewidth of the experimental spectrum depend on the relative concentration of these complexes, their lifetimes, and their tumbling times. If a study does not involve frozen solutions [104–106], interpreting the experimental spectra in terms of the structure and geometry of the potential complexes will require evaluating numerous model structures, the relative average concentration of which, in solution, cannot always be reliably predicted based on general assumptions [107]. Ideally, additional experimental measurements at very low concentrations should also be performed. In any case, for TEMPO, the list of possible model structures must include complexes featuring two conjugated hydrogen bonds.

This work examines the influence of various non-covalent interactions on $A_{iso}(^{14}N)$ without a thorough analysis of the changes in the electron density distribution caused by these interactions, which, in turn, determines the observed changes in the $A_{iso}(^{14}N)$ value. However, conducting such an analysis at a reliable level would require a separate, specialized study based on the implementation of appropriate experimental studies. Such studies are beyond the scope of this paper, although the results obtained here can help in planning the practical part of such studies.

Supplementary Materials: The following supporting information can be downloaded at: https://www.mdpi.com/article/10.3390/molecules29215032/s1, Table S1: The optimized molecular structure of TEMPO in toluene; Table S2: The optimized molecular structure of TEMPO in DMSO; Table S3: The optimized molecular structure of TEMPO in n-Hexane; Table S4: The optimized molecular structure of 1:1 complex of TEMPO and water; Table S5: The optimized molecular structure of 1:1 complex of TEMPO and methanol; Table S6: The optimized molecular structure of 1:1 complex of TEMPO and ethanol; Table S7: The optimized molecular structure of 1:1 complex of TEMPO and Butan-2-ol; Table S8: The optimized molecular structure of 1:1 complex of TEMPO and 2-methylpropan-2-ol; Table S9: The optimized molecular structure of 1:1 complex of TEMPO and phenol; Table S10: The optimized molecular structure of 1:1 complex of TEMPO and 4-nitrophenol; Table S11: The optimized molecular structure of 1:1 complex of TEMPO and 2,5-dinitrophenol; Table S12: The optimized molecular structure of 1:1 complex of TEMPO and 2,4,6-trinitrophenol; Table S13: The optimized molecular structure of $[TEMPO\text{-}H]^+$; Table S14: The optimized molecular structure of $[TEMPO\text{-}2H]^{2+}$; Table S15: The optimized molecular structure of 1:1 complex of TEMPO and HF; Table S16: The optimized molecular structure of 1:2 complex of TEMPO and HF; Table S17: The optimized molecular structure of 1:1 complex of TEMPO and $[H\text{-pyridine}]^+$; Table S18: The optimized molecular structure of 1:1 complex of TEMPO and $[H\text{-BiPy}]^+$; Table S19: The optimized molecular structure of 1:1 complex of TEMPO and [H-Phen]$^+$; Table S20: The optimized molecular structure of 1:1 complex of TEMPO and iodo 2,2,2-trifluoroacetate; Table S21: The isotropic Fermi contact couplings in TEMPO in toluene; Table S22: The isotropic Fermi contact couplings in TEMPO in DMSO; Table S23: The isotropic Fermi contact couplings in TEMPO in n-hexane; and Table S24: The isotropic Fermi contact couplings in 1:1 complex of TEMPO and water.

Funding: This research received no external funding.

Institutional Review Board Statement: Not applicable.

Informed Consent Statement: Not applicable.

Data Availability Statement: Data are contained within the article and Supplementary Materials.

Acknowledgments: The author thanks Anton Savitsky for assistance with the literature searches and motivation to write this article, and Gábor Balázs for help with conducting model EPR measurements.

Conflicts of Interest: The author declares no conflict of interest.

References

1. Vega, S. Dynamic Nuclear Polarization with or without Spin Diffusion? *Online Lect.* 2011. Available online: https://www.youtube.com/watch?v=-7xu7a0YQy4 (accessed on 7 October 2024).
2. Asanbaeva, N.B.; Dobrynin, S.A.; Morozov, D.A.; Haro-Mares, N.; Gutmann, T.; Buntkowsky, G.; Bagryanskaya, E.G. An EPR Study on Highly Stable Nitroxyl-Nitroxyl Biradicals for Dynamic Nuclear Polarization Applications at High Magnetic Fields. *Molecules* **2023**, *28*, 1926. [CrossRef] [PubMed]
3. Herr, K.; Fleckenstein, M.; Brodrecht, M.; Höfler, M.V.; Heise, H.; Aussenac, F.; Gutmann, T.; Reggelin, M.; Buntkowsky, G. A novel strategy for site selective spin-labeling to investigate bioactive entities by DNP and EPR spectroscopy. *Sci. Rep.* **2021**, *11*, 13714. [CrossRef] [PubMed]
4. van der Heijden, G.H.A.; Kentgens, A.P.M.; van Bentum, P.J.M. Liquid state dynamic nuclear polarization of ethanol at 3.4 T (95 GHz). *Phys. Chem. Chem. Phys.* **2014**, *16*, 8493–8502. [CrossRef] [PubMed]
5. Russ, J.L.; Gu, J.; Tsai, K.-H.; Glass, T.; Duchamp, J.C.; Dorn, H.C. Nitroxide/Substrate Weak Hydrogen Bonding: Attitude and Dynamics of Collisions in Solution. *J. Am. Chem. Soc.* **2007**, *129*, 7018–7027. [CrossRef] [PubMed]
6. Sert, I.; Sünnetçioğlu, M.M.; Sungur, R.; Bingöl, G. Dynamic Nuclear Polarization Studies of TANOL/water-glycerol Solutions. *Z. Für Naturforschung A* **2000**, *55*, 682–686. [CrossRef]
7. Waterton, J.C.; Sanders, J.K.M. Radical-induced pseudocontact shifts. *J. Am. Chem. Soc.* **1978**, *100*, 1295–1296. [CrossRef]
8. Qiu, Z.W.; Grant, D.M.; Pugmire, R.J. Paramagnetic Carbon- 13 Shifts Induced by the Free Radical 2,2,6,6-Tetramethylpiperidinyl-1-oxy. 1. Simple Aromatic and Paraffinic Hydrocarbons. *J. Am. Chem. Soc.* **1982**, *104*, 2747–2753. [CrossRef]
9. Qiu, Z.W.; Grant, D.M.; Pugmire, R.J. Paramagnetic carbon-13 shifts induced by the free radical Tempo. 2. Nitrogen heterocycles. *J. Am. Chem. Soc.* **1984**, *106*, 557–563. [CrossRef]
10. Sysoeva, N.A.; Pekhk, T.I.; Lippmaa, É.T.; Buchachenko, A.L. NMR in paramagnetic radical complexes with organic ligands. III. Paramagnetic shifts of C^{13}. *J. Struct. Chem.* **1972**, *13*, 392–395. [CrossRef]
11. Draney, D.; Kingsbury, C.A. Free radical induced nuclear magnetic resonance shifts: Comments on contact shift mechanisms. *J. Am. Chem. Soc.* **1981**, *103*, 1041–1047. [CrossRef]
12. Winnik, F.M.; Ottaviani, M.F.; Bossman, S.H.; Pan, W.; Garcia-Garibay, M.; Turro, N.J. Phase separation of poly (N-isopropylacrylamide) in water: A spectroscopic study of a polymer tagged with a fluorescent dye and a spin label. *J. Phys. Chem.* **1993**, *97*, 12998–13005. [CrossRef]
13. Wawer, I.; Ströhl, D.; Kleinpeter, E. Nitroxide radical as a probe for molecular conformation. Part I. ^{13}C NMR study of the interaction of N,N-dimethylamides with nitroxide radical. *Magn. Reason. Chem.* **1993**, *31*, 758–762. [CrossRef]
14. Endo, K.; Hazama, Y.; Okabayashi, K.; Tonoike, I.; Suzuki, K. Pressure dependence of NMR spectroscopy for studying intermolecular interactions ^{1}H NMR in organic solvents involving the nitroxide radical. *Chem. Phys. Lett.* **1979**, *61*, 336–338. [CrossRef]
15. Sysoeva, N.A.; Buchachenko, A.L. NMR of paramagnetic complexes of radicals with organic ligands. I. Complexes with Zydroxyl-Containing Ligands. *J. Struct. Chem.* **1972**, *13*, 35–41. [CrossRef]
16. Sysoeva, N.A.; Buchachenko, A.L. NMR of paramagnetic complexes of radicals with organic ligands. I. Method of determination of lifetimes and distances in complexes. *Chem. Phys.* **1975**, *7*, 123–129. [CrossRef]
17. Sysoeva, N.A.; Karmilov, A.Y.; Buchachenko, A.L. NMR in paramagnetic complexes of radicals with organic ligands. II. Method of identification of electronic structure of complexes—Theory. *Chem. Phys.* **1976**, *15*, 313–319. [CrossRef]
18. Sysoeva, N.A.; Karmilov, A.Y.; Buchachenko, A.L. NMR in paramagnetic complexes of radicals with organic ligands: III. Method of identification of electronic structure of complexes—Experiment. *Chem. Phys.* **1976**, *15*, 321–330. [CrossRef]
19. Borah, B.; Bates, R.D., Jr. Interspecies hydrogen bonding between a nitroxide spin label and perfluoro-t-butanol studied by DNP and ^{1}H and ^{19}F relaxation measurements. *Chem. Phys. Lett.* **1980**, *76*, 101–104. [CrossRef]

20. Borah, B.; Bates, R.D., Jr. Dynamics of interspecies hydrogen bonding of a fluorinated alcohol and a nitroxide free radical. *J. Chem. Phys.* **1981**, *74*, 1538–1545. [CrossRef]
21. Bundfuss, K.; Meise-Gresch, K.; Müller-Warmuth, W. Hydrogen bond formation and molecular motion in solutions of chlorofluoromethanes containing nitroxide free radicals. *J. Magn. Res.* **1983**, *55*, 408–420. [CrossRef]
22. Darges, G.; Müller-Warmuth, W. NMR T_1 study of dipolar and scalar spectral densities for intermolecular relaxation in solutions of free radicals over a large frequency and temperature range. *J. Magn. Res.* **1985**, *65*, 444–458. [CrossRef]
23. Nientiedt, H.-W.; Bundfuss, K.; Müller-Warmuth, W. Frequency and temperature dependence of NMR relaxation for protic solvents with complexes of nitroxide radicals. *J. Magn. Res.* **1981**, *43*, 154–166. [CrossRef]
24. Sysoeva, N.A.; Stepanyants, A.U.; Buchachenko, A.L. Paramagentic shift of solvent protons in the presence of organic radicals. *J. Struct. Chem.* **1968**, *9*, 248–249. [CrossRef]
25. Sysoeva, N.A.; Buchachenko, A.L. NMR in complexes of radicals with organic ligands II. Amines, halogen-containing compounds and unsaturated compounds. *J. Struct. Chem.* **1972**, *13*, 205–208. [CrossRef]
26. Grabowski, S.J. What Is the Covalency of Hydrogen Bonding? *Chem. Rew.* **2011**, *111*, 2597–2625. [CrossRef]
27. Alkorta, I.; Rozas, I.; Elguero, J. Non-conventional hydrogen bonds. *Chem. Soc. Rev.* **1998**, *27*, 163170. [CrossRef]
28. Samitov, Y.Y.; Ibraimov, D. NMR study of molecular interaction in solutions of 2,2,6,6-tetramethylpentamethylenenitroxyl. *Theor. Exp. Chem.* **1965**, *1*, 251–254. [CrossRef]
29. Koodziejski, W.; Kęcki, Z. ^{13}C NMR study of alcohols complexed with nitroxide radical. *Chem. Phys. Lett.* **1978**, *54*, 286–288. [CrossRef]
30. Bonesteel, J.-A.K.; Borah, B.; Bates, R.D., Jr. Effects of nitroxide free radicals on proton donor chemical shifts. *J. Mang. Reason.* **1992**, *98*, 475–482. [CrossRef]
31. Dally, E.; Müller-Warmuth, W. NMR-Untersuchungen an Lösungen mit Nitroxidradikalkomplexen. II. Hyperfeinkopplungskonstanten, Linienbreiten und Relaxation von Komplexen mit n-Propanol, Isopropanol und tert.-Butanol. *Ber. Bunsenges. Phys. Chem.* **1978**, *82*, 792–798. [CrossRef]
32. Dally, E.; Müller-Warmuth, W. NMR-Untersuchungen an Lösungen mit Nitroxidradikalkomplexen. III. Komplexe mit Äthanol, Isoamylalkohol, n-Dekanol, Glykol, Cyclohexanol und Triäthylcarbinol. *Ber. Bunsenges. Phys. Chem.* **1980**, *84*, 260–265. [CrossRef]
33. Endo, K.; Morishima, I.; Yonezawa, T. Use of a stable free radical as a NMR spin probe for studying intermolecular interactions. XIV. A proton relaxation study of the hydrogen bond involving a stable free radical. *J. Chem. Phys.* **1977**, *67*, 4760–4767. [CrossRef]
34. Morishima, I.; Endo, K.; Yonezawa, T. Nuclear magnetic resonance contact shifts induced by hydrogen bonding with organic radicals. I. ^{1}H and ^{13}C contact shifts of protic molecules in the presence of the nitroxide radical. *J. Am. Chem. Soc.* **1971**, *93*, 2048–2050. [CrossRef]
35. Wawer, I.; Krysiak, T.; Kęcki, Z. NMR and EPR study of hydrogen bonding in chloroanilines. *J. Mol. Struct.* **1994**, *326*, 163–170. [CrossRef]
36. Stark, U.; Müller-Warmuth, W. Hydrogen Bond Formation and Molecular Motion in Solutions of Aniline with Nitroxide Free Radicals. *Ber. Bunsenges. Phys. Chem.* **1990**, *94*, 168–172. [CrossRef]
37. Dally, E.; Müller-Warmuth, W. Hochauflösende NMR an Lösungen mit Nitroxidradikalkomplexen. *Ber. Bunsenges. Phys. Chem.* **1977**, *81*, 1133–1137. [CrossRef]
38. Endo, K.; Knuettel, B.; Morishima, I.; Inubushi, T.; Yonezawa, T. Interaction between closed-shell and open-shell molecules. ^{1}H relaxation studies of the hydrogen bond between proton-donor molecules and a nitroxide radical. *Chem. Phys. Lett.* **1975**, *31*, 387–391. [CrossRef]
39. Meise, K.; Müller-Warmuth, W.; Nientiedt, H.-W. Magnetische Resonanzuntersuchungen zur Struktur, Kinetik und Hyperfeinwechselwirkung des Systems Nitroxidradikal-Wasser. *Ber. Bunsenges. Phys. Chem.* **1976**, *80*, 584–590. [CrossRef]
40. Pokhodenko, V.D.; Slipenyuk, T.S.; Kuts, V.S. Effect of the electronic structure of a diamagnetic molecule on the formation of complexes with free radicals. *Theor. Exp. Chem.* **1976**, *11*, 39–43. [CrossRef]
41. Cramer, R.E.; Dahlstrom, P.L.; Heya, H. Fluorine-19 and hydrogen contact shifts of Lewis acids hydrogen bonded to a stable free radical. Non-Curie law behavior. *J. Phys. Chem.* **1975**, *79*, 376–380. [CrossRef]
42. Hatch, G.F.; Kreilick, R.W. NMR of Some Nitroxide Radicals: ^{13}C Coupling Constants. *J. Chem. Phys.* **1972**, *57*, 3696–3699. [CrossRef]
43. Kuts, V.S.; Radchenko, N.F.; Pokhodenko, V.D. Electronic structure of free-nitroxide-radical—Aromatic-molecule π complexes. *Theor. Exp. Chem.* **1981**, *16*, 447–453. [CrossRef]
44. Milton Franklin Benial, A.; Kumara Dhas, M.; Jawahar, A. Rotational Correlation Time Studies on Nitroxyl Radicals Using 300 MHz ESR Spectrometer in High Viscous Liquid. *Appl. Magn. Reason.* **2011**, *40*, 311–319. [CrossRef]
45. Strehmel, V.; Berdzinski, S.; Rexhausen, H. Interactions between ionic liquids and radicals. *J. Mol. Liq.* **2014**, *192*, 153–170. [CrossRef]
46. Noël, M.A.M.; Allendoerfer, R.D.; Osteryoung, R.A. Solvation in Ionic Liquids: An EPR Study. *J. Phys. Chem.* **1992**, *96*, 2391–2394. [CrossRef]
47. Švajdlenková, H.; Šauša, O.; Adichtchev, S.V.; Surovtsev, N.V.; Novikov, V.N.; Bartoš, J. On the Mutual Relationships between Molecular Probe Mobility and Free Volume and Polymer Dynamics in Organic Glass Formers: Cis-1,4-poly(isoprene). *Polymers* **2021**, *13*, 294. [CrossRef]

48. Walski, T.; Gałecka, T.; Grzeszczuk-Kuć, K.; Komorowska, M. Modification of the hydrogen bonding network at the reversed micelles interface by near infrared radiation. *Nukleonika* **2013**, *58*, 435–438. Available online: https://yadda.icm.edu.pl/baztech/element/bwmeta1.element.baztech-4e16d03d-8cc6-4fef-b50e-d6b60865d291/c/Modification_of_the_hydrogen_bonding_network.pdf (accessed on 30 June 2023).
49. Tedeschi, A.M.; D'Errico, G.; Busi, E.; Basosi, R.; Barone, V. Micellar aggregation of sulfonate surfactants studied by electron paramagnetic resonance of a cationic nitroxide: An experimental and computational approach. *Phys. Chem. Chem. Phys.* **2002**, *4*, 2180–2188. [CrossRef]
50. Liu, F.; Karoui, H.; Rockenbauer, A.; Liu, S.; Ouari, O.; Bardelang, D. EPR Spectroscopy: A Powerful Tool to Analyze Supramolecular Host•Guest Complexes of Stable Radicals with Cucurbiturils. *Molecules* **2020**, *25*, 776. [CrossRef]
51. Griffith, O.H.; Dehlinger, P.J.; Van, S.P. Shape of the hydrophobic barrier of phospholipid bilayers (Evidence for water penetration in biological membranes). *J. Membrain Biol.* **1974**, *15*, 159–192. [CrossRef]
52. Shenderovich, I.G.; Wawer, I.; Denisov, G.S.; Kecki, Z. NMR and EPR study of the nitroxide radical TEMPO interaction with phenols. *Spectrosc. Lett.* **1997**, *30*, 1515–1523. [CrossRef]
53. Kuts, V.S.; Slipenyuk, T.S.; Pokhodenko, V.D. Effect of the steric structure of molecules on complex formation in the free radical -2,6-dialkylphenol system. *Theor. Exp. Chem.* **1976**, *12*, 614–619. [CrossRef]
54. Kuts, V.S.; Slipenyuk, T.S.; Platonova, É.P.; Pokhodenko, V.D. Influence of the electronic structure of a free radical on complex-formation with a diamagnetic molecule. *Theor. Exp. Chem.* **1978**, *13*, 268–271. [CrossRef]
55. Morishima, I.; Ishihara, K.; Tomishima, K.; Inubushi, T.; Yonezawa, T. Nitroxide radical induced nuclear magnetic resonance contact shift studies. Potential utility of specific downfield ^{1}H contact shifts induced by hydrogen bonding with di-*tert*-butyl nitroxide radical. *J. Am. Chem. Soc.* **1975**, *97*, 2749–2756. [CrossRef]
56. Morishima, I.; Endo, K.; Yonezawa, T. Interaction between closed- and open-shell molecules. VI. ^{1}H and ^{13}C contact shifts and molecular orbital studies on the hydrogen bond of nitroxide radical. *J. Chem. Phys.* **1973**, *58*, 3146–3154. [CrossRef]
57. Knauer, B.R.; Napier, J.J. The nitrogen hyperfine splitting constant of the nitroxide functional group as a solvent polarity parameter. The relative importance for a solvent polarity parameter of its being a cybotactic probe vs. its being a model process. *J. Am. Chem. Soc.* **1976**, *98*, 4395–4400. [CrossRef]
58. Marsh, D. Reaction fields and solvent dependence of the EPR parameters of nitroxides: The microenvironment of spin labels. *J. Magn. Res.* **2008**, *190*, 60–67. [CrossRef]
59. Vogler, A.; Shenderovich, I.G. Photochemistry of deprotonated rhenium(I) (3,3′-dihydroxy-2,2′-bipyridine) tricarbonyl chloride. Photoisomerization at the chelate in basic solution. *Inorg. Chim. Acta* **2014**, *421*, 496–499. [CrossRef]
60. Shenderovich, I.G. Actual Symmetry of Symmetric Molecular Adducts in the Gas Phase, Solution and in the Solid State. *Symmetry* **2021**, *13*, 756. [CrossRef]
61. Gedat, E.; Schreiber, A.; Findenegg, G.H.; Shenderovich, I.; Limbach, H.-H.; Buntkowsky, G. Stray field gradient NMR reveals effects of hydrogen bonding on diffusion coefficients of pyridine in mesoporous silica. *Magn. Res. Chem.* **2001**, *39*, S149–S157. [CrossRef]
62. Shenderovich, I.G.; Denisov, G.S. Modeling of Solute-Solvent Interactions Using an External Electric Field—From Tautomeric Equilibrium in Nonpolar Solvents to the Dissociation of Alkali Metal Halides. *Molecules* **2021**, *26*, 1283. [CrossRef] [PubMed]
63. Morishima, I.; Inubushi, T.; Endo, K.; Yonezawa, T. Interaction between closed and open-shell molecules. V. ^{13}C contact shift and molecular orbital studies on the interaction between halogenated molecules and nitroxide radical. *Chem. Phys. Lett.* **1972**, *14*, 372–376. [CrossRef]
64. Morishima, I.; Inubushi, T.; Endo, K.; Yonezawa, T.; Goto, K. Interaction between Closed- and Open-Shell Molecules. VII. Carbon-13 Contact Shift and Molecular Orbital Studies on the Charge-Transfer Interaction between Halogenated Molecules and Nitroxide Radical. *J. Am. Chem. Soc.* **1972**, *94*, 4812–4818. [CrossRef]
65. Cavallo, G.; Metrangolo, P.; Milani, R.; Pilati, T.; Priimagi, A.; Resnati, G.; Terraneo, G. The Halogen Bond. *Chem. Rev.* **2016**, *116*, 2478–2601. [CrossRef]
66. Grabowski, S.J. Hydrogen and halogen bonds are ruled by the same mechanisms. *Phys. Chem. Chem. Phys.* **2013**, *15*, 7249–7259. [CrossRef]
67. Gleeson, R.; Andersen, C.L.; Rapta, P.; Machata, P.; Christensen, J.B.; Hammerich, O.; Sauer, S.P.A. A Combined Experimental and Theoretical Study of ESR Hyperfine Coupling Constants for N,N,N′,N′-Tetrasubstituted p-Phenylenediamine Radical Cations. *Int. J. Mol. Sci.* **2023**, *24*, 3447. [CrossRef]
68. Khafizov, N.R.; Mukhametgaleev, R.; Madzhidov, T.I.; Kadkin, O.N.; Antipin, I.S. Effect of core substituents on the intramolecular exchange interaction in N,N′-dioxy-2,6-diazaadamantane biradical: DFT studies. *Int. J. Quantum Chem.* **2018**, *118*, e25568. [CrossRef]
69. Shil, S. Effect of deprotonation on the magnetic exchange coupling constant of fluorene-based verdazyl diradical: A computational study. *Int. J. Quantum Chem.* **2018**, *118*, e25543. [CrossRef]
70. Wawer, I.; Kęcki, Z.; Denisow, G.S. Hydrogen bonding in fluoroanilines studied radical probe. *J. Mol. Struct.* **1994**, *327*, 313–319. [CrossRef]
71. Shenderovich, I.G. Electric field effect on ^{31}P NMR magnetic shielding. *J. Chem. Phys.* **2020**, *153*, 184501. [CrossRef]
72. Shenderovich, I.G.; Denisov, G.S. Modeling of the Response of Hydrogen Bond Properties on an External Electric Field: Geometry, NMR Chemical Shift, Spin-Spin Scalar Coupling. *Molecules* **2021**, *26*, 4967. [CrossRef] [PubMed]

73. Shenderovich, I.G.; Denisov, G.S. Adduct under Field—A Qualitative Approach to Account for Solvent Effect on Hydrogen Bonding. *Molecules* **2020**, *25*, 436. [CrossRef] [PubMed]
74. Kolodziejski, W.; Kęcki, Z. ESR and PMR Studies of Alcohols Hydrogen Bonded with Nitroxide Radical. *Ber. Bunsenges. Phys. Chem.* **1978**, *82*, 1312–1314. [CrossRef]
75. Windle, J.J. Hyperfine coupling constants for nitroxide spin probes in water and carbon tetrachloride. *J. Magn. Res.* **1981**, *45*, 432–439. [CrossRef]
76. Malatesta, V.; Ingold, K.U. Protonated nitroxide free radical. *J. Am. Chem. Soc.* **1973**, *95*, 6404–6407. [CrossRef]
77. Uçar, E.; Dadayli, D.; Sünnetçioğlu, M.M.; Sungur, R. An EPR Analysis of Nitroxide Spin Probes in Ethyl Alcohol and Tetrahydrofuran. *Phys. Chem. Liq.* **1999**, *37*, 323–331. [CrossRef]
78. Hoffman, B.M.; Thomas, B.; Eames, T.B. Protonated nitroxide free radical. *J. Am. Chem. Soc.* **1969**, *91*, 2169–2170. [CrossRef]
79. Tachikawa, T.; Akiyama, K.; Yokoyama, C.; Tero-Kubota, S. Local density effects on the hyperfine splitting constant and line width of TEMPO radical in gaseous and supercritical carbon dioxide. *Chem. Phys. Lett.* **2003**, *376*, 350–357. [CrossRef]
80. Sharif, S.; Shenderovich, I.G.; González, L.; Denisov, G.S.; Silverman, D.N.; Limbach, H.-H. NMR and Ab initio Studies of Small Complexes Formed between Water and Pyridine Derivatives in Solid and Liquid Phase. *J. Phys. Chem. A* **2007**, *111*, 6084–6093. [CrossRef]
81. Andreeva, D.V.; Ip, B.; Gurinov, A.A.; Tolstoy, P.M.; Denisov, G.S.; Shenderovich, I.G.; Limbach, H.-H. Geometrical Features of Hydrogen Bonded Complexes Involving Sterically Hindered Pyridines. *J. Phys. Chem. A* **2006**, *110*, 10872–10879. [CrossRef]
82. Ma, Y.; Loyns, C.; Price, P.; Chechik, V. Thermal decay of TEMPO in acidic media via an N-oxoammonium salt intermediate. *Org. Biomol. Chem.* **2011**, *9*, 5573–5578. [CrossRef] [PubMed]
83. Sen′, V.D.; Golubev, V.A. Kinetics and mechanism for acid-catalyzed disproportionation of 2,2,6,6-tetramethylpiperidine-1-oxyl. *J. Phys. Org. Chem.* **2009**, *22*, 138–143. [CrossRef]
84. Golubev, V.A.; Rozantsev, É.G.; Neiman, M.B. Some reactions of free iminoxyl radicals with the participation of the unpaired electron. *Russ. Chem. Bull.* **1965**, *14*, 1898–1904. [CrossRef]
85. Hoffman, A.K.; Henderson, A.T. A New Stable Free Radical: DI-t-Butylnitroxide. *J. Am. Chem. Soc.* **1961**, *83*, 4671–4672. [CrossRef]
86. Tupikina, E.Y.; Bodensteiner, M.; Tolstoy, P.M.; Denisov, G.S.; Shenderovich, I.G. P=O Moiety as an Ambidextrous Hydrogen Bond Acceptor. *J. Phys. Chem. C* **2018**, *122*, 1711–1720. [CrossRef]
87. Lyssenko, K.A.; Grintselev-Knyazev, G.V.; Antipin, M.Y. Nature of the P–O Bond in Diphenylphosphonic Acid: Experimental Charge Density and Electron Localization Function Analysis. *Mendeleev Commun.* **2002**, *12*, 128–130. [CrossRef]
88. Chernyshov, I.Y.; Vener, M.V.; Shenderovich, I.G. Local-structure effects on ^{31}P NMR chemical shift tensors in solid state. *J. Chem. Phys.* **2019**, *150*, 144706. [CrossRef]
89. Arp, F.F.; Bhuvanesh, N.; Blümel, J. Hydrogen peroxide adducts of triarylphosphine oxides. *Dalton Trans.* **2019**, *48*, 14312–14325. [CrossRef]
90. Byrne, E.L.; O'Donnell, R.; Gilmore, M.; Artioli, N.; Holbrey, J.D.; Swadźba-Kwaśny, M. Hydrophobic functional liquids based on trioctylphosphine oxide (TOPO) and carboxylic acids. *Phys. Chem. Chem. Phys.* **2020**, *22*, 24744–24763. [CrossRef]
91. Ivshin, K.; Metlushka, K.; Zinnatullin, R.; Nikitina, K.; Pashagin, A.; Zakharychev, D.V.; Efimova, A.; Kiiamov, A.; Latypov, S.; Kataeva, O. Competitive Hydrogen Bonding and Unprecedented Polymorphism in Selected Chiral Phosphorylated Thioureas. *Cryst. Growth Des.* **2021**, *21*, 5460–5471. [CrossRef]
92. Hoefler, J.C.; Vu, A.; Perez, A.J.; Blümel, J. Immobilized di(hydroperoxy)propane adducts of phosphine oxides as traceless and recyclable oxidizing agents. *Appl. Surf. Sci.* **2023**, *629*, 157333. [CrossRef]
93. Borissova, A.O.; Lyssenko, K.A.; Gurinov, A.A.; Shenderovich, I.G. Energy Analysis of Competing Non-Covalent Interaction in 1:1 and 1:2 Adducts of Collidine with Benzoic Acids by Means of X-Ray Diffraction. *Z. Phys. Chem.* **2013**, *227*, 775–790. [CrossRef]
94. Shenderovich, I.G. Weak, Broken, but Working—Intramolecular Hydrogen Bond in 2,2′-bipyridine. *Int. J. Mol. Sci.* **2023**, *24*, 10390. [CrossRef] [PubMed]
95. Lesnichin, S.B.; Tolstoy, P.M.; Limbach, H.-H.; Shenderovich, I.G. Counteranion-Dependent Mechanisms of Intramolecular Proton Transfer in Aprotic Solution. *Phys. Chem. Chem. Phys.* **2010**, *12*, 10373–10379. [CrossRef] [PubMed]
96. Yu, S.; Ward, J.S.; Truong, K.-N.; Rissanen, K. Carbonyl Hypoiodites as Extremely Strong Halogen Bond Donors. *Angew. Chem. Int. Ed.* **2021**, *60*, 20739–20743. [CrossRef]
97. Frisch, M.J.; Trucks, G.W.; Schlegel, H.B.; Scuseria, G.E.; Robb, M.A.; Cheeseman, J.R.; Scalmani, G.; Barone, V.; Petersson, G.A.; Nakatsuji, H.; et al. *Gaussian 09; Revision D.01*; Gaussian, Inc.: Wallingford, CT, USA, 2013.
98. Chai, J.-D.; Head-Gordon, M. Long-range corrected hybrid density functionals with damped atom–atom dispersion corrections. *Phys. Chem. Chem. Phys.* **2008**, *10*, 6615–6620. [CrossRef]
99. Weigend, F.; Ahlrichs, R. Balanced basis sets of split valence, triple zeta valence and quadruple zeta valence quality for H to Rn: Design and assessment of accuracy. *Phys. Chem. Chem. Phys.* **2005**, *7*, 3297–3305. [CrossRef]
100. Dunning, T.H., Jr.; Hay, P.J. Gaussian basis sets for molecular calculations. In *Methods of Electronic Structure Theory*; Schaefer, H.F., Ed.; Springer: New York, NY, USA, 1977; Volume 3, pp. 1–27. [CrossRef]
101. Ernzerhof, M.; Perdew, J.P. Generalized gradient approximation to the angle- and system-averaged exchange hole. *J. Chem. Phys.* **1998**, *109*, 3313–3320. [CrossRef]
102. Shenderovich, I.G. Simplified Calculation Approaches Designed to Reproduce the Geometry of Hydrogen Bonds in Molecular Complexes in Aprotic Solvents. *J. Chem. Phys.* **2018**, *148*, 124313. [CrossRef]

103. Vogler, S.; Dietschreit, J.C.B.; Peters, L.D.M.; Ochsenfeld, C. Important components for accurate pyperfine coupling constants: Electron correlation, dynamic contributions, and solvation effects. *Mol. Phys.* **2020**, *118*, e1772515. [CrossRef]
104. Savitsky, A.; Nalepa, A.; Petrenko, T.; Plato, M.; Möbius, K.; Lubitz, W. Hydrogen-Bonded Complexes of Neutral Nitroxide Radicals with 2-Propanol Studied by Multifrequency EPR/ENDOR. *Appl. Magn. Reason.* **2022**, *53*, 1239–1263. [CrossRef]
105. Möbius, K.; Savitsky, A.; Malferrari, M.; Francia, F.; Mamedov, M.D.; Semenov, A.Y.; Lubitz, W.; Venturoli, G. Soft Dynamic Confinement of Membrane Proteins by Dehydrated Trehalose Matrices: High-Field EPR and Fast-Laser Studies. *Appl. Magn. Reason.* **2020**, *51*, 773–850. [CrossRef]
106. Nalepa, A.; Möbius, K.; Plato, M.; Lubitz, W.; Savitsky, A. Nitroxide Spin Labels—Magnetic Parameters and Hydrogen-Bond Formation: A High-Field EPR and EDNMR Study. *Appl. Magn. Reason.* **2019**, *50*, 1–16. [CrossRef]
107. Wang, K.; He, X.; Rong, C.; Zhong, A.; Liu, S.; Zhao, D. On the origin and nature of internal methyl rotation barriers: An information-theoretic approach study. *Theor. Chem. Acc.* **2022**, *141*, 68. [CrossRef]

molecules

MDPI

Article

Intermolecular Interactions between Aldehydes and Alcohols: Conformational Equilibrium and Rotational Spectra of Acrolein-Methanol Complex

Dingding Lv [1], David Sundelin [2], Assimo Maris [1], Luca Evangelisti [1], Wolf Dietrich Geppert [2,*] and Sonia Melandri [1,*]

[1] Dipartimento di Chimica "G. Ciamician" Università di Bologna, Via Selmi 2, I-40126 Bologna, Italy
[2] Fysikum, Stockholm University, Roslagstullsbacken 21, 106 91 Stockholm, Sweden; david.sundelin@fysik.su.se
* Correspondence: wgeppert@fysik.su.se (W.D.G.); sonia.melandri@unibo.it (S.M.)

Abstract: The rotational spectra of the 1:1 complex formed by acrolein and methanol and its deuterated isotopologues have been analyzed. Two stable conformations in which two hydrogen bonds between the two moieties are formed were detected. The rotational lines show a hyperfine structure due to the methyl group internal rotation in the complex and the V_3 barriers hindering the motion were determined as 2.629(5) kJ mol^{-1} and 2.722(5) kJ mol^{-1} for the two conformations, respectively. Quantum mechanical calculations at the MP2/aug-cc-pVTZ level and comprehensive analysis of the intermolecular interactions, utilizing NCI and SAPT approaches, highlight the driving forces of the interactions and allow the determination of the binding energies of complex formation.

Keywords: acrolein; methanol; hydrogen bond; microwave spectroscopy; molecular structure; methyl internal rotation

1. Introduction

Hydrogen bonds (HBs) have long been of paramount interest to the chemical community. Some types of molecules (for example, alcohols [1], acids [2] and amides [3]) are very prone to form HBs, both as donors and acceptors. The conformational preference in forming HBs affects the energetics and kinetics of chemical reactions and the structure of biological macromolecules [4,5]. Complexes involving aldehydes and alcohols are particularly interesting systems in HB studies because they often contain a combination of stronger HBs and weaker secondary interactions. This study of clusters of aldehydes and alcohols aims to determine the interaction sites and relative arrangement of monomers and determine the intermolecular forces at play, and it lays the foundation for studies of larger systems.

In the past few decades, the scientific community has made great progress in the understanding of the nature and properties of the hydrogen bond. Hydrogen-bonded clusters have enjoyed considerable attention from the scientific community due to their importance in many fields of chemistry, e.g., in atmospheric and biological processes. They also can serve as simple model systems to aid the understanding of the structures and intermolecular interactions in bulk solutions. In this respect, clusters by water and simple organic molecules like alcohols, which are often used as solvents in synthetic processes (both in the lab and industries), play an eminent role. The cluster bonding between water and alcohols with organic molecules can be very strong. This especially holds for species in which the water or the alcohol can act both as a hydrogen acceptor and a hydrogen donor, like in the case of 2-aminopyridine-water [6]. The formation of clusters of organic compounds with small molecules can also change the geometry of the species. For example, it was found that the clustering of benzaldehyde with a larger number of water molecules distorts the planar structure of the benzaldehyde moiety [7]. Very recently, a similar behavior was detected for water clusters of salicylic acid [8]. Also, in crown ethers, a change in the conformation of the organic molecule upon clustering with water was observed [9]. Very often, several isomers with similar energies exist for organic water clusters, so high-level ab initio calculations combined with rotational spectroscopy are necessary to pin down the correct geometries of the clusters [10]. In a number of instances, more than one isomer has been observed, as in the cases of 2- and 3-furonitrile [11], and for the formamide-water cluster, even three different isomers could be detected [12]. Furthermore, the clustering of organic molecules, especially α-unsaturated aldehydes like acroleine, can affect their reactivity and the product distributions of the reactions they undergo [13–15]. Since acrolein is a major pollutant in the terrestrial atmosphere, this is of high relevance.

Moving from water to methanol clusters increases the complexity of the spectra by introducing the possibility of internal rotation of the methyl group. This phenomenon and the associated rotational barrier have since long been a subject of great attention from the spectroscopic community. For example, the cluster of thioacetaldehyde was investigated by Judge et al. already in 1987 [16]. Clustering of methanol tends to lower the internal rotational barrier. Earlier, this was attributed to the complexation, but more recent research efforts have interpreted it as being due to an internal motion coupled to the internal rotation of the methyl group, namely a librational motion of the entire MeOH moiety, the methyl group or even the hydroxyl group [17,18]. Nevertheless, further studies of different methanol clusters are necessary to fully understand this phenomenon, since such endeavors have implications for understanding the stability and interactions of complexes involving aldehydes and alcohols, contributing to the broader field of chemical and biological molecular research. Expanding the number of systems studied also makes it possible to identify the influence of changes in the nature of substituents on the nature and energy of non-covalent binding and the existence of various conformers of adducts.

Acrolein (from here on ACR) is also one of the most important environmental pollutants and plays an important role in the formation of atmospheric aerosols [19–21]. It is also a very important synthetic intermediate in the chemical industry and is widely used in resin production and organic synthesis [22]. Two conformations of ACR, s-*trans* and s-*cis*, have been characterized by microwave spectroscopy [23–29], pointing out that the s-*trans* form is more stable than the s-*cis* one by 9.2(5) kJ mol^{-1} [30]. The presence of those two different conformers of the ACR monomer can make the investigation of the complex formation more interesting, since the relative energy of the conformations can change upon complexation. The rotational spectra of ACR involved in binary and ternary water complexes and some of their isotopologues have been analyzed [31]. All the complexes show a planar heavy atom skeleton and are stabilized by ring-like structures. In the binary complex ACR-water [31], the conformation where the aldehyde oxygen forms an HB with the water hydrogen and the water oxygen forms an HB with the methylene group are most stable and the ACR adopts the s-*trans* form. The electron pairs located at the aldehyde oxygen are quite efficient proton acceptors that can easily form an HB with proton donors, like water, alcohols or other protonated compounds. At the same time, the oxygen atom of methanol (from here on abbreviated as MeOH) can easily form another interaction with the methylene hydrogen or aldehydic hydrogen atoms of ACR. In this case, it is possible that the 1:1 complex of ACR and MeOH will also show a ring-like structure where aldehyde oxygen interacts with hydroxyl hydrogen and hydroxyl oxygen interacts with methylenic hydrogen or aldehydic hydrogen.

The rotational spectra of the MeOH monomer and those of many of its clusters with other molecules have been studied [17,18,21,32–44]. The splitting of the rotational lines caused by the internal rotation of the methyl group was always observed and the determined barriers of the methyl group's internal rotation in the complexes were lower than the one determined for the MeOH monomer (4.463(4) kJ mol^{-1} [44]). For example, in the 1:1 complex of MeOH-SO_2 [37], the observed methyl top internal rotation barrier is

$V_3 = 1.54(1)$ kJ mol^{-1} and in the more loosely bound MeOH-Ar complex [39], an even more striking drop in the methyl torsional barrier height to about 0.82(2) kJ mol^{-1} was observed in the experiment. The model used to deduce the barrier height is based on the assumption that the methyl group rotates against a heavy frame, but the experimental data confirm that the barrier reductions are not a result of complexation effects, but rather are related to another large amplitude internal motion coupled to the internal rotation of the methyl group. In the note by Fraser et al. [17], the decrease in the MeOH internal rotation barrier observed in several methanol complexes with light partners is attributed to the effects of a librational motion either of the whole MeOH moiety or of the methyl group taking place in the complex. This might not be the case when a heavier moiety is complexed by MeOH. For example, in a more recent study of the 1:1 complex of acrylonitrile−MeOH in the gas phase using rotational spectroscopy [18], a consistent change was not observed in the determined internal rotation parameters upon deuteration of the whole methyl group, which led the authors to exclude that the decrease in the MeOH internal rotation barrier after complexation is related to the coupling of another internal motion of the whole MeOH moiety or the methyl group.

The study of the driving forces stabilizing the molecular complexes and the effects of complexation on the structure and the internal motions are the aims of this study.

Rotational spectroscopy provides high resolution and exceptional sensitivity to the molecular mass distribution and geometry, making it ideal for obtaining precise experimental evidence of tautomeric equilibria. In the present study, a COBRA-type pulsed supersonic-jet) spectrometer was employed to measure the radiofrequency spectrum of the ACR-methanol cluster. Deuterated isotopologues of methanol were used to validate the assignment of spectra, to assess the effects of deuteration on the A-E line splitting and the influence of deuteration upon the methyl top internal rotation barrier of the cluster and to compare those parameters with previously studied systems. The high spectral resolution of state-of-the-art Fourier transform microwave spectrometers allows the full resolution of closely spaced rotational spectra of isotopically labeled clusters. Ab initio at the MP2/aug-cc-pVTZ were employed to identify the structures of the different clusters and to determine their relative energies. Moreover, such calculations yield rotational constants and electric dipole moment components of the different conformations, which can aid the assignment of lines, identify experimentally observed isomers, and be used as benchmarks for the experimentally found parameters (and vice versa). The combination of rotational spectroscopy analysis and theoretical methods thus provides a synergistic method for studying the structure and internal dynamics of isolated molecules and weakly bound complexes.

2. Results and Discussion

2.1. Computational Results

For simplicity, the s-*trans* and s-*cis* conformations of ACR are labeled just *t* and *c*. Distinct conformers of ACR-MeOH, which originate from the same conformer of ACR have been further distinguished with a progressive number (1, 2 or 3). The choice of the initial geometries was based on the previous study on ACR-water [31]. In that case, four planar conformers, in which the water forms an HB with the ACR oxygen atom and a secondary interaction involving the water oxygen with either the methylenic or the aldehydic ACR hydrogen, were identified. In the ACR-MeOH case, the same kind of C_s arrangement is found, where the methanol frame lies in the plane of acrolein, but also, two additional non-planar conformers (one from each of ACR's conformation and indicated by the number 3) were characterized. In both cases, the methyl group is located almost perpendicular to the ACR's plane. The theoretical (MP2/aug-cc-pVTZ) spectroscopic parameters and relative energies of all conformers are reported in Table 1.

Table 1. Ab initio (MP2/aug-cc-pVTZ) structures, rotational constants, electric dipole moment components and relative energies of the different conformations of ACR-MeOH.

	***t*-ACR-MeOH-1**	***t*-ACR-MeOH-2**	***t*-ACR-MeOH-3**
	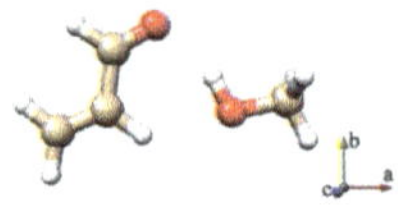	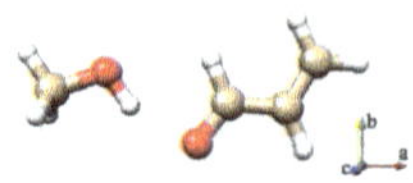	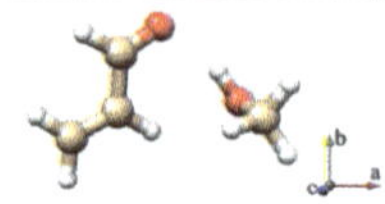
A, B, C/MHz	6233, 1390, 1145	11158, 1036, 954	4905, 1692, 1373
μ_a, μ_b, μ_c/D	−2.4, −0.5, 0.0	2.7, 0.6, 0.0	−3.2, −1.4, 1.6
ΔE_e/kJ mol^{-1}	0	3.7	1.4
ΔE_0/kJ mol^{-1}	0	3.2	1.3
	***c*-ACR-MeOH-1**	***c*-ACR-MeOH-2**	***c*-ACR-MeOH-3**
	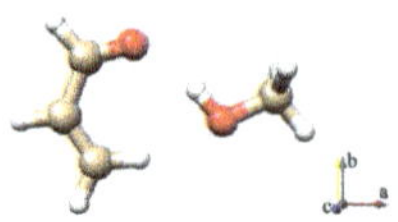	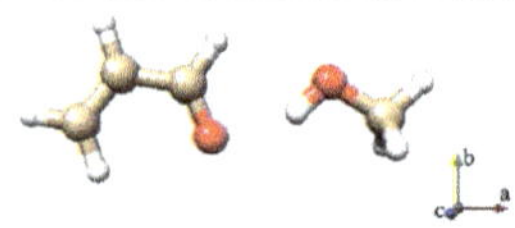	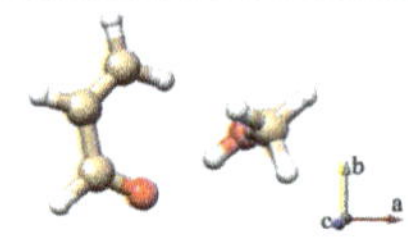
A, B, C/MHz	5974, 1505, 1211	16485, 1007, 955	4958, 1824, 1481
μ_a, μ_b, μ_c/D	−2.4, 0.0, 0.0	−2.5, 0.2, 0.0	−3.0, −1.2, −1.8
ΔE_e/kJ mol^{-1}	10.5	13.5	11.0
ΔE_0/kJ mol^{-1}	10.1	12.7	10.9

The complexes of MeOH with *t*-ACR show, in general, lower energies than those with *c*-ACR. In fact, the three lowest energy structures are the complexes of MeOH with *t*-ACR and span about 3.7 kJ mol^{-1}, and the lowest energy structure formed by *c*-ACR (*c*-ACR-MeOH-1) is located 10.5 kJ mol^{-1} higher. This energy difference is very close to the energy difference (9.2(5) kJ mol^{-1}) determined for the two forms of the ACR monomer [30] and the energy difference (9.9 kJ mol^{-1}) for the *t*- and *c*-ACR complexes with water [31].

The six lowest energy structures (three originating from *t*-ACR and three from *c*-ACR) all exhibit a strong OH···O HB and a weak C–H···O interaction. The hydroxyl group of MeOH acts as a proton donor for the OH···O interaction and as a proton acceptor for the C–H···O interaction. The conformations 1 and 3, in which the MeOH oxygen forms a weak interaction with the methylenic hydrogen, are more stable than the conformations labeled 2 where the oxygen is bound to the aldehydic hydrogen. For example, the conformer *t*-ACR-MeOH-2 is 3.7 kJ mol^{-1} higher in energy than *t*-ACR-MeOH-1 and the conformer *c*-ACR-MeOH-2 is 3.0 kJ mol^{-1} higher in energy than *c*-ACR-MeOH-1.

Considering conformations 1 and 3, where the MeOH oxygen is bound to the methylenic hydrogen, the planar conformers show lower energy than the non-planar ones, although, in general, those differences are smaller than the ones between 1 and 2. The energy difference between *t*-ACR-MeOH-3 and *t*-ACR-MeOH-1 is 1.4 kJ mol^{-1}, and the energy difference between *c*-ACR-MeOH-3 and *c*-ACR-MeOH-1 is 0.5 kJ mol^{-1}.

2.2. Spectroscopic Results

According to the computational results, *t*-ACR-MeOH-1 is the global minimum, and it shows the most intense spectrum originating from the μ_a electric dipole moment component, which is much higher in value with respect to μ_b and μ_c (the latter one being zero by symmetry). We started our experiment with the search for the $J' \leftarrow J''$: 3←2 μ_a-R rotational transition of *t*-ACR-MeOH-1, and after the detection and fit of one group of rotational transitions, the measurements were extended following new predictions to 42 rotational transitions with the rotational quantum number J ranging from 1 to 6 and K_a ranging from 0 to 2, including μ_b type transition lines.

After finding all detectable rotational transition lines of the conformer *t*-ACR-MeOH-1, the measurements were extended to detect those of conformer *t*-ACR-MeOH-2. From the calculated relative energy and considering the temperature prior to the expansion (273 K), the population of *t*-ACR-MeOH-2 was predicted to be 19.6% of the global minimum, and taking into account the values of the dipole moment components, the most intense μ_a lines of *t*-ACR-MeOH-2 were predicted to be 22.0% of the most intense ones of *t*-ACR-MeOH-1. Indeed, 42 rotational transitions with the rotational quantum number J ranging from 1 to 7 and K_a ranging from 0 to 2 were observed for *t*-ACR-MeOH-2, including μ_b type transition lines.

The list of frequencies for both conformers is reported in the Supporting Information (Tables S1–S4), while the spectral parameters of the observed conformations of *t*-ACR-MeOH-1 and *t*-ACR-MeOH-2 obtained from a direct fit of the lines through the XIAM program [45] are summarized in Table 2. The fitted rotational constants are within 3% of the ones predicted from theory for both conformers. The detected μ_a-type and μ_b-type rotational transitions and the non-observation of the μ_c-type lines confirm the assignment to the planar conformers, which are in agreement with the zero value of the μ_c dipole moment components for both two conformers.

Table 2. Experimental spectroscopic parameters of the parent and deuterated species of *t*-ACR-MeOH-1 and *t*-ACR-MeOH-2 and the deuterated species.

	***t*-ACR-MeOH-1**	***t*-ACR-MeOD-1-**	***t*-ACR-MeOH-2**	***t*-ACR-MeOD-2-**
A/MHz	6180.152(3) [a]	6193.285(9)	11263.02(4)	11312(12)
B/MHz	1350.2493(7)	1345.401(2)	1009.515(3)	1004.652(2)
C/MHz	1116.9532(7)	1114.199(1)	933.289(3)	929.742(2)
D_J/kHz	0.905(6)	0.905(6)	0.724(9)	0.71(1)
D_{JK}/kHz	−4.30(9)	−4.7(1)	−37.1(2)	−38(1)
d_1/kHz	−0.219(6)	−0.217(9)	−0.09(1)	−0.082(9)
V_3/kJ mol^{-1}	2.686(3)	2.825(3)	2.722(6)	2.856(1)
δ/deg	0.089(7)	0.09(2)	2.896(5)	2.881(4)
F_0/GHz	158.6(1)	159.3(2)	156.1(3)	[156.1] [b]
$Dc3J$/kHz	−1.25(5)	−1.4(1)	-	
σ [c]/kHz	5.8	12.0	15.9	8.9
N [d]	42	46	42	24
M_{cc}/uÅ^2	1.7991	1.8275	1.9915	2.0731

[a] Error in parentheses in units of the last digit. [b] The data in brackets were fixed at the corresponding normal species values because they were not determined in the fit. [c] RMS error of the fit. [d] Number of lines in the fit.

As an example, the $3_{03}\leftarrow 2_{02}$ transition for the *t*-ACR-MeOH-1 conformer is shown in Figure 1 where the splitting of the rotational lines due to the internal rotation of the methyl group of MeOH can be observed; A and E are methyl torsional state symmetry labels.

2.3. Determination of Spectroscopic Parameters and Torsion Barriers

From the fitting procedure performed on the experimental transition frequencies, in addition to the rotational constants (A, B, C) and the quartic centrifugal distortion constants (D_J, D_{JK} and d_1), internal rotation parameters are also obtained, as reported in Table 2. The determined internal rotation parameters are V_3, the three-fold barrier to internal rotation of the methyl group, F_0, the rotational constant of the internal rotor, and δ, the angle between the internal rotor axis and the a-principal axis of inertia, while D_{c3J} is an empirical internal-overall rotation distortion constant. The determined V_3 is smaller in *t*-ACR-MeOH-1 (2.686(3) kJ mol^{-1}) than in *t*-ACR-MeOH-2 (2.723(6) kJ mol^{-1}), but both conformers have a low internal rotational barrier compared to the MeOH monomer (4.463(4) kJ mol^{-1} [44]).

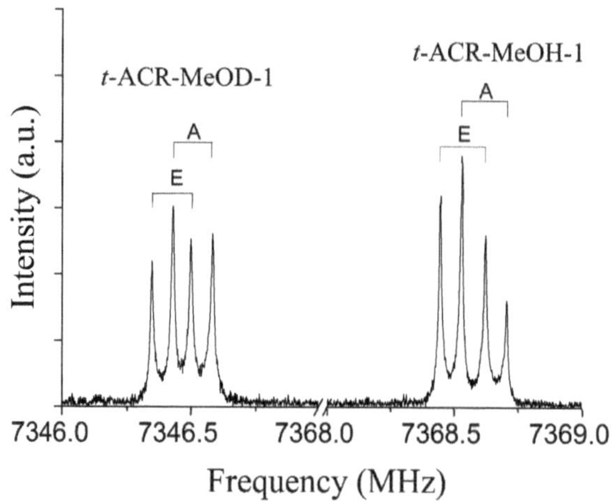

Figure 1. The $3_{03} \leftarrow 2_{02}$ transitions of *t*-ACR-MeOH-1 and *t*-ACR-MeOD-1 enriched isotopologue showing the methyl internal rotation A-E doublets. Each transition appears as a doublet (indicated by square brackets) due to the instrumental Doppler effect.

In order to understand the reason for the lowering of the V_3 barrier in the complex compared to the MeOH monomer, the potential energy surface of the MeOH methyl torsion was calculated by changing the dihedral angle of τ = HC-OH on the regular grid with $\Delta\tau = 10°$. The results are shown in Figure 2 and the calculated data, expressed in red bullets and dots, are well reproduced by the threefold function: $V(\tau) = \frac{1}{2}V_3[1 + \cos(3\tau)]$, which is shown as a green line for *t*-ACR-MeOH-1 and a blue line for *t*-ACR-MeOH-2. The maximum value of *t*-ACR-MeOH-1 (3.32 kJ mol^{-1}) represents the theoretical barrier hindering the methyl group internal rotation in the complex, and it can be noted that it is larger (about 26%) than the experimental values obtained for the normal species, V_3 = 2.686(3) kJ mol^{-1} for *t*-ACR-MeOH-1 (see Table 2). The difference between the theoretical (3.45 kJ mol^{-1}) and experimental (2.723(6) kJ mol^{-1}) V_3 values was also observed in *t*-ACR-MeOH-2. It is possible to attribute these differences to the accuracy of the theoretical method. This can be tested by calculating the barrier to internal rotation for free MeOH with the same theoretical method (4.16 kJ mol^{-1}) and comparing it to the experimental barrier (4.50 kJ mol^{-1}) [46]. The calculated barrier for free MeOH is indeed lower than the experimental one but only by about 7%, which is much smaller than the 26% difference found for *t*-ACR-MeOH-1 and the 27% for *t*-ACR-MeOH-2. This difference could be due to a lower accuracy of the method in calculating methyl torsional barriers in molecular complexes or to a possible large amplitude motion coupled to the torsion for which no evidence could be found in the experimentally measured rotational spectrum.

2.4. Deuterated Isotopologues

The rotational spectra of the deuterated isotope of methanol, abbreviated as *t*-ACR-MeOD-1 and *t*-ACR-MeOD-2, were also measured. Similarly, to the spectra of the parent species, the peaks of the deuterated isotopologues show up as doublets due to the internal rotation of the methyl group. The spectral parameters obtained with the same fitting procedure used for the undeuterated species are summarized in Table 2, and the observed transition frequencies of all measured isotopes are provided in the Supporting Information (Tables S3 and S4). As shown in Table 2, the deuterated species *t*-ACR-MeOD-1 (V_3 = 2.825(3) kJ mol^{-1}) and *t*-ACR-MeOD-2 (V_3 = 2.856(1f) kJ mol^{-1}) shows a higher barrier of V_3 than the corresponding parent species *t*-ACR-MeOH-1 (V_3 = 2.686(3) kJ mol^{-1}) and *t*-ACR-MeOH-2 (V_3 = 2.723(6) kJ mol^{-1}). One would therefore expect to find a larger A–E splitting of the spectrum of the deuterated species *t*-ACR-MeOD-1. On the contrary, in the $3_{03} \leftarrow 2_{02}$ transition (Figure 1), the deuterium-bonded dimer *t*-ACR-MeOD-1 shows a splitting of only ~0.16 MHz, while the hydrogen-bonded dimer *t*-ACR-MeOH-1 is split by ~0.18 MHz. The smaller A-E splittings in the deuterium-bound complexes are likely due to the Ubbelohde effect, which is the shortening of the HBs upon deuteration [47]. The effect of the shortening of the HB on the A-E splitting in deuterated complexes was also observed in the ethanol–methanol dimer [43]. The A-E splittings for the deuterated species *t*-ACR-MeOD-1 and *t*-ACR-MeOD-2 are summarized in Table S6.

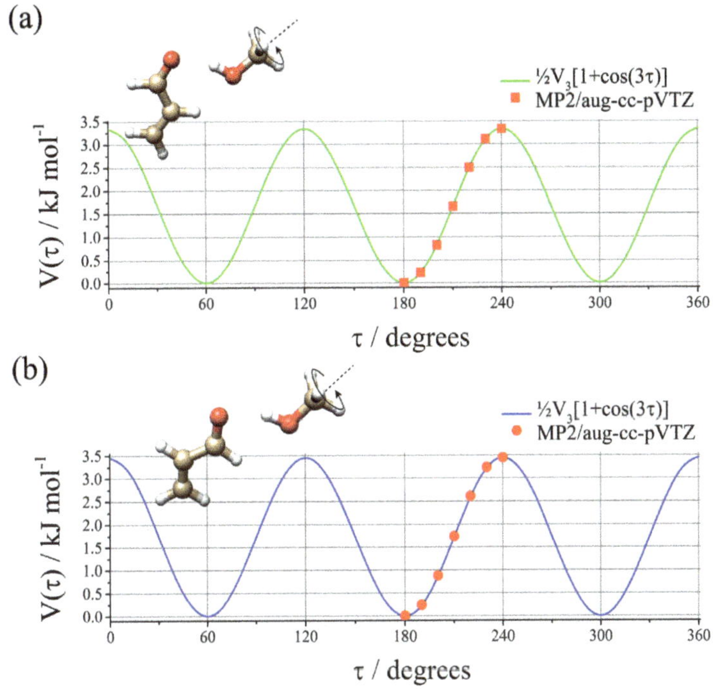

Figure 2. Ab initio (MP2/aug-cc-pVTZ) methyl internal rotational potential energy surface for (**a**) *t*-ACR-MeOH-1 and (**b**) *t*-ACR-MeOH-2.

In Table 2, the planar moments of inertia M_{cc} for the two species are also reported. Their values are consistent with an overall planar structure of the complex where only the methyl hydrogen atoms are out of the plane. The small differences between the values of the two conformers and those of the deuterated species with respect to the parent species can be attributed to small contributions of the low-frequency vibrational modes to the experimentally determined moments of inertia.

2.5. Possible Presence of Other Isomers

Considering the relative energy value, the temperature prior to the expansion and the dipole moment component's values, the most intense μ_a lines of *t*-ACR-MeOH-3, were predicted to be 72.0% of the most intense ones of *t*-ACR-MeOH-1. However, the spectrum originating from conformer *t*-ACR-MeOH-3 was not found in the experiment. This can be attributed to a relaxation of the population of this conformation onto that of the global minimum during adiabatic expansion, and in order to confirm this, the potential energy surface for the interconversion was calculated. This was performed by changing the dihedral angle τ = CO-HO on a regular grid with a step of 5° for both conformers, and the results are shown in Figure 3. In Figure 3a, one can see the conformations in which the MeOH oxygen is bound to the methylenic hydrogen atom (conformers *t*-ACR-MeOH-1 and *t*-ACR-MeOH-3), while in Figure 3b, are those where the MeOH oxygen is bound to the aldehydic hydrogen (conformer *t*-ACR-MeOH-2). In the first case, the higher energy conformer *t*-ACR-MeOH-3 gives rise to two equivalent minima, with respect to the ACR plane. The population's relaxation of this conformation to the global minimum *t*-ACR-MeOH-1 during adiabatic expansion is demonstrated by the low barrier to interconversion between the two conformations (approximately 0.1 kJ mol^{-1}). On the contrary, when the MeOH oxygen is bound to the aldehydic hydrogen (conformer *t*-ACR-MeOH-2), only one minimum is found, as shown in Figure 3b, and the non-planar arrangement does not correspond to a minimum. This can be ascribed to two reasons: the non-planar conformer exists but was not optimized due to the accuracy of the calculation method, or the non-planar conformer is not stable considering the weak intermolecular interaction between MeOH oxygen and aldehydic hydrogen.

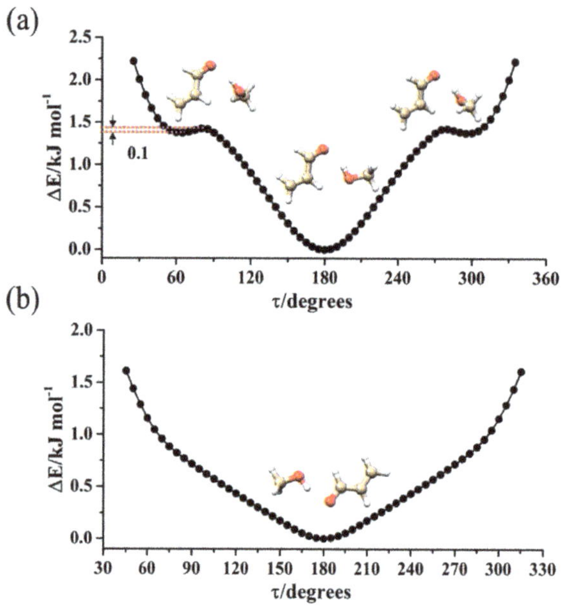

Figure 3. Ab initio (MP2/aug-cc-pVTZ) potential energy surface for the torsion of MeOH's methyl group around the OH bond for the (**a**) *t*-ACR-MeOH-1and (**b**) *t*-ACR-MeOH-2. In the upper panel, this motion interconverts the in-plane conformer (absolute minimum) with the out-of-plane ones (relative equivalent minima).

2.6. Analysis and General Discussion of the Clustering Interaction

The structure and the PES shown in Figure 3a are similar to the ones observed in the acrylonitrile-MeOH complex [18] where the water molecular binds to acrylonitrile forming a cyclic planar structure with a OH···N HB and a CH···O_W (global minimum), while a non-planar structure stabilized by the same interactions, but with the methyl group perpendicular to the acrylonitrile plane, is at a relative minimum with two equivalent forms. Also, in that case, the barrier to interconversion was found to be quite low (about 0.6 kJ mol^{-1}), preventing the observation of the non-planar conformer in the jet expansion.

The HBs in the six optimized conformers can be better visualized by using Johnson's NCI method [48]. Figure 4 displays the NCI plots, revealing the pattern of intermolecular interactions for all optimized conformations of ACR-MeOH. The gradient isosurfaces are colored according to the corresponding values of sign(λ_2) ρ. A negative sign of sign(λ_2)ρ (corresponding to blue regions) and small values of the reduced density gradient (RGD <0.5 a.u.) suggest stronger OH···O HB, while green regions indicate a weaker

C–H···O attractive interaction. Positive values of sign(λ_2)ρ and small values of the RGD (<0.5 a.u.), corresponding to orange regions, suggest a weak repulsive interaction.

One can observe the relatively large orange regions between oxygen and hydrogen in the *c*-ACR frame (Figure 4d–f), suggesting a weak repulsive intramolecular interaction accounting for the lower stability of *c*-ACR compared to *t*-ACR where this repulsion is absent (Figure 4a–c). Focusing on the most stable conformations of each series, one can see that in conformations 1 (Figure 4a,d) and 2 (Figure 4b,e) the OH···O HB is of similar strength, while the CH···O interactions in conformations 1 (CCH···O_W) are stronger than those in conformations 2 (O=CH···O_W); this is in agreement with the results of the theoretical calculation where conformations 1 have lower energy than conformations 2.

For comparison, Johnson's NCI method is also applied to the complexes of ACR-water, as shown in Figure 5. Two hydrogen bonds are formed between ACR and water in all four conformations (Figure 5a–d): a stronger OH···O HB (the blue regions) and a weaker CH···O interaction (the green regions), which correspond to the planar conformations of ACR-MeOH described above (Figure 4a,b,d,e), respectively. The strengths of hydrogen bonds are of the same level in view of the area of the NCI plots from ACR-water to ACR-MeOH.

A careful analysis of the geometrical parameters can also be performed. The OH···O/CH···O distances are 1.914/2.353 Å and 1.922/2.608 Å in the ACR-MeOH conformers 1 and 2, respectively, and 1.923/2.385 Å and 1.928/2.636 Å in the ACR-W conformers 1 and 2. These values confirm the relative strength of the interactions discussed previously, with the OH···O being the stronger interaction compared to the CH···O one and the latter being strongest in conformer 1 compared to to conformer 2.

(a)

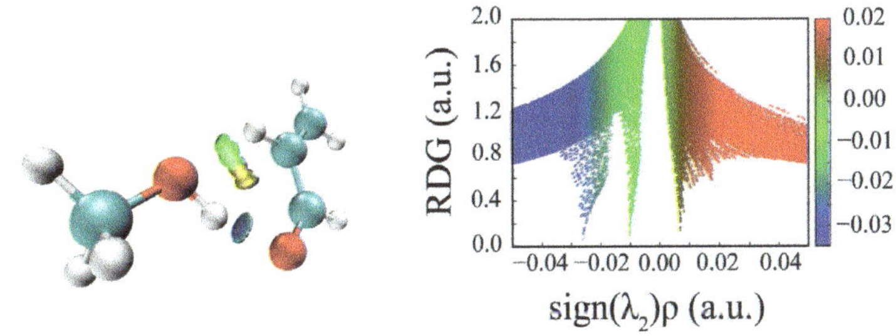

(b)

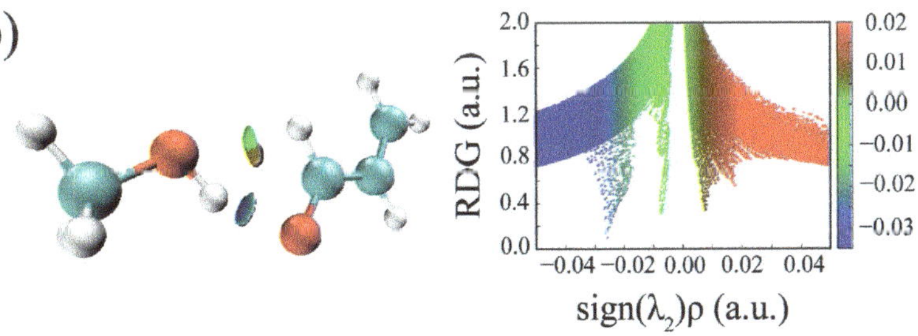

Figure 4. *Cont.*

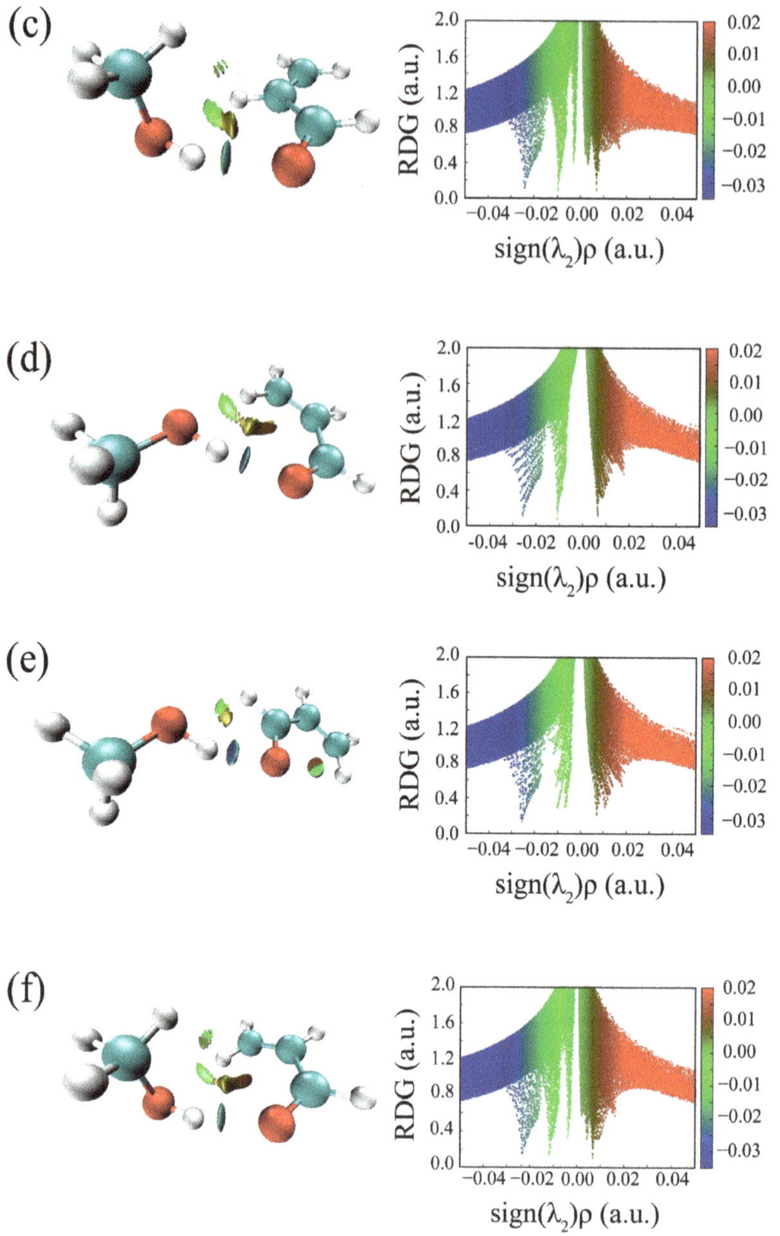

Figure 4. The NCI plots from the ab initio (MP2/aug-cc-pVTZ) outputs for conformers: (**a**) *t*-ACR-MeOH-1, (**b**) *t*-ACR-MeOH-2, (**c**) *t*-ACR-MeOH-3, (**d**) *c*-ACR-MeOH-1, (**e**) *c*-ACR-MeOH-2 and (**f**) *c*-ACR-MeOH-3. Left panel: gradient isosurfaces according to the values of the sign(λ_2)ρ (−0.04~0.04 a.u.). Color coding is blue (stronger attractive interactions), green (weaker attractive interactions) and orange-red (repulsive interaction). Right panel: The reduced density gradient (RDG) versus sign(λ_2)ρ. Positive values of the sign(λ_2)ρ indicate repulsive interactions, and negative values of the sign(λ_2)ρ indicate attractive interactions.

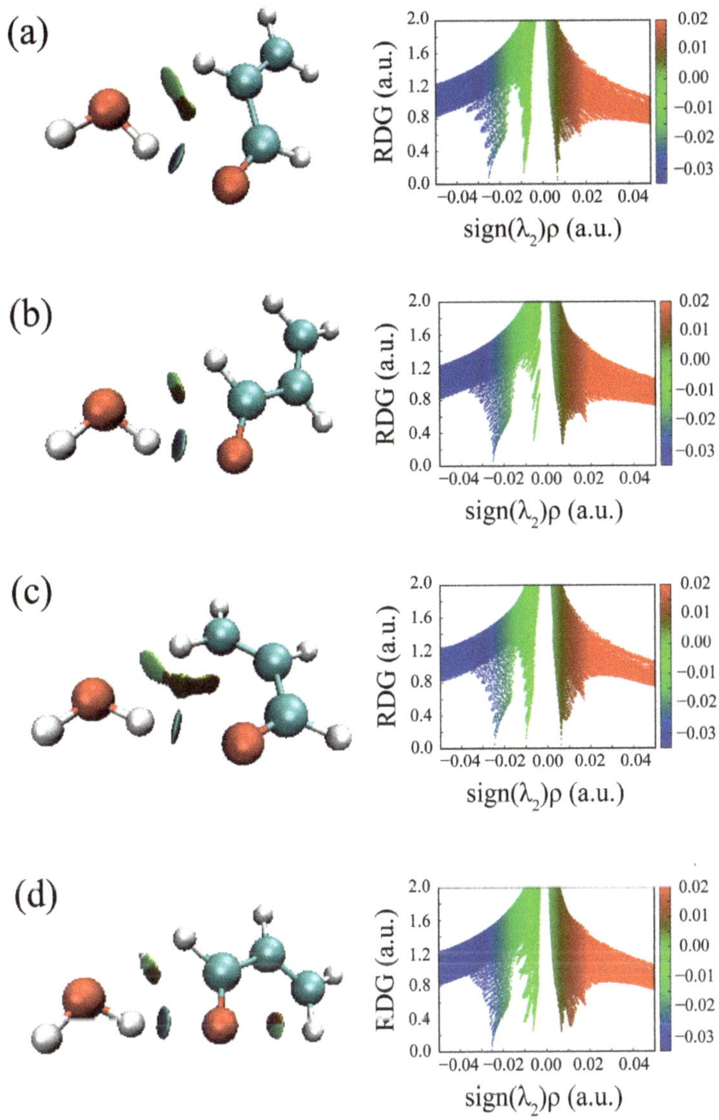

Figure 5. The NCI plots from the ab initio (MP2/aug-cc-pVTZ) outputs for three conformers: (**a**) *t*-ACR-W-1, (**b**) *t*-ACR-W-2, (**c**) *c*-ACR-W-1 and (**d**) *c*-ACR-W-2. (**Left panel**): gradient isosurfaces according to the values of the sign(λ_2)ρ (−0.04~0.04 a.u.). Color coding is blue (stronger attractive interactions), green (weaker attractive interactions) and orange-red (repulsive interaction). (**Right panel**): The reduced density gradient (RDG) versus sign(λ_2)ρ. Positive values of the sign(λ_2)ρ indicate repulsive interactions, and negative values of the sign(λ_2)ρ indicate attractive interactions.

Comparative analysis of the computational results of ACR-MeOH complexes and reported experimental geometries of MeOH and ACR reveals several key structural changes upon complexation. The O-H bond length in MeOH tends to be longer, and the ∠C5O6H11 angle is smaller after complexation. When compared to the ACR monomer, the complexation leads to a longer C3=O4 bond and an extended O4-H10 distance. Additionally, the ∠C2C3H10 angle becomes larger, whereas the angles ∠C1C2C3 and ∠O4=C3H10 decrease. For the ∠C2C3O4 angle, it tends to increase in *t*-ACR-MeOH-1 but decreases in *t*-ACR-MeOH-2. The determined barriers for the internal rotation of the methyl group in these complexes are lower than those found for the MeOH monomer. According to Fraser et al. [34], this is attributed to the effects of the librational motion of either the entire MeOH moiety or the methyl group within the complex. Schmitt [40] points out that the small absolute value of the barrier compared to the monomer is likely an artifact due to the one-dimensional description of the large amplitude motion. In MeOH-HCl complexes [38], the apparent barrier reduction arises from neglecting the large amplitude wagging motion of the hydroxyl hydrogen relative to HCl. In the acrylonitrile-MeOH complex [18], the results exclude the librational motion of the entire methanol moiety or the methyl group, but the motion of the hydroxyl group cannot be excluded. In summary, the decreased barrier is not due to complexation effects but is related to another large amplitude internal motion, such as the motion of the hydroxyl group coupled with the internal rotation of the methyl group.

Other complexes of small organic molecules with methanol have been characterized by rotational spectroscopy and, in all cases, the global minimum is stabilized by similar cyclic structures. In formamide-MeOH [41], there is an HB from the MeOH hydroxyl hydrogen to the oxygen atom of formamide and one from the N-H group of formamide to the MeOH oxygen. The corresponding OH···O and NH···O distances are quite short: 2.01(1) and 1.97(1) Å, respectively. In formaldehyde-MeOH [35], the HBs are from the MeOH hydroxyl hydrogen to the oxygen of formaldehyde (2.097(6) Å) and from a CH group of the second moiety to the MeOH oxygen, while in acrylonitrile-MeOH [18] the HB is from MeOH to the cyano group and from the methylenic hydrogen to the MeOH oxygen; the OH···N and CH···O distances being 2.257(1) and 2.484(1) Å (MP2/aug-cc-pVTZ values). The shorter distances in formamide-MeOH seem to imply a larger binding energy, which is indeed confirmed by the ab initio (MP2/aug-cc-pVTZ) value of 44.5 kJ mol^{-1} for this system compared to those calculated for formaldehyde-MeOH (24.1 kJ mol^{-1}) and acrylonitrile-MeOH (26.3 kJ mol^{-1}). For the ACR-MeOH system, the HB distances and energy binding values (30.4 kJ mol^{-1}, MP2/aug-cc-pVTZ, Table 3) are found to be halfway between the three systems cited above. The HB structure for ACR-MeOH has a lower binding energy than formamide-MeOH and higher than formaldehyde-MeOH and acrylonitrile-MeOH. Notably, the binding energy value for ACR-MeOH is slightly larger than that of ACR–water (28.8 kJ mol^{-1} MP2/aug-cc-pVTZ, Table 3) [31].

Table 3. Ab initio binding energies calculated by MP2/aug-cc-pVTZ method (BE MP2), SAPT2+(3)dMP2/aug-cc-pVTZ analysis and total binding energies (BE SAPT) for the complexes of methanol and water with acrolein (in bold the observed conformations), all values in kJ mol^{-1}.

Energies	Electrostatic	Induction	Dispersion	Exchange	BE SAPT	BE MP2
Acrolein-Methanol						
***t*-ACR-MeOH-1**	−46.15	−16.16	−20.45	54.08	−28.68	−30.4
***t*-ACR-MeOH-2**	−38.67	−14.24	−17.15	44.22	−25.84	−26.8
t-ACR-MeOH-3	−43.84	−14.07	−22.36	53.60	−26.68	−29.1
c-ACR-MeOH-1	−45.38	−16.44	−20.75	54.23	−28.34	−29.4
c-ACR-MeOH-2	−38.39	−14.20	−17.04	44.08	−25.55	−26.3
c-ACR-MeOH-3	−43.96	−14.24	−23.41	54.59	−27.02	−28.9
Acrolein-Water						
***t*-ACR-W-1**	−44.43	−15.15	−17.96	49.78	−27.76	−28.8
***t*-ACR-W-2**	−37.85	−13.69	−15.27	41.47	−25.35	−25.8
c-ACR-W-1	−42.90	−14.94	−17.94	48.63	−27.14	−27.6
c-ACR-W-2	−37.57	−13.66	−15.17	41.34	−25.05	−25.4

A quantitative understanding of the chemical nature of the NCIs can be achieved by energy decomposition analysis, using the Symmetry-Adapted Perturbation theory (SAPT). According to SAPT [49], the energy of the intermolecular interaction can be interpreted as the sum of different terms with defined physical meanings: electrostatic, induction, dispersion and exchange-repulsion terms. The results, summarized in Table 3, show that the intermolecular interactions become slightly stronger following the substitution of the hydrogen atom in water with the methyl group in MeOH. These results are confirmed by the ab initio binding energies (MP2-aug-cc-pVTZ) and are also reported in Table 3. The dispersion interactions are mainly responsible for the increased total interactions and are related to electron correlations in the whole molecules, and they become stronger as the size of the molecules and the number of electrons increase.

3. Methods

The experiment was carried out with the COBRA-type pulsed supersonic-jet FTMW spectrometer, which was previously described [50–53]. ACR and MeOH were acquired from Sigma-Aldrich (St. Louis, MO, USA) (purity > 99%), while D_2O was acquired from Cambridge Isotope Laboratories, Inc. (Tewksbury, MA, USA) (purity 99.9%). All compounds were used without further purification. For the preparation of the complexes, samples of ACR (cooled to 273 K) and MeOH (at 298 K) were prepared in two separate containers, and helium, at a stagnation pressure of about 0.6 MPa, was made to flow over them creating a 1% mixture of both ACR and MeOH in the carrier gas. The hydroxyl-enriched deuterated isotope of methanol was prepared by directly mixing methanol with the D_2O sample in a 1:2 volume ratio.

The molecular beam was expanded via a solenoid valve (General Valve, Series 9, nozzle diameter 0.5 mm) into a Fabry–Pérot cavity. This expansion allowed the molecules and their complexes to reach low rotational temperatures (1–2 K), trapping the most stable forms at their energy minimum. Spectral line positions were identified using Fourier transformation of the time-domain signal with 8 k data points recorded at a 100 ns sampling interval.

Geometry optimizations and subsequent harmonic frequency calculations were performed at the MP2/aug-cc-pVTZ level using the GAUSSIAN 16 program [54]. This method is known to provide very accurate results regarding both the geometries and energies. The small size of the system allowed the use of this method, which is more computationally expensive than DFT. The analysis of the spectral data was performed with the program XIAM [45], which is based on the combined axis method [55] and directly supplies the V_3 barrier to internal rotation, the angles between the internal rotation axis and the principal axes and the moment of inertia of the internal top, while a set of rotational constants common to both the A- and E-states is provided (corresponding to the values for the infinite barrier limit).

Johnson's NCI method [48], which can visualize and quantify non-covalent interactions through the reduced density gradient (RDG) of the electron distribution, was used to study the non-covalent interactions between ACR and MeOH.

4. Conclusions

The rotational spectrum of the 1:1 complex of ACR and MeOH was studied to obtain information on its structure, on the HBs and on the barrier hindering the methyl group's internal rotation in the complex. Two stable conformations of ACR-MeOH and their deuterated istopologues have been measured. The determined rotational constants are coherent with a cyclic structure (MP2/aug-cc-pVTZ) stabilized by a primary OH···O HB between MeOH and the aldehyde group and a secondary one between the terminal CH group and the MeOH oxygen, which were also characterized and visualized using NCI plots. Splitting of the rotational lines due to the hindered internal rotation of the methyl group was observed and the global analysis of the spectrum led to the determination of a V_3 value of 2.686(3) kJ mol^{-1} and 2.723(6) kJ mol^{-1} for the two conformations, respectively. These values are about 40% lower than those experimentally determined for free MeOH, revealing a significant discrepancy between the theoretical and experimental rotational barriers. This demonstrates the importance of continuous improvement in computational chemistry techniques to achieve more accurate predictions and better alignment with experimental observations. The finding that the reduced mass of the internal rotation motion is extremely close to the calculated one allowed us to exclude that the lowering of the barrier could be ascribed to a librational motion of the methyl group. The present findings thus have implications for understanding large-amplitude motions within molecular complexes. Such insights provide valuable empirical evidence that can enhance the accuracy of theoretical models of internal dynamics in molecular chemistry, offering a deeper comprehension of the factors influencing molecular behavior.

Supplementary Materials: The following supporting information can be downloaded at: https://www.mdpi.com/article/10.3390/molecules29153444/s1: Table S1: Experimental transition frequencies (ν/MHz) and observed minus calculated values (Δν/MHz) of t-ACR-MeOH-1; Table S2: Experimental transition frequencies (ν/MHz) and observed minus calculated values (Δν/MHz) of t-ACR-MeOH-2; Table S3: Experimental transition frequencies (ν/MHz) and observed minus calculated values (Δν/MHz) of t-ACR-MeOD-1; Table S4: Experimental transition frequencies (ν/MHz) and observed minus calculated values (Δν/MHz) of t-ACR-MeOD-2; Table S5: MP2/aug-cc-pVTZ geometries of the two conformers of the ACR-MeOH complexes; Table S6: The different splitting (ΔSplit/MHz) of A-E lines between parent species and deuterated species; Figure S1: MP2/aug-cc-pVTZ geometries of the two conformers of the ACR-MeOH complexes; Table S7: MP2/aug-cc-pVTZ geometries of the two conformers of the ACR-MeOH complexes and experimental geometries of MeOH and ACR; References [28,44].

Author Contributions: The manuscript was written with contributions from all authors. All authors have provided approval to the final version of the manuscript. All authors have read and agreed to the published version of the manuscript.

Funding: This research was funded by the Italian Ministry of Research (2020AFB3FX_003 PRIN; grant n. J43C21000050001) and the Swedish Research Council (grant n. 2019-04332).

Institutional Review Board Statement: Not applicable.

Informed Consent Statement: Not applicable.

Data Availability Statement: No new data were created or analyzed in this study.

Acknowledgments: The authors thank the University of Bologna for financial support (RFO), and the CINECA award under the ISCRA initiative, for the availability of high-performance computing resources and support.

Conflicts of Interest: The authors declare no conflict of interest.

References

1. Bellamy, L.J.; Pace, R.J. Hydrogen Bonding by Alcohols and Phenols—I. The Nature of the Hydrogen Bond in Alcohol Dimers and Polymers. *Spectrochim. Acta* **1966**, *22*, 525–533. [CrossRef]
2. Li, W.; Evangelisti, L.; Gou, Q.; Caminati, W.; Meyer, R. The Barrier to Proton Transfer in the Dimer of Formic Acid: A Pure Rotational Study. *Angew. Chem. Int. Ed.* **2019**, *58*, 859–865. [CrossRef]
3. Dixon, D.A.; Dobbs, K.D.; Valentini, J.J. Amide-Water and Amide-Amide Hydrogen Bond Strengths. *J. Phys. Chem.* **1994**, *98*, 13435–13439. [CrossRef]
4. Rose, G.D.; Wolfenden, R. Hydrogen Bonding, Hydrophobicity, Packing, and Protein Folding. *Annu. Rev. Biophys. Biomol. Struct.* **1993**, *22*, 381–415. [CrossRef]
5. Desiraju, G.R.; Steiner, T. *The Weak Hydrogen Bond: In Structural Chemistry and Biology*; International Union of Crystallography; Monograpg on crystallography Volume 9; Oxford University Press Inc.: New York, Ny, USA, 1999.
6. Kraśnicki, A.; Kisiel, Z.; Guillemin, J.-C. From Molecular to Cluster Properties: Rotational Spectroscopy of 2-Aminopyridine and of Its Biomimetic Cluster with Water. *Molecules* **2021**, *26*, 6870. [CrossRef]
7. Li, W.; Pérez, C.; Steber, A.L.; Schnell, M.; Lv, D.; Wang, G.; Zeng, X.; Zhou, M. Evolution of Solute–Water Interactions in the Benzaldehyde-(H_2O) 1–6 Clusters by Rotational Spectroscopy. *J. Am. Chem. Soc.* **2023**, *145*, 4119–4128. [CrossRef]
8. Macario, A.; López, J.C.; Blanco, S. Molecular Structure of Salicylic Acid and Its Hydrates: A Rotational Spectroscopy Study. *Int. J. Mol. Sci.* **2024**, *25*, 4074. [CrossRef]
9. Pérez, C.; López, J.C.; Blanco, S.; Schnell, M. Water-Induced Structural Changes in Crown Ethers from Broadband Rotational Spectroscopy. *J. Phys. Chem. Lett.* **2016**, *7*, 4053–4058. [CrossRef]

10. Murugachandran, S.I.; Sanz, M.E. Interactions of Limonene with the Water Dimer. *Phys. Chem. Chem. Phys.* **2022**, *24*, 26529–26538. [CrossRef]
11. Melosso, M.; Alessandrini, S.; Spada, L.; Melli, A.; Wang, X.; Zheng, Y.; Duan, C.; Li, J.; Du, W.; Gou, Q. Rotational Spectra and Semi-Experimental Structures of Furonitrile and Its Water Cluster. *Phys. Chem. Chem. Phys.* **2023**, *25*, 31281–31291. [CrossRef]
12. Blanco, S.; López, J.C.; Lesarri, A.; Alonso, J.L. Microsolvation of Formamide: A Rotational Study. *J. Am. Chem. Soc.* **2006**, *128*, 12111–12121. [CrossRef] [PubMed]
13. Magneron, I.; Thevenet, R.; Mellouki, A.; Le Bras, G.; Moortgat, G.K.; Wirtz, K. A Study of the Photolysis and OH-Initiated Oxidation of Acrolein and Trans-Crotonaldehyde. *J. Phys. Chem. A* **2002**, *106*, 2526–2537. [CrossRef]
14. Atkinson, R. A Structure-activity Relationship for the Estimation of Rate Constants for the Gas-phase Reactions of OH Radicals with Organic Compounds. *Int. J. Chem. Kinet.* **1987**, *19*, 799–828. [CrossRef]
15. Kwok, E.S.C.; Atkinson, R. Estimation of Hydroxyl Radical Reaction Rate Constants for Gas-Phase Organic Compounds Using a Structure-Reactivity Relationship: An Update. *Atmos. Environ.* **1995**, *29*, 1685–1695. [CrossRef]
16. Judge, R.H.; Moule, D.C. Thiocarbonyl Spectroscopy: Methyl Torsional Vibrations and Internal. *J. Chem. Phys.* **1987**, *87*, 60–67. [CrossRef]
17. Fraser, G.T.; Lovas, F.J.; Suenram, R.D. On the Apparent Methyl Internal-Rotation Barrier Decrease in Weakly Bound Methanol Complexes. *J. Mol. Spectrosc.* **1994**, *167*, 231–235. [CrossRef]
18. Calabrese, C.; Maris, A.; Vigorito, A.; Mariotti, S.; Fathi, P.; Geppert, W.D.; Melandri, S. Structure, Dynamics, and Accurate Laboratory Rotational Frequencies of the Acrylonitrile--Methanol Complex. *J. Phys. Chem. A* **2020**, *124*, 3601–3608. [CrossRef]
19. Li, Z.; Nizkorodov, S.A.; Chen, H.; Lu, X.; Yang, X.; Chen, J. Nitrogen-Containing Secondary Organic Aerosol Formation by Acrolein Reaction with Ammonia/Ammonium. *Atmos. Chem. Phys.* **2019**, *19*, 1343–1356. [CrossRef]
20. Borchers, M.T.; Wesselkamper, S.; Wert, S.E.; Shapiro, S.D.; Leikauf, G.D. Monocyte Inflammation Augments Acrolein-Induced Muc5ac Expression in Mouse Lung. *Am. J. Physiol. Lung Cell. Mol. Physiol.* **1999**, *277*, L489–L497. [CrossRef] [PubMed]
21. Ayer, H.E.; Yeager, D.W. Irritants in Cigarette Smoke Plumes. *Am. J. Public Health* **1982**, *72*, 1283–1285. [CrossRef]
22. Pradipta, A.R.; Tanaka, K. Application of Acrolein Imines to Organic Synthesis, Biofunctional Studies, and Clinical Practice. *Chem. Rec.* **2021**, *21*, 646–662. [CrossRef] [PubMed]
23. Evangelisti, L.; Maris, A.; Grieco, F.; Calabrese, C.; Melandri, S. Millimeter Wave Free-Jet Spectrum of Acrolein and Several Isotopologues. *Can. J. Phys.* **2020**, *98*, 555–559. [CrossRef]
24. Cherniak, E.A.; Costain, C.C. Microwave Spectrum and Molecular Structure of Trans-Acrolein. *J. Chem. Phys.* **1966**, *45*, 104–110. [CrossRef]
25. Wagner, R.; Fine, J.; Simmons, J.W.; Goldstein, J.H. Microwave Spectrum, Structure, and Dipole Moment of s-Trans Acrolein. *J. Chem. Phys.* **1957**, *26*, 634–637. [CrossRef]
26. Winnewisser, M.; Winnewisser, G.; Honda, T.; Hirota, E. Ground State Centrifugal Distortion Constants of Trans-Acrolein, CH_2 = CH–CHO from the Microwave and Millimeter Wave Rotational Spectra 1, 2. *Z. Für Naturforsch. A* **1975**, *30*, 1001–1014. [CrossRef]
27. Blom, C.E.; Bauder, A. Microwave Spectrum, Rotational Constants and Dipole Moment of s-Cis Acrolein. *Chem. Phys. Lett.* **1982**, *88*, 55–58. [CrossRef]
28. Blom, C.E.; Grassi, G.; Bauder, A. Molecular Structure of S-Cis- and s-Trans-Acrolein Determined by Microwave Spectroscopy. *J. Am. Chem. Soc.* **1984**, *106*, 7427–7431. [CrossRef]
29. Jaman, A.I.; Bhattacharya, R. Millimeter-Wave Rotational Spectra of *Trans*-Acrolein (Propenal) (CH_2CHCOH): A DC Discharge Product of Allyl Alcohol (CH_2CHCH_2OH) Vapor and DFT Calculation. *J. At. Mol. Opt. Phys.* **2012**, *2012*, 363247. [CrossRef]
30. Alves, A.C.P.; Christoffersen, J.; Hollas, J.M. Near Ultra-Violet Spectra of the s-Trans and a Second Rotamer of Acrolein Vapour. *Mol. Phys.* **1971**, *20*, 625–644. [CrossRef]
31. Li, W.; Maris, A.; Calabrese, C.; Usabiaga, I.; Geppert, W.D.; Evangelisti, L.; Melandri, S. Atmospherically Relevant Acrolein--Water Complexes: Spectroscopic Evidence of Aldehyde Hydration and Oxygen Atom Exchange. *Phys. Chem. Chem. Phys.* **2019**, *21*, 23559–23566. [CrossRef]
32. Stockman, P.A.; Blake, G.A.; Lovas, F.J.; Suenram, R.D. Microwave Rotation-Tunneling Spectroscopy of the Water--Methanol Dimer: Direct Structural Proof for the Strongest Bound Conformation. *J. Chem. Phys.* **1997**, *107*, 3782–3790. [CrossRef]
33. Haeckel, M.; Stahl, W. The Microwave Spectrum and Molecular Structure of the Hydrogen-Bonded Aniline–Methanol Complex. *J. Mol. Spectrosc.* **1999**, *198*, 263–277. [CrossRef] [PubMed]
34. Lovas, F.J.; Hartwig, H. The Microwave Spectrum of the Methanol Dimer ForK= 0 and 1 States. *J. Mol. Spectrosc.* **1997**, *185*, 98–109. [CrossRef] [PubMed]
35. Ilyushin, V.V.; Lovas, F.J.; Plusquellic, D.F. Microwave Spectrum of the Heterodimers: CH_3OH–CO_2 and CH_3OH–H_2CO. *J. Mol. Spectrosc.* **2006**, *239*, 94–100. [CrossRef]
36. Westphal, A.; Jacoby, C.; Ratzer, C.; Reichelt, A.; Schmitt, M. Determination of the Intermolecular Geometry of the Phenol–Methanol Cluster. *Phys. Chem. Chem. Phys.* **2003**, *5*, 4114–4122. [CrossRef]
37. Sun, L.; Tan, X.; Oh, J.J.; Kuczkowski, R.L. The Microwave Spectrum and Structure of the Methanol·SO_2 Complex. *J. Chem. Phys.* **1995**, *103*, 6440–6449. [CrossRef]
38. Tan, X.-Q.; Ioannou, I.I.; Kuczkowski, R.L. The Methanol·HCl Complex: Structure and Methyl Group Internal Rotation Barrier. *J. Mol. Struct.* **1995**, *356*, 105–115. [CrossRef]

39. Tan, X.Q.; Sun, L.H.; Kuczkowski, R.L. The Methanol · Ar Complex: Apparent Reduction of the Methyl Group Internal Rotation Barrier. *J. Mol. Spectrosc.* **1995**, *171*, 248–264. [CrossRef]
40. Schmitt, M.; Küpper, J.; Spangenberg, D.; Westphal, A. Determination of the Structures and Barriers to Hindered Internal Rotation of the Phenol–Methanol Cluster in the S0 and S1 States. *Chem. Phys.* **2000**, *254*, 349–361. [CrossRef]
41. Lovas, F.J.; Suenram, R.D.; Fraser, G.T.; Gillies, C.W.; Zozom, J. The Microwave Spectrum of Formamide–Water and Formamide–Methanol Complexes. *J. Chem. Phys.* **1988**, *88*, 722–729. [CrossRef]
42. Lovas, F.J.; Belov, S.P.; Tretyakov, M.Y.; Ortigoso, J.; Suenram, R.D. The Microwave Spectrum and Structure of the CH_3OH CO Dimer. *J. Mol. Spectrosc.* **1994**, *167*, 191–204. [CrossRef]
43. Finneran, I.A.; Carroll, P.B.; Mead, G.J.; Blake, G.A. Hydrogen Bond Competition in the Ethanol–Methanol Dimer. *Phys. Chem. Chem. Phys.* **2016**, *18*, 22565–22572. [CrossRef] [PubMed]
44. Lees, R.M.; Lovas, F.J.; Kirchhoff, W.H.; Johnson, D. Microwave Spectra of Molecules of Astrophysical Interest: III. Methanol. *J. Phys. Chem. Ref. Data* **1973**, *2*, 205–214. [CrossRef]
45. Hartwig, H.; Dreizler, H. The Microwave Spectrum of Trans-2,3-Dimethyloxirane in Torsional Excited States. *Z. Für Naturforsch. A* **1996**, *51*, 923–932. [CrossRef]
46. Xu, L.-H.; Lovas, F.J. Microwave Spectra of Molecules of Astrophysical Interest. XXIV. Methanol (CH_3OH and $^{13}CH_3OH$). *J. Phys. Chem. Ref. Data* **1997**, *26*, 17–156. [CrossRef]
47. Ubbelohde, A.R.; Gallagher, K.J. Acid-Base Effects in Hydrogen Bonds in Crystals. *Acta Crystallogr.* **1955**, *8*, 71–83. [CrossRef]
48. Johnson, E.R.; Keinan, S.; Mori-Sánchez, P.; Contreras-García, J.; Cohen, A.J.; Yang, W. Revealing Noncovalent Interactions. *J. Am. Chem. Soc.* **2010**, *132*, 6498–6506. [CrossRef] [PubMed]
49. Jeziorski, B.; Moszynski, R.; Szalewicz, K. Perturbation Theory Approach to Intermolecular Potential Energy Surfaces of van Der Waals Complexes. *Chem. Rev.* **1994**, *94*, 1887–1930. [CrossRef]
50. Grabow, J.-U.; Stahl, W.; Dreizler, H. A Multioctave Coaxially Oriented Beam-Resonator Arrangement Fourier-Transform Microwave Spectrometer. *Rev. Sci. Instrum.* **1996**, *67*, 4072–4084. [CrossRef]
51. Balle, T.J.; Flygare, W.H. Fabry—Perot Cavity Pulsed Fourier Transform Microwave Spectrometer with a Pulsed Nozzle Particle Source. *Rev. Sci. Instrum.* **1981**, *52*, 33–45. [CrossRef]
52. Caminati, W.; Millemaggi, A.; Alonso, J.L.; Lesarri, A.; López, J.C.; Mata, S. Molecular Beam Fourier Transform Microwave Spectrum of the Dimethylether--Xenon Complex: Tunnelling Splitting and ^{131}Xe Quadrupole Coupling Constants. *Chem. Phys. Lett.* **2004**, *392*, 1–6. [CrossRef]
53. Caminati, W.; Evangelisti, L.; Feng, G.; Giuliano, B.M.; Gou, Q.; Melandri, S.; Grabow, J.-U. On the Cl···C Halogen Bond: A Rotational Study of CF_3Cl-CO. *Phys. Chem. Chem. Phys.* **2016**, *18*, 17851–17855. [CrossRef] [PubMed]
54. Frisch, M.J.; Trucks, G.W.; Schlegel, H.B.; Scuseria, G.E.; Robb, M.A.; Cheeseman, J.R.; Scalmani, G.; Barone, V.; Petersson, G.A.; Nakatsuji, H.; et al. *Gaussian 16 Revision B. 01, 2016*; Gaussian Inc.: Wallingford, CT, USA, 2016; Volume 1.
55. Woods, R.C. A General Program for the Calculation of Internal Rotation Splittings in Microwave Spectroscopy. *J. Mol. Spectrosc.* **1966**, *21*, 4–24. [CrossRef]

molecules

MDPI

Article

The Role of Hydrogen Bonding in the Raman Spectral Signals of Caffeine in Aqueous Solution

Sara Gómez * and Chiara Cappelli *

Scuola Normale Superiore, Classe di Scienze, Piazza dei Cavalieri 7, 56126 Pisa, Italy
* Correspondence: sara.gomezmaya@sns.it (S.G.); chiara.cappelli@sns.it (C.C.)

Abstract: The identification and quantification of caffeine is a common need in the food and pharmaceutical industries and lately also in the field of environmental science. For that purpose, Raman spectroscopy has been used as an analytical technique, but the interpretation of the spectra requires reliable and accurate computational protocols, especially as regards the Resonance Raman (RR) variant. Herein, caffeine solutions are sampled using Molecular Dynamics simulations. Upon quantification of the strength of the non-covalent intermolecular interactions such as hydrogen bonding between caffeine and water, UV-Vis, Raman, and RR spectra are computed. The results provide general insights into the hydrogen bonding role in mediating the Raman spectral signals of caffeine in aqueous solution. Also, by analyzing the dependence of RR enhancement on the absorption spectrum of caffeine, it is proposed that the sensitivity of the RR technique could be exploited at excitation wavelengths moderately far from 266 nm, yet achieving very low detection limits in the quantification caffeine content.

Keywords: Raman spectroscopy; Resonance Raman; caffeine; simulations; Molecular Dynamics; hydrogen bonding; UV-Vis; Quantum Mechanics/Fluctuating Charges; non-covalent interactions

1. Introduction

Caffeine (1,3,7-trimethylxanthine) is a purine alkaloid that acts as a psychoactive drug and stimulant agent [1]. It can be consumed from natural sources like coffee, chocolate, and tea, but it can be found in other foods and beverages, and, in recent decades, also in medications, because of the enhanced effect it provides in a mix with certain analgesics [2]. The effects of caffeine on cognitive or physical performance have been the subject of several reviews [3–6].

Structurally, caffeine is a combination of two fused rings: pyrimidinedione, a six-membered pyrimidine ring with two ketone groups linked to it at meta positions, and imidazole, a five-membered ring with two nitrogen atoms. Caffeine's structure is shown in the innermost layer of the drop in Figure 1.

Due to the high level of caffeine consumption worldwide, controlling the caffeine content of coffee, tea samples, and caffeine-containing food products is indispensable [7]. Moreover, this methyl xanthine, as well as diclofenac, ibuprofen, and acetaminophen, is currently considered an emerging contaminant that poses health risks to aquatic life [8] and poses a danger to the environment because of its persistence [9,10]. In fact, caffeine and paraxanthine pollution in aqueous environments are declared to be ubiquitous [11]. Hence, the determination of caffeine is relevant in chemical analysis and it is desirable to have practical methods for carrying it out in various matrices. Many analytical techniques have been employed for that goal, among them Raman spectroscopy [7,12]. As a matter of fact, the application of vibrational spectroscopy techniques to quantify nutraceuticals in fruits and plants has proven successful [13], and Raman spectroscopy has been used for quantitative analysis in the pharmaceutical industry [14,15].

The evolution of Raman as an extremely sensitive analytical tool was further boosted by the discovery of several other Raman phenomena, including Resonance Raman (RR), coherent anti-Stokes Raman scattering (CARS), and surface-enhanced Raman scattering (SERS) [16,17]. All of them can improve the intensity of the Raman signal, thus overcoming the low sensitivity caused by the inherently weak Raman scattering, and in most cases, they can minimize the interference from fluorescence. In contrast to SERS, which is a method where the sample is adsorbed onto a metallic substrate (gold, silver, etc.) [18], RR benefits from choosing an excitation wavelength that couples to an electronic transition or photon absorption of the target analytes and chromophore segments of material or macromolecules, affording a much more intense Raman spectrum. Thus, some modes absent in normal Raman spectra can be resonantly enhanced to appear in RR spectra. The combination of the two techniques, RR and SERS, gives rise to surface resonance-enhanced Raman spectroscopy (SERRS) and can greatly increase sensitivity by up to 10 orders of magnitude compared to just using RR [7]. In the context of determining caffeine content, the low solubility of the compound in water at room temperature has been a limitation to employing conventional Raman methods; nonetheless, it has been investigated with the SERS modality in various works [19,20]. SERS has afforded a simultaneous multiplexed quantification of caffeine and its major metabolites theobromine and paraxanthine [21], and SERS-coupled multivariate calibration has been recently used in the rapid prediction of caffeine in tea [22]. In addition, Frosch et al. [23] presented a fiber-array-based Raman hyperspectral imaging technique for direct simultaneous in situ monitoring of different active pharmaceutical ingredients, such as caffeine, in analgesic tablets. Without requiring metallic surfaces or further agents, diluents, or matrices, RR represents an excellent technique for single or simultaneous identification and quantification of multiple components of interest in a sample thanks to its good selectivity and sensitivity [24–28]. To our knowledge, there is only one experimental study in which RR has been used for the caffeine spectral analysis of extremely diluted samples, such as 0.0022 M [29].

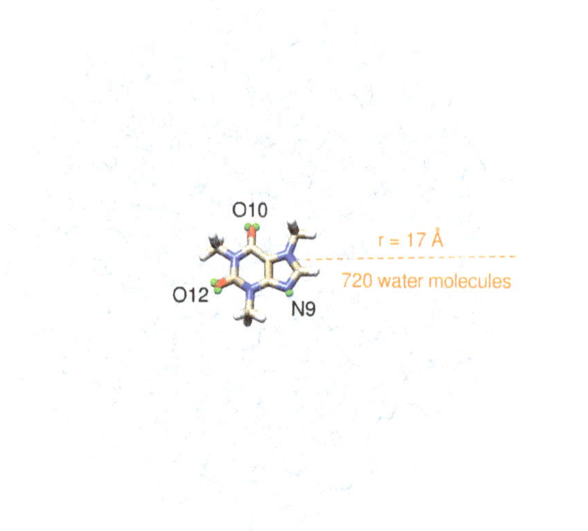

Figure 1. Pictorial view of caffeine dissolved in aqueous solution as treated in QM/FQ calculations. Virtual sites for the atoms that are more prone to hydrogen bonding with the solvent, labeled N9, O10, and O12, are shown in small green spheres.

Quantum chemical calculations are useful tools to properly describe and analyze Raman and in particular RR spectra [30–35]. Some modeling with isolated caffeine in the gas phase or describing the condensed phase using the Polarizable Continuum Model (PCM) [36] or a microsolvation approach has been used in several works to assist the peak assignments of Raman spectra [19,29,37–42]. Nevertheless, understanding the properties and spectral behavior of caffeine in solution requires a computational strategy that includes conformational sampling, environmental effects with a proper description of hydrogen bonding (HB) interactions, the dynamical aspects of the solvation, and mutual solute–solvent polarization [43]. Such features are all integrated into Quantum Mechanics/Fluctuating Charges (QM/FQ) computational protocols developed during the last years in our research group [44–46]. By following the protocol, we have previously studied the absorption spectra of caffeine and similar xanthines [47,48] and so far we have also computed RR spectroscopies on solvated amides [49], dipeptides [50], anionic ibuprofen [51], contaminants [52], doxorubicin (DOX) and DOX-DNA [53] complexes. As a general finding, we have seen that the multiscale QM/FQ protocols outperform implicit and non-polarizable QM/Molecular Mechanics approaches, and this is due to the atomistic treatment of the solvent molecules and thus of the solute–solvent interactions. On the subject of intermolecular interactions, Natural Bond Orbitals (NBO) [54–56] and the topological analysis of electron density as formulated in the Quantum Theory of Atoms in Molecules (QTAIM) by Bader [57] are well-established tools for studying these cooperative, non-covalent contacts and have been applied to diverse chemical systems [58].

The purpose of this paper is to use computational tools to analyze intermolecular interactions and provide an overview of the chemical physics influence of the environment when recognizing resonance enhancement in the Raman spectrum of caffeine in an aqueous solution. Since the careful selection of a laser source that is appropriate for a specific compound and the geometry of the spectrometer is essential to the efficient use of RR spectroscopy as an analytical tool [59], several excitation wavelengths are proposed to quantify the caffeine content in very diluted samples.

The next section outlines the computational steps, while the modeled electronic absorption, Raman, and RR spectral profiles for solvated caffeine are then discussed in light of the solute–solvent interactions.

2. Results and Discussion

This section provides a characterization of the main interactions taking place when caffeine is dissolved in water as well as a description of the multiscale QM/FQ simulated electronic absorption, Raman, and RR spectra, whose quality and interpretation are improved by an atomistic description of the environment in the modeling.

Geometrical and energetic aspects of isolated and hydrated forms of caffeine have been the subject of investigation in earlier studies [38,40,60–62]. Structural analysis of caffeine has revealed small barriers associated with the rotation of the methyl groups [47] which are reported to be affected by non-covalent interaction with their neighboring functional groups and by different degrees of hyperconjugation [39,63]. Despite its limited solubility in water that is certainly overcome at moderate temperatures, caffeine solutions can be seen as a series of non-covalent interactions, mainly HBs.

2.1. *Caffeine· · · Water Interactions*

The hydrogen-bonding patterns of caffeine in an aqueous solution may be examined from the MD trajectories. Caffeine has three specific solvation sites and five hydrogen bond acceptor sites: four acceptor sites are provided by the carbonyl oxygen atoms and one by a nitrogen atom in the imidazole ring. The solvent structure around those atoms is shown in Figure 2a. The two carbonyl oxygen atoms, O10 and O12, are better solvated than the N9 site, as reflected by the higher peak value in the RDFs associated with the former. The corresponding running coordination numbers (RCNs) resulting from integrating the RDFs over the first solvation shell give approximately 3.9 water molecules around the carbonyl oxygens and 1.6 for the nitrogen, yielding a total of 5.5 water molecules close to caffeine. A similar number was obtained in our previous work [47], with the difference being in the incorporation of virtual sites (see Figure 1) in the lone pairs of O10, O12, and N9. Here, it is clear that as a consequence of a more specific/directional interaction when adding virtual sites, the peaks of the RDFs are higher and shifted to smaller distances (continuous vs. dashed curves in Figure 2a). After studying the influence of HB directionality on vibrational cooling dynamics of methyl xanthines, Zhang et al. [64] suggested that HB acceptors do not play a primary role in mediating vibrational energy flow from solute to solvent. Unsurprisingly, the five water molecules in the first solvation shell coincide with the average number of HBs between the caffeine and the solvent based on geometric criteria ($d_{X\cdots O_w} \leq 3.5$ Å and $\theta_{H_w-O_w-X} \leq 30°$, X being N9, O10, and O12 atoms in caffeine), as can be inferred from Figure 2b. The hydration structure of caffeine in aqueous solution has been studied in earlier [65] and recent [66] works using classical MD simulations, and the same hydration patterns were found.

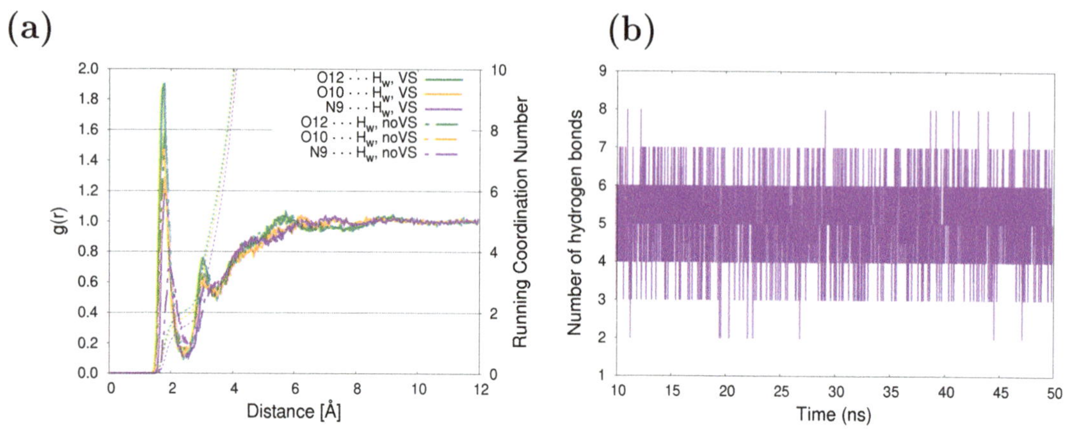

Figure 2. (**a**) Radial distribution function between selected sites of caffeine and water molecules. Running coordination numbers are also included (dashed lines). (**b**) Evolution in time of the number of hydrogen bonds between caffeine and its surrounding water molecules.

To further pinpoint the source of the stabilization when caffeine is dissolved in an aqueous solution, more selective scrutiny is included by performing a quantitative analysis of intermolecular interactions with NBO and QTAIM approaches. Given that determining explicit interactions of caffeine atoms with environmental molecules requires the explicit presence of water molecules, the QM region was expanded on each snapshot to accommodate the first solvation shell in addition to caffeine, but still keeping the other FQ water molecules. First, we resort to NBO to study NBO donor–acceptor interactions.

According to the NBO results, caffeine and its neighboring water molecules interact through delocalizations from lone pairs (n) in the solvent to antibonding π^* orbitals in the solute, or from π Lewis type orbitals in caffeine to antibonding $\sigma^*_{H_w-O_w}$ in the water molecules. However, the dominant specific solute–solvent contacts involve HBs with orbital interactions from lone pairs (n) in the O10, O12, and N9 caffeine atoms to antibonding (σ^*) orbitals in the water molecules. They are of the $n_{O_c} \rightarrow \sigma^*_{H_w-O}$ $O_c = O10, O12$, $n_{N9} \rightarrow \sigma^*_{H_w-O}$ type. Their interaction energies on each frame, obtained via second-order perturbation corrections to the Fock matrix, are presented as stacked histograms in Figure 3a. The corresponding orbital overlap representation for one of the MD configurations where caffeine more heavily interacts with five water molecules is depicted in Figure 3b.

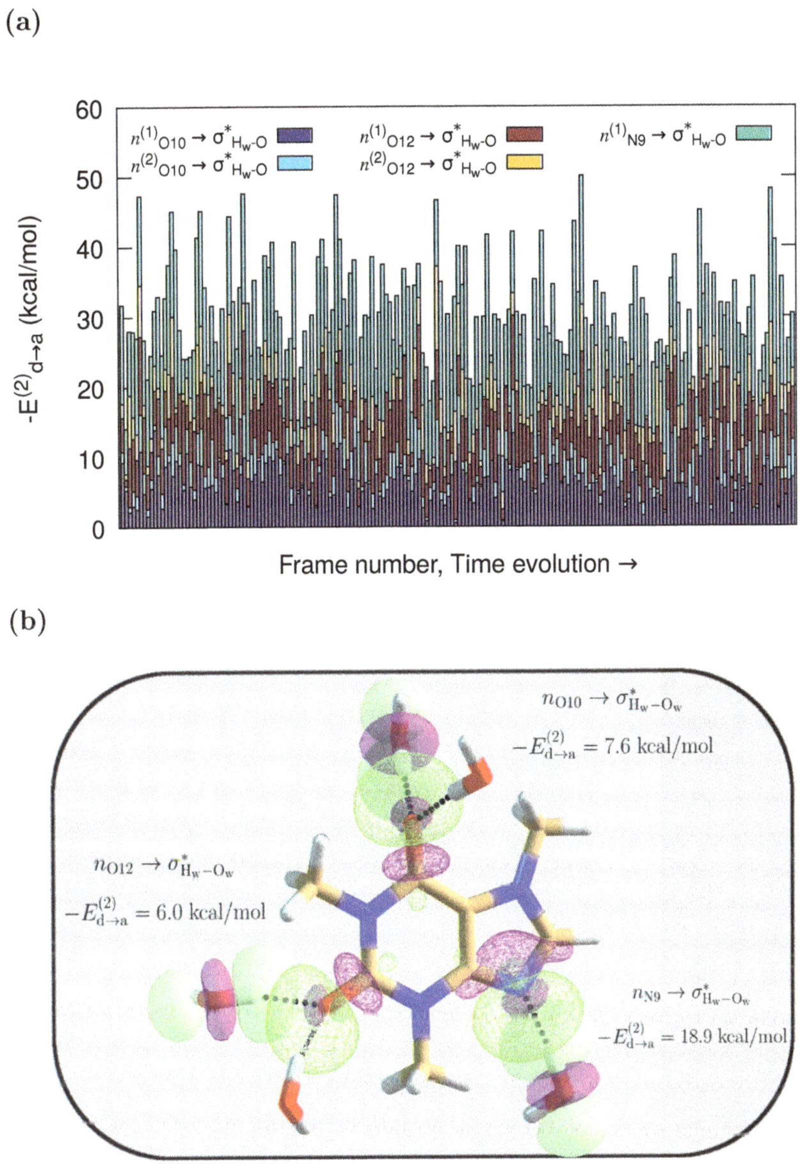

Figure 3. QM/FQ NBO stabilization energies. (**a**) Cumulative quantification of the strength of the orbital interactions that keep caffeine in contact with adjacent water molecules. All stabilization energies, $E^{(2)}_{\text{d}\to\text{a}}$, are associated with the $n_{\text{O}_\text{c}} \to \sigma^*_{\text{H}_\text{w}-\text{O}}$ $\text{O}_\text{c} = \text{O10, O12}$ charge transfers, and $n_{\text{N9}} \to \sigma^*_{\text{H}_\text{w}-\text{O}}$ charge transfer. (**b**) Orbital representation within the NBO picture for the intermolecular interactions in solvated caffeine for one randomly chosen configuration from the MD trajectory. The QM region includes five explicit waters for that configuration, while the remaining solvent molecules are treated at the FQ level.

The average values of stabilization energies in Figure 3 are around $-E^{(2)}_{\text{d}\to\text{a}} = 7$ kcal/mol and $-E^{(2)}_{\text{d}\to\text{a}} = 11$ kcal/mol, for $n_{\text{O}_\text{c}} \to \sigma^*_{\text{H}_\text{w}-\text{O}}$ $\text{O}_\text{c} = \text{O10, O12}$, $n_{\text{N9}} \to \sigma^*_{\text{H}_\text{w}-\text{O}}$, respectively, which are associated with moderate or mild interactions, unlike the strong interaction found for charge-assisted HBs [51,67]. As is also true for the reported UV-Vis and ECD spectra of captopril and naproxen [67], configurations with a large cumulative $|E^{(2)}_{\mathrm{d}\to\mathrm{a}}|$ have spectra that better match the experiments. The individual stabilization that those charge transfers provide to the system is quantified in Figure S1 in the Supplementary Materials.

Two observations can be made from the cumulative plot in Figure 3a. First, the different interaction strengths seen for each snapshot underline the variability in the solvent arrangements around the caffeine molecule. This, in turn, directly affects the spectra because as a result of such dynamical variability, assorted oscillator strengths and Raman cross-sections are obtained for each snapshot. Second, given that the higher the $|E^{(2)}_{\mathrm{d}\to\mathrm{a}}|$ value, the stronger the interaction, the water contacts with the lone pair in the N9 atom are the strongest ones most of the time ($-E^{(2)}_{\mathrm{d}\to\mathrm{a}}$ up to 26 kcal/mol, Figure S1), even if all the interactions have the same nature. Interestingly, in the crystal structure of caffeine hydrate, the only existing hydrogen bond between the crystallization water and the caffeine molecule is noted to occur through the hydrophilic N9 center [68].

Further characterization of the hydration patterns of solvated caffeine is carried out by analyzing the topological descriptors, particularly the electron density, $\rho(\mathbf{r}_c)$, accumulated at the BCPs that correspond to HBs. Figure 4 displays the distribution of the $\rho(\mathbf{r}_c)$ values for located $N9 \cdots H_w$ contacts in the whole group of configurations. Electron densities at those points cover the $[1.0 \times 10^{-2}, 5.0 \times 10^{-2}]$ a.u. interval. Similar to what happens for the interaction with the other atoms susceptible to H-bonding in caffeine (Figure S2 in the Supplementary Materials), the distributions are centered around 3.0×10^{-2} a.u., thus exhibiting larger densities than the usual reference of the water dimer, 2.3×10^{-2} a.u. [69]. For the interactions studied in this work, the accumulations of electron densities at the BCPs do not present any direct correlation with the $-E^{(2)}_{\mathrm{d}\to\mathrm{a}}$ stabilization energies.

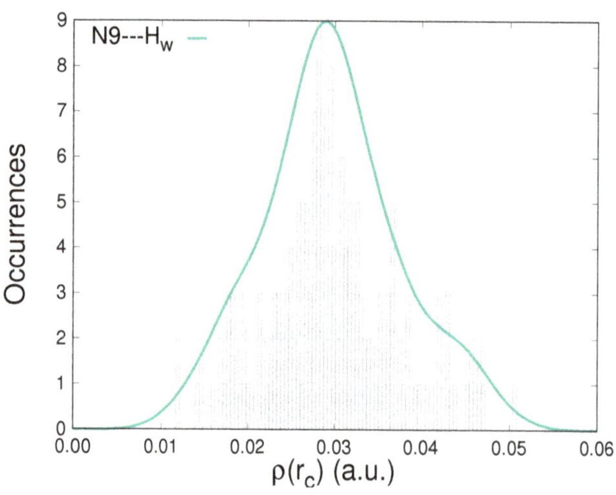

Figure 4. Distribution of the electron densities at the bond critical points for the $N9 \cdots H_w$ intermolecular contacts between caffeine and water molecules along an MD trajectory.

2.2. Spectra

After sampling the caffeine$\cdots$water phase space through MD simulations and characterizing the dynamic nature of their interactions, this section reports on the simulated UV-Vis, Raman, and RR spectra of solvated caffeine, making a comparison with experimental data and offering a detailed view of the enhanced signals.

Convergence tests displayed in Figures S9–S11 in the Supplementary Materials indicate that UV-Vis, Raman, and RR spectra remain unchanged when more than $\approx$100 uncorrelated snapshots are included. Next, the results and discussion built upon converged spectra averaged on 200 frames in each case. We recall here that for all of the spectral calculations, the caffeine moiety is the only one treated at the QM level, whereas all the water molecules are described with FQ.

2.2.1. UV-Vis Spectrum

The QM/FQ calculated absorption spectrum of solvated caffeine is shown in Figure 5 along with the experimental curve (dashed line) and the stick-like spectrum colored according to every specific excited state. Two main bands characterize the UV-Vis absorption spectrum. Excitations to S_1 (black sticks) are mainly responsible for the appearance of the band centered at around $\lambda_{max,1}$ = 267 nm. In contrast, the band with the highest absorption is located at $\lambda_{max,2}$ = 206 nm and results from a more complex combination of excitations. Experimental observations placed such bands at 273 nm and 205 nm, respectively, [70,71] or at very similar values [38,39,61,72]. We have previously concluded [47] that the vertical absorption in the lowest energy band corresponds to a $\pi \rightarrow \pi^*$ electronic transition involving the HOMO and LUMO orbitals, based on the canonical molecular orbital decomposition [56]. In line with that assignment, Table S1 in the Supplementary Materials reports, for a single snapshot, the nature of the electronic states giving rise to the absorption spectrum. Beyond the fair agreement with the experimental spectrum, the computed UV-Vis absorption spectrum helps recognize potential excitation wavelengths for irradiating a solvated caffeine sample in RR spectroscopy (see below).

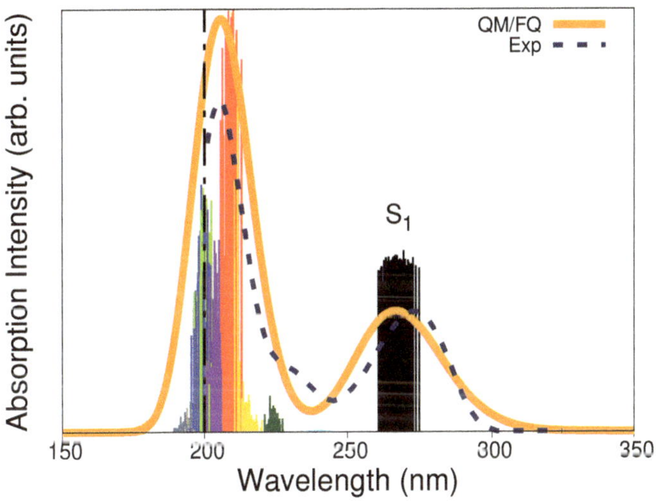

Figure 5. QM/FQ and experimental UV–Vis absorption spectra of caffeine in aqueous solution. Ten excited states, each one associated with a different stick color, were converged in the TD-DFT calculations. The spectrum was convoluted with Gaussian functions using a FWHM of 0.6 eV. The label S_1 indicates the first excited states. Experimental reports collected from refs. [70,71] determine the first maximum at 273 nm. The dashed vertical line indicates that there is no experimental information below 200 nm.

Multiscale methodologies have also been employed by Skarmoutsos et al. [66] to calculate absorption and emission spectra of caffeine in aqueous solution and the authors emphasized the need for an explicit inclusion of the solvent for the correct reproduction of the emission spectrum.

2.2.2. Spontaneous Raman and UV Resonance Raman Spectra

Despite having noticed in our previous publication that specific solute–solvent interactions are not essential for the reproduction of excitation energies of caffeine in aqueous solution, they are found to be significant in the modeling of relative intensities of the two main bands of the spectrum [47] and become fundamental when going to vibrational spectroscopies as we have already exemplified for amides [49], dipeptides [50], and anionic ibuprofen [51] cases.

Multiple works in the literature deal with the Raman spectra of caffeine at different excitation sources. Some of the experiments refer to the hydrated and anhydrous crystalline caffeine [42,68,73,74], whereas others cover caffeine solutions [29,75] and the effect of different pH values on the Raman spectra [19,42]. Calculations have also been used as guidelines for a reliable and complete vibrational assignment [19,40,42,60,76,77]. Since caffeine does not have tautomers in water, only small changes are appreciated in the peak assignments for the crystalline and solvated forms. The most relevant vibrational frequencies in the 1000–2000 cm^{-1} range of the caffeine Raman spectra are ascribed to the following:

- Carbonyl stretching frequencies, $\nu_{C=O}$, experimentally reported to appear at 1647 and 1692 cm^{-1} (theoretically located at 1728 and 1670 cm^{-1}, respectively) and due to the two C=O groups couple into an in-phase and an out-of-phase stretching vibration. The in-phase one is predicted to have less intensity (sometimes a half) than the out-of-phase carbonyl vibration [40].
- C=C and C=N stretching modes in the purine ring system, $\nu_{C=C}$ and $\nu_{C=N}$, at 1598 cm^{-1}.
- Imidazole ring stretching plus ν_{C-N} + $\nu_{C=C}$ + C-H bend appearing as a peak at 1549 cm^{-1}.
- Symmetric CH-bending vibrations in the methyl groups appearing as a broad band or a set of peaks [40] at 1488, 1470, 1454, and 1431 cm^{-1}. These are collectively labeled as δ_{CM} in ref. [29] and centered at approximately 1497 cm^{-1}. Importantly, some of them also include a contribution of the ν_{C-N} in the imidazole ring.
- Bending of C8-H atoms in the imidazole group combined with CH_3 bending of methyl groups, located at about 1437 cm^{-1}. This peak is labeled as δ_{C8-H} in ref. [29].
- Joint quadratal $\nu_{(imidazolering)}$, $\nu_{C=N}$ and bending of methyl groups at 1406 cm^{-1}.
- Stretching in the imidazole ring ν_{C-N} plus some bending of methyl groups at 1362 cm^{-1}. It will be called $\nu_{i-ring,1}$ in what follows.
- Trigonal C-N stretching vibrations in the imidazole ring combined with N-CH_3 (M1) stretching vibration at about 1335 cm^{-1}. In some reports, this is the most intense Raman peak and is labeled as $\nu i-$ ring in ref. [29]. It will be termed as $\nu_{i-ring,2}$ in the next discussion.
- Mixed $\nu_{N-CH_3(M2)}$, ν_{C-N} and ν_{C-C} in both rings expressed as a band centered at 1291 cm^{-1}. The general ν_{rings} nomenclature proposed in ref. [29] for this peak will be maintained here.
- CH_3 rocking (in plane), $\rho_{CH_3,M1}$ and $\rho_{CH_3,M3}$ located at 1034 and 1079 cm^{-1}, respectively.

The theoretical positions of those peaks in one specific frame and after averaging the convoluted spectra along all MD snapshots are listed in Table 1. From the standard deviations, it is noticed that the positions of some peaks are particularly sensitive to the solvent arrangements, thus indicating either close contact with the solvent or limited vs. more flexible vibrations due to the presence/absence of water molecules in the vicinities. To offer a better description of the normal modes, Figure 6 depicts some selected ones.

Figure 7a shows the QM/FQ spontaneous (far from Resonance) Raman and the experimental spectra [68] at ω_0 = 1064 nm. We have chosen data from ref. [68] as a reference because of the larger measurement region available to compare, but spectral features are common to all experimental works. The calculated frequencies agree very well with the experimental ones, and the computed intensities accurately capture the relative intensities of the main peaks. Consistent with previous findings, the configurational wealth translated in a diversity of sticks (Figure S5a in the Supplementary Materials) confers the modeled spectrum with a natural inhomogeneous broadening that brings it close to the measurements.

Table 1. Raman and RR cross-sections in cm^2 cm mol^{-1} sr^{-1} for selected vibrations of solvated caffeine. Two peak positions are given: in one specific frame and after averaging the spectra along all MD snapshots. StdDev stands for the standard deviation of the average position of the peak. The experimental RR scattering condition is fully achieved with an incident wavelength of 266 nm, but there is another more intense absorption maximum at about 205 nm (Figure 5). Enhancement factors computed with respect to the Raman cross-sections at 1064 nm.

Vibration	Position cm^{-1}	StdDev cm^{-1}	Raman @1064 nm	RR @266 nm	RR [1] @210 nm	Enhancement Factor @266 nm	Enhancement Factor @210 nm
$\rho_{CH_3,M1}$	1050/1049	5	5.07×10^{-10}	0.0023	0.0084	4.54×10^{6}	1.66×10^{7}
$\rho_{CH_3,M3}$	1101/1097	7	9.73×10^{-10}	0.0026	0.0164	2.67×10^{6}	1.69×10^{7}
ν_{rings}	1308/1313	6	2.13×10^{-9}	0.0026	0.0120	1.22×10^{6}	5.63×10^{6}
$\nu_{i-ring,2}$	1360/1362	4	3.71×10^{-9}	0.0011	0.0383	2.96×10^{5}	1.03×10^{7}
$\nu_{i-ring,1}$	1396/1394	4	4.15×10^{-9}	0.0034	0.0074	8.19×10^{5}	1.78×10^{6}
δ_{C8-H}	1455, 1469/1469	7	1.38×10^{-9}	0.0014	0.0110	1.01×10^{6}	7.97×10^{6}
δ_{CM}	1517, 1538/1536	10	1.58×10^{-9}	0.0013	0.0104	8.23×10^{5}	6.58×10^{6}
ν_{C-C}	1583/1583	4	5.90×10^{-10}	0.0002	0.0140	3.39×10^{5}	2.37×10^{7}
$\nu_{C=C} + \nu_{C=N}$	1618/1624	6	2.12×10^{-9}	0.0003	0.0265	1.42×10^{5}	1.25×10^{7}

[1] Among the ω_0 list used to compute RR cross-sections, this value is the closest one to the absorption maximum at 205 nm.

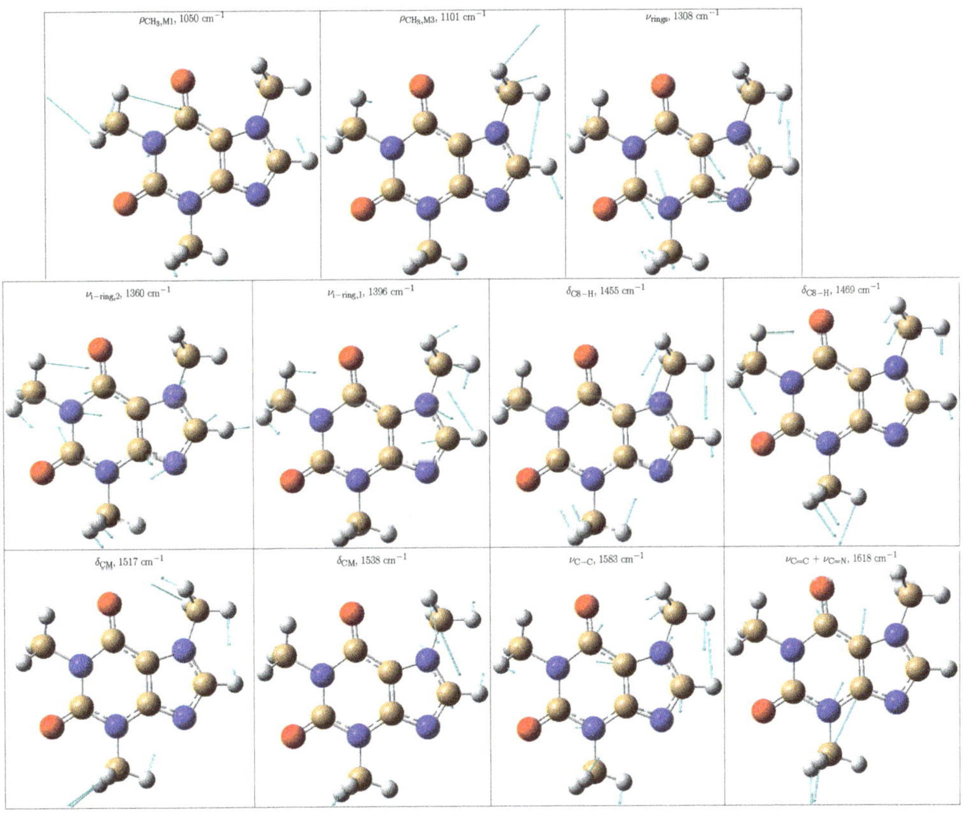

Figure 6. Selected vibrational modes giving rise to the most enhanced peaks in the Resonance Raman spectrum of caffeine in aqueous solution. Color code: C atoms in tan, O atoms in red, N atoms in blue, and H atoms in light gray.

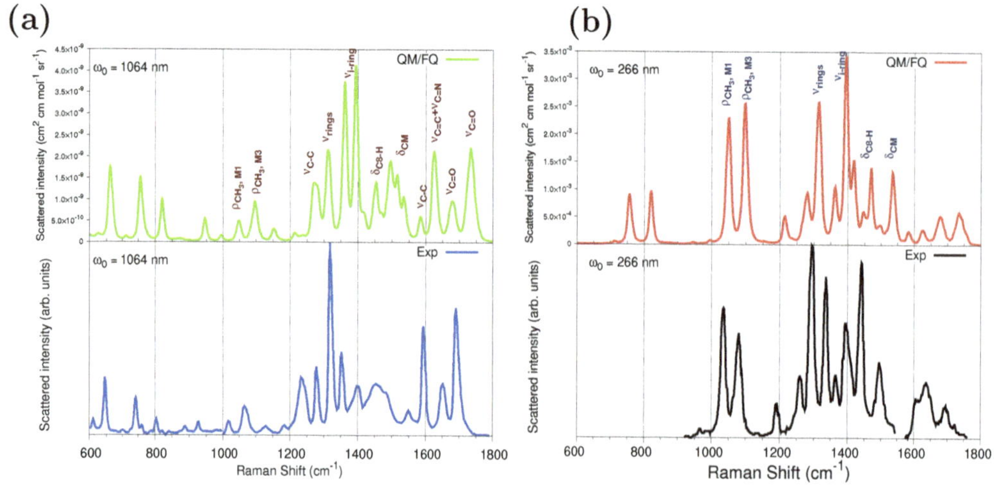

Figure 7. Convoluted QM/FQ (top) and experimental (bottom) Raman spectra of caffeine in aqueous solution. (**a**) Spontaneous (far from Resonance) Raman spectra simulated and experimental measurement from ref. [68] using 1064 nm as excitation wavelength. (**b**) UVRR spectra. RR intensities were calculated with a 200 cm^{-1} damping factor. In the UVRR experimental spectrum [29] (bottom panel), the authors reported that data at about 1550 cm^{-1} are affected by an experimental laser satellite artifact and therefore were omitted. Raman and RR intensities were broadened using Lorentzian functions with FWHM = 8 cm^{-1}. No scaling factors were applied to the frequencies. See also Figure S1a for atom numbering.

To investigate the RR enhancement pattern, Figure 7b compares the computed QM/FQ RR spectrum of caffeine aqueous solution with UVRR scattering measurements reported in ref. [29] and acquired at $\omega_0 = 266$ nm. Figure S5b in the Supplementary Materials depicts raw data plotted as a stick-like RR spectrum. The convoluted RR spectrum follows a similar pattern as that seen in the experiment and most experimental relative intensities are well reproduced. The position of one of the computed absorption maxima in Figure 5 matches the experimental RR condition; thus, simulating at 266 nm preserves the resonance enhancement observed in the experiment. Under this excitation, of particular spectroscopic interest are changes involving $\rho_{CH_3,M1}$, $\rho_{CH_3,M3}$, ν_{rings}, $\nu_{i-ring,2}$, δ_{C8-H} and δ_{CM} signals. This result agrees with the earlier experimental RR studies of caffeine in solution by Tavagnacco et al. [29]. The authors pointed out that this quite unselective peak enhancement obeys the fact that the orbitals mainly contributing to the transition at 273 nm are highly delocalized along the entire molecule. The Raman and RR intensities for those vibrations are summarized in Table 1, for calculations performed at 1064 nm and 266 nm, respectively, and the enhancement goes up to six orders of magnitude for some vibrations.

It was mentioned before that the strongest absorption band of solvated caffeine occurs at 267 nm (273 nm in the experiments) and is merely owing to the contribution of the S_1 excited state. Calculated Raman spectra at a 266 nm excitation wavelength using the ten excited states and only S_1 are shown in Figure S6a of the Supplementary Materials and are virtually identical. Instead, using excited state gradients for states going from S_2 (sticks in cyan in Figure 5) to S_{10} (sticks in gray in Figure 5), which are the cause of the appearance of the other more intense band (at 205 nm), produces an RR spectrum with weaker signals and introduces new spectral features.

As is to be expected, the RR enhancement of some normal modes of caffeine in water is highly dependent on the excitation wavelength. Based upon the results of the UV-Vis spectra in Figure 5, the effect of using excitation wavelengths near the center of the highest absorption band on the RR spectrum was also explored. As illustrated in Figure S6b of the Supplementary Materials, choosing ω_0 =210 nm leads to a quite different spectral profile that alters the intensity of the peaks revealing another enhancement pattern when compared against the spectrum at $\omega_0 = 266$ nm. In that scenario, Raman signals around 1600 cm^{-1}, $\nu_{C=C}$, $\nu_{C=N}$, and ν_{C-C}, as well as $\nu_{i-ring,1}$ appear to be substantially enhanced. Enhancement factors at 210 nm are tabulated in Table 1 and indicate that it is possible to track the presence of caffeine by analyzing further signals that are selectively enhanced at shorter wavelengths.

The observed signal enhancement in the RR spectrum can also be rationalized in terms of the displacement of the excited states and S_i geometries with respect to the ground state. In the (VG | FC) approximation, such displacements are related to the shift vectors (widely denoted as **K**) since the Duschinsky rotation is ignored. As noted in Figure 5, S_1 dominantly contributes to the resonance when exciting caffeine in water at 266 nm. For that particular excited state, a graphical representation of the shift vector, and in turn, of the displacements, Δ_1 obtained by projecting the excited state gradient onto the normal modes is reported in Figure S7 in the Supplementary Materials for one of the snapshots extracted from the trajectory. Noticeably, the most intense peaks in the RR spectrum (see Table 1 for their positions) are associated with the normal modes exhibiting the largest shifts/displacement values which is exactly the spirit of the short-time approximation, although that one is better suited for the pre-resonance conditions [78,79].

To further investigate the relationship between the RR enhancement and the absorption spectrum of caffeine, we split the 65,000–30,000 cm^{-1} range (154–333 nm) into 15 excitation wavelengths and computed RR cross-sections. The resulting RREP is shown in Figure S8 in the Supplementary Materials. Figure 8 displays an RREP portion for excitation wavelengths in the vicinity of 266 nm, namely, 222, 235, 250, 267, 285, 307, and 333 nm, which are all associated with the lowest energy band of the absorption spectra in Figure 5. RREPs support the observation that the strongest Raman signals suggested to detect or quantify caffeine come from enhancements with ω_0 values around 210 and 266 nm.

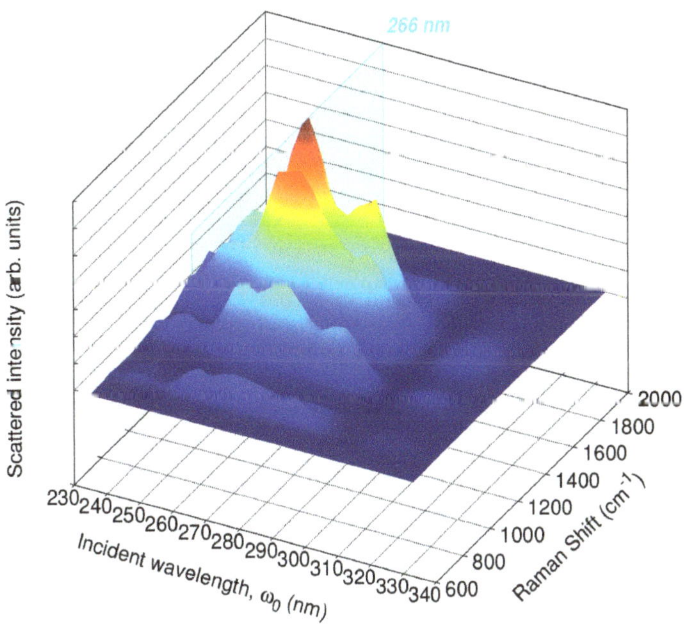

Figure 8. Calculated QM/FQ Resonance Raman Excitation Profiles (RREP) of caffeine in aqueous solution. In total, 200 structures were considered to achieve convergence at the QM/FQ level, with B3LYP/6–311++G(d, p).

3. Materials and Methods

To analyze the potential effect of the explicit water molecules on the spectra of solvated caffeine, we initially computed all the spectra of interest both in the gas phase and in a mimicked bulk solvent environment with the implicit model PCM. Results for Raman and RR are shown in Figure S4 in the Supplementary Materials. Although the absorption spectrum is less sensitive, vibrational spectra exhibit erroneous positions of the peaks and overestimate the enhancement of some signals when compared against experiments. Due to the discrepancies between the QM/PCM and the Raman and RR experimental data, we opted to utilize an atomistic representation of the solvent in what follows.

There is a robust computational protocol, described by Giovannini et al. [44], to compute spectra of molecular systems. Such a methodology relies on a fully atomistic and polarizable classical modeling of the solvent coupled with a QM description of the solute. Recently, it was successfully applied to simulate the UV-Vis spectra of a series of methylated xanthines [47], including caffeine, in an aqueous solution. Under the assumption that virtual sites could recover the correct directionality of hydrogen bonds between the solute and solvent [49,80–83], we followed the same procedure outlined in ref. [47] but placed virtual sites at the nitrogen atom, N9, of the imidazole ring, and at the oxygen atoms, O10, and O12, of the pyrimidinedione ring, specifically at their centroid positions, determined by the Boys localization procedure [84]. Figure 1 depicts the position of those dummy atoms. MD simulations were performed in the GROMACS software, version 2020.3. Ref. [85] using the General Amber Force Field (GAFF) [86] and TIP3P [87] force fields to describe inter and intra-molecular interactions for caffeine and water molecules, respectively. From the 50 ns trajectory, we obtain the average number of solute–solvent HBs and extract the radial distribution functions (RDFs) between water molecules and the carboxylic groups and the nitrogen atom, N9, of the imidazole ring of caffeine, using the TRAVIS package [88,89]. We also determine the number of waters in the first solvation shell by integrating the RDFs.

After extracting 200 uncorrelated snapshots from the trajectory and cutting them into a solute-centered sphere with a radius of 17 Å (Figure 1), QM/FQ calculations were carried out at the B3LYP/6-311++G(d, p)/FQ level in the Gaussian16 package [90] exploiting the FQ parametrization proposed by Rick et al. [91]. The same QM level of theory has been employed in other computational works concerning caffeine and xanthines [38–40,47,61,63,70,77]. In spectral calculations, the caffeine molecule is the only part of the system treated at the QM level. However, we expanded the QM region to include the solvent molecules in the first solvation shell to estimate the strength of the interactions between caffeine and water molecules in the closest contact. This is carried out using the NBO analysis [54–56] in the NBO7 program [92]. Similar QM/FQ NBO analyses have been conducted for common pharmaceuticals [51,67] and hypoxanthine [48] in solution. Conventional bond critical points (BCP) of the electron density and localized Natural BCPs topological descriptors are computed with the Natural BCP (NBCP) analysis as implemented in NBO7 [92].

The caffeine geometry was optimized in each snapshot at the QM/FQ level by keeping the solvent molecules fixed. Vertical excitation energies were then estimated by converging 10 excited states again exploiting the B3LYP/6-311++G(d, p)/FQ level in the linear response time-dependent density functional theory (TD-DFT) framework. As a side note, optimizing the solute, or including explicit water molecules during the TD-DFT calculation did not have any effect on the UV-Vis absorption spectrum, as can be seen in Figure S3 in the Supplementary Material (SM). The reported averaged absorption spectrum was obtained by convoluting peak intensities with Gaussian functions and a full width at half maximum (FWHM) of 0.6 eV.

On the optimized geometries, frequencies and excited state gradients were also computed to model vibrational spectroscopies. Using analytical response theory as implemented for QM/FQ [93], the spontaneous Raman spectrum was calculated in the dynamic regime by setting the incident frequency (ω_0) to match the experimental value of 1064 nm. We used the QM/FQ approach adapted to RR spectroscopy in ref. [49] and computed the RR cross-sections by resorting to the time-independent sum-over-state formulation detailed in refs. [79,94]. Among the diverse ways for modeling the excited-state Potential Energy Surface, we chose the Vertical Gradient, Franck–Condon (VG | FC) approximation, which assumes that the vibrational frequencies and normal modes of the excited state are the same as those of the ground state, while the transition dipole moments are thought to be unaffected by the geometry of the molecules. An RR excitation profile (RREP) was also built by choosing several incident wavelengths and recomputing RR cross-sections. Sticks for spontaneous Raman and RR were convoluted with Lorentzian profiles and FWHM values of 8 cm^{-1}. Final UV-Vis, Raman, and RR spectra were obtained by averaging over the spectra of the MD snapshots. The convergence of QM/MM computed spectra was closely examined by considering increasing numbers of snapshots (see Figures S9–S11 in the Supplementary Materials).

4. Conclusions

In this study, interactions and spectral signatures of caffeine in aqueous solution have been investigated from a computational perspective to understand the impact of the dynamical variability of the solute–solvent interactions on modeled spectroscopies.

The configurational landscape of neutral caffeine in water was explored with MD simulations. From the trajectory, the hydrogen bond formation between the caffeine and surrounding waters was the dominating intermolecular interaction. The non-covalent interactions in the caffeine–water complexes were thoroughly examined using NBO and QTAIM. Among the three H-bonding sites, the caffeine interactions through the nitrogen atom of the imidazole ring turned out to be stronger than those with the two C=O groups.

Building on the fact that several analytical methods—including UV-Vis and Raman—have been proposed to determine the caffeine content in coffee samples or to detect if this emerging and persistent micropollutant is present in the environment even at very low concentrations, electronic absorption and mixed electronic–vibrational spectra were also modeled. Computed spectral profiles obtained with a multiscale QM/FQ approach were compared with experimental UV-Vis, Raman, and RR data of solvated caffeine, and a good agreement was found between simulated and experimental results. The quality of the spectra upon atomistically treating the solvent confirmed that HB interactions play an essential role in the caffeine spectral behavior. This underscores the robustness of a computational protocol able to prioritize the explicit intermolecular interactions [62] and the dynamical aspect of solvation, both important characteristics to be considered in the reproduction of any spectra, but crucial when the presence of solvent molecules influences the vibrational modes as also seen in the case of the RR spectrum of cytosine in water [35]. For caffeine, these modes are the carbonyl stretchings and C=N stretchings in the imidazole ring, the latter being involved in one of the enhanced vibrations.

From the RREPs, the excitation wavelengths found to provide an intense enhancement are those that match the maxima in the absorption spectrum, namely the close vicinity of 205 and 273 nm, mostly enhancing $\nu_{\mathrm{i-ring}}$ and $\nu_{\mathrm{C=C}}$ signals with enhancement factors up to six orders of magnitude when compared to their intensities in conventional Raman. These results suggest that RR can be employed as a quantitative technique regarding caffeine. Indeed, experiments to monitor vibrational modes at specific excitation wavelengths, shorter than the earlier tested ω_0 = 266 nm, can be useful to more thoroughly detect and quantify caffeine in different samples, which is a growing need because of the everyday use of this stimulating agent.

Supplementary Materials: The following supporting information can be downloaded at: https://www.mdpi.com/article/10.3390/molecules29133035/s1, NBO and QTAIM analyses; Additional plot concerning UV-Vis spectra; Analysis of electronic transitions; Raman and RR spectra in the gas phase and PCM; Stick-spectra; Effect of specific excited states; Shift vector; RREP; Convergence plots.

Author Contributions: Conceptualization, S.G. and C.C.; methodology, S.G.; validation, S.G. and C.C.; formal analysis, S.G. and C.C.; investigation, S.G. and C.C.; resources, C.C.; data curation, S.G.; writing—original draft preparation, S.G.; writing—review and editing, S.G. and C.C.; visualization, S.G.; supervision, C.C.; project administration, C.C.; funding acquisition, C.C. All authors have read and agreed to the published version of the manuscript.

Funding: CC acknowledges the support of the European Union by the Next Generation EU project ECS00000017 'Ecosistema dell'Innovazione' Tuscany Health Ecosystem (THE, PNRR, Spoke 4: Nanotechnologies for diagnosis and therapy).

Institutional Review Board Statement: Not applicable.

Informed Consent Statement: Not applicable.

Data Availability Statement: The original contributions presented in the study are included in the article/Supplementary Material, further inquiries can be directed to the corresponding authors.

Acknowledgments: We gratefully acknowledge the Center for High-Performance Computing (CHPC) at SNS for providing the computational infrastructure.

Conflicts of Interest: The authors declare no conflicts of interest. The funders had no role in the design of the study; in the collection, analyses, or interpretation of data; in the writing of the manuscript; or in the decision to publish the results.

Abbreviations

The following abbreviations are used in this manuscript:

(UV)RR	(Ultra–Violet) Resonance Raman
QM	Quantum Mechanics
FQ	Fluctuating Charges
RREP	Resonance Raman Excitation Profile
MD	Molecular Dynamics
SERS	Surface-Enhanced Raman Scattering
RDF	Radial Distribution Function
NBO	Natural Bond Orbitals
BCP	Bond Critical Point
QTAIM	Quantum Theory of Atoms in Molecules
FWHM	Full Width at Half Maximum
PCM	Polarizable Continuum Model

References

1. Temple , J.L.; Bernard, C.; Lipshultz, S.E.; Czachor, J.D.; Westphal, J.A.; Mestre, M.A. The safety of ingested caffeine: A comprehensive review. *Front. Psychiatry* **2017**, *8*, 80. [CrossRef]
2. Derry, C.J.; Derry, S.; Moore, R.A. Caffeine as an analgesic adjuvant for acute pain in adults. *Cochrane Database Syst. Rev.* **2014** , *12*. [CrossRef] [PubMed]
3. Grgic, J.; Trexler, E.T.; Lazinica, B.; Pedisic, Z. Effects of caffeine intake on muscle strength and power: A systematic review and meta-analysis. *J. Int. Soc. Sports Nutr.* **2018**, *15*, 11. [CrossRef]
4. de Souza, J.G.; Del Coso, J.; Fonseca, F.d.S.; Silva, B.V.C.; de Souza, D.B.; da Silva Gianoni, R.L.; Filip-Stachnik, A.; Serrão, J.C.; Claudino, J.G. Risk or benefit? Side effects of caffeine supplementation in sport: A systematic review. *Eur. J. Nutr.* **2022**, *61*, 3823–3834. [CrossRef]
5. McLellan, T.M.; Caldwell, J.A.; Lieberman, H.R. A review of caffeine's effects on cognitive, physical and occupational performance. *Neurosci. Biobehav. Rev.* **2016**, *71*, 294–312. [CrossRef] [PubMed]
6. Dong, X.; Li, S.; Sun, J.; Li, Y.; Zhang, D. Association of Coffee, Decaffeinated Coffee and Caffeine Intake from Coffee with Cognitive Performance in Older Adults: National Health and Nutrition Examination Survey (NHANES) 2011–2014. *Nutrients* **2020**, *12*, 840. [CrossRef]
7. Zaukuu, J.L.Z.; Benes, E.; Bázár, G.; Kovács, Z.; Fodor, M. Agricultural potentials of molecular spectroscopy and advances for food authentication: An overview. *Processes* **2022**, *10*, 214. [CrossRef]
8. Vieira, L.; Soares, A.; Freitas, R. Caffeine as a contaminant of concern: A review on concentrations and impacts in marine coastal systems. *Chemosphere* **2022**, *286*, 131675. [CrossRef] [PubMed]

9. Rigueto, C.V.T.; Nazari, M.T.; De Souza, C.F.; Cadore, J.S.; Brião, V.B.; Piccin, J.S. Alternative techniques for caffeine removal from wastewater: An overview of opportunities and challenges. *J. Water Process Eng.* **2020**, *35*, 101231. [CrossRef]
10. Korekar, G.; Kumar, A.; Ugale, C. Occurrence, fate, persistence and remediation of caffeine: A review. *Environ. Sci. Pollution Res.* **2020**, *27*, 34715–34733. [CrossRef]
11. Rodríguez-Gil, J.; Cáceres, N.; Dafouz, R.; Valcárcel, Y. Caffeine and paraxanthine in aquatic systems: Global exposure distributions and probabilistic risk assessment. *Sci. Total Environ.* **2018**, *612*, 1058–1071. [CrossRef] [PubMed]
12. Schulz, H. Spectroscopic technique: Raman spectroscopy. In *Modern Techniques for Food Authentication*; Elsevier: Amsterdam, The Netherlands, 2018; pp. 139–191.
13. Cozzolino, D. An overview of the successful application of vibrational spectroscopy techniques to quantify nutraceuticals in fruits and plants. *Foods* **2022**, *11*, 315. [CrossRef] [PubMed]
14. Jesus, J.I.S.d.S.d.; Löbenberg, R.; Bou-Chacra, N.A. Raman spectroscopy for quantitative analysis in the pharmaceutical industry. *J. Pharm. Pharm. Sci.* **2020**, *23*, 24–46. [CrossRef] [PubMed]
15. Shah, K.C.; Shah, M.B.; Solanki, S.J.; Makwana, V.D.; Sureja, D.K.; Gajjar, A.K.; Bodiwala, K.B.; Dhameliya, T.M. Recent advancements and applications of Raman spectroscopy in pharmaceutical analysis. *J. Mol. Struct.* **2023**, *1278*, 134914. [CrossRef]
16. Li, Y.S.; Church, J.S. Raman spectroscopy in the analysis of food and pharmaceutical nanomaterials. *J. Food Drug Anal.* **2014**, *22*, 29–48. [CrossRef] [PubMed]
17. Silge, A.; Weber, K.; Cialla-May, D.; Müller-Bötticher, L.; Fischer, D.; Popp, J. Trends in pharmaceutical analysis and quality control by modern Raman spectroscopic techniques. *TrAC Trends Anal. Chem.* **2022**, *153*, 116623. [CrossRef]
18. Norton, R.D.; Phan, H.T.; Gibbons, S.N.; Haes, A.J. Quantitative surface-enhanced spectroscopy. *Annu. Rev. Phys. Chem.* **2022**, *73*, 141–162. [CrossRef] [PubMed]
19. Kang, J.; Gu, H.; Zhong, L.; Hu, Y.; Liu, F. The pH dependent Raman spectroscopic study of caffeine. *Spectrochim. Acta A Mol. Biomol. Spectrosc.* **2011**, *78*, 757–762. [CrossRef] [PubMed]
20. Zheng, H.; Ni, D.; Yu, Z.; Liang, P.; Chen, H. Fabrication of flower-like silver nanostructures for rapid detection of caffeine using surface enhanced Raman spectroscopy. *Sens. Actuators B Chem.* **2016**, *231*, 423–430. [CrossRef]
21. Alharbi, O.; Xu, Y.; Goodacre, R. Simultaneous multiplexed quantification of caffeine and its major metabolites theobromine and paraxanthine using surface-enhanced Raman scattering. *Anal. Bioanal. Chem.* **2015**, *407*, 8253–8261. [CrossRef]
22. Zareef, M.; Hassan, M.M.; Arslan, M.; Ahmad, W.; Ali, S.; Ouyang, Q.; Li, H.; Wu, X.; Chen, Q. Rapid prediction of caffeine in tea based on surface-enhanced Raman spectroscopy coupled multivariate calibration. *Microchem. J.* **2020**, *159*, 105431. [CrossRef]
23. Frosch, T.; Wyrwich, E.; Yan, D.; Popp, J.; Frosch, T. Fiber-Array-Based Raman hyperspectral imaging for simultaneous, Chemically Selective monitoring of particle size and shape of active ingredients in analgesic tablets. *Molecules* **2019**, *24*, 4381. [CrossRef]
24. Chi, Z.; Chen, X.; Holtz, J.S.; Asher, S.A. UV Resonance Raman-Selective Amide Vibrational Enhancement: Quantitative Methodology for Determining Protein Secondary Structure. *Biochemistry* **1998**, *37*, 2854–2864. [CrossRef]
25. Efremov, E.V.; Ariese, F.; Gooijer, C. Achievements in Resonance Raman Spectroscopy: Review of a Technique with a Distinct Analytical Chemistry Potential. *Anal. Chim. Acta* **2008**, *606*, 119–134. [CrossRef]
26. Xiong, K. UV Resonance Raman Spectroscopy: A Highly Sensitive, Selective and Fast Technique for Environmental Analysis. *J. Environ. Anal. Chem.* **2014**, *2*, e107. [CrossRef]
27. Chowdhury, J. Chapter 7—Resonance Raman Spectroscopy: Principles and Applications. In *Molecular and Laser Spectroscopy*; Gupta, V., Ed.; Elsevier: Amsterdam, The Netherlands, 2018; pp. 147–164. [CrossRef]
28. Rossi, B.; Bottari, C.; Catalini, S.; D'Amico, F.; Gessini, A.; Masciovecchio, C. Chapter 13—Synchrotron-based Ultraviolet Resonance Raman Scattering for Material Science. In *Molecular and Laser Spectroscopy*; Gupta, V., Ozaki, Y., Eds.; Elsevier: Amsterdam, The Netherlands, 2020; pp. 447–482. [CrossRef]
29. Tavagnacco, L.; Di Fonzo, S.; D'Amico, F.; Masciovecchio, C.; Brady, J.; Cesàro, A. Stacking of purines in water: The role of dipolar interactions in caffeine. *Phys. Chem. Chem. Phys.* **2016**, *18*, 13478–13486. [CrossRef]
30. Guthmuller, J.; Champagne, B. Time dependent density functional theory investigation of the resonance Raman properties of the julolidinemalononitrile push-pull chromophore in various solvents. *J. Chem. Phys.* **2007**, *127*, 164507. [CrossRef]
31. Kane, K.A.; Jensen, L. Calculation of absolute resonance Raman intensities: Vibronic theory vs. short-time approximation. *J. Phys. Chem. C* **2010**, *114*, 5540–5546. [CrossRef]
32. Avila Ferrer, F.J.; Barone, V.; Cappelli, C.; Santoro, F. Duschinsky, Herzberg–Teller, and multiple electronic resonance interferential effects in resonance Raman spectra and excitation profiles. The case of pyrene. *J. Chem. Theory Comput.* **2013**, *9*, 3597–3611. [CrossRef]
33. Ozaki, Y.; Wójcik, M.J.; Popp, J. *Molecular Spectroscopy: A Quantum Chemistry Approach*; John Wiley & Sons: Hoboken, NJ, USA, 2019.
34. Liang, W.; Ma, H.; Zang, H.; Ye, C. Generalized time-dependent approaches to vibrationally resolved electronic and Raman spectra: theory and applications. *Int. J. Quantum Chem.* **2015**, *115*, 550–563. [CrossRef]
35. Xu, Q.; Liu, Y.; Wang, M.; Cerezo, J.; Improta, R.; Santoro, F. The Resonance Raman Spectrum of Cytosine in Water: Analysis of the Effect of Specific Solute–Solvent Interactions and Non-Adiabatic Couplings. *Molecules* **2023**, *28*, 2286. [CrossRef]
36. Mennucci, B. Polarizable continuum model. *Wiley Interdiscip. Rev. Comput. Mol. Sci.* **2012**, *2*, 386–404.[CrossRef]

37. Ucun, F.; Sağlam, A.; Güçlü, V. Molecular structures and vibrational frequencies of xanthine and its methyl derivatives (caffeine and theobromine) by ab initio Hartree–Fock and density functional theory calculations. *Spectrochim. Acta A Mol. Biomol. Spectrosc.* **2007**, *67*, 342–349. [CrossRef]
38. Kim, D.; Kim, H.M.; Yang, K.Y.; Kim, S.K.; Kim, N.J. Molecular beam resonant two-photon ionization study of caffeine and its hydrated clusters. *J. Chem. Phys.* **2008**, *128*, 13. [CrossRef]
39. Kim, D.; Yang, K.Y.; Kim, H.M.; Kim, T.R.; Kim, N.J.; Shin, S.; Kim, S.K. Site-dependent effects of methylation on the electronic spectra of jet-cooled methylated xanthine compounds. *Phys. Chem. Chem. Phys.* **2017**, *19*, 22375–22384. [CrossRef]
40. Srivastava, S.K.; Singh, V.B. Ab initio and DFT studies of the structure and vibrational spectra of anhydrous caffeine. *Spectrochim. Acta A Mol. Biomol. Spectrosc.* **2013**, *115*, 45–50. [CrossRef]
41. Chenouf, J.; Boutahir, M.; Mejía-López, J.; Rahmani, A.; Fakrach, B.; Chadli, H.; Rahmani, A. Predicting the structure configuration and Raman analysis of caffeine molecules encapsulated into single-walled carbon nanotubes: Evidence for charge transfer. *Sol. Energy* **2022**, *232*, 204–211. [CrossRef]
42. Pavel, I.; Szeghalmi, A.; Moigno, D.; Cîntă, S.; Kiefer, W. Theoretical and pH dependent surface enhanced Raman spectroscopy study on caffeine. *Biopolym. Orig. Res. Biomol.* **2003**, *72*, 25–37. [CrossRef]
43. Cappelli, C. Integrated QM/polarizable MM/continuum approaches to model chiroptical properties of strongly interacting solute–solvent systems. *Int. J. Quantum Chem.* **2016**, *116*, 1532–1542. [CrossRef]
44. Giovannini, T.; Egidi, F.; Cappelli, C. Molecular spectroscopy of aqueous solutions: a theoretical perspective. *Chem. Soc. Rev.* **2020**, *49*, 5664–5677. [CrossRef]
45. Giovannini, T.; Cappelli, C. Continuum vs. atomistic approaches to computational spectroscopy of solvated systems. *Chem. Commun.* **2023**, *59*, 5644–5660. [CrossRef]
46. Gómez, S.; Giovannini, T.; Cappelli, C. Multiple Facets of Modeling Electronic Absorption Spectra of Systems in Solution. *ACS Phys. Chem. Au.* **2023**, *3*, 1–16. [CrossRef]
47. Gómez, S.; Giovannini, T.; Cappelli, C. Absorption spectra of xanthines in aqueous solution: A computational study. *Phys. Chem. Chem. Phys.* **2020**, *22*, 5929–5941. [CrossRef]
48. Gomez, S.; Cappelli, C. When Tautomers Matter: UV-Vis Absorption Spectra of Hypoxanthine in Aqueous Solution from Fully Atomistic Simulations. *ChemPhysChem* **2024**, *2024*, e202400107. [CrossRef]
49. Gómez, S.; Egidi, F.; Puglisi, A.; Giovannini, T.; Rossi, B.; Cappelli, C. Unlocking the power of resonance Raman spectroscopy: The case of amides in aqueous solution. *J. Mol. Liq.* **2021**, *346*, 117841. [CrossRef]
50. Gómez, S.; Bottari, C.; Egidi, F.; Giovannini, T.; Rossi, B.; Cappelli, C. Amide Spectral Fingerprints are Hydrogen Bonding-Mediated. *J. Phys. Chem. Lett.* **2022**, *13*, 6200–6207. [CrossRef]
51. Gómez, S.; Rojas-Valencia, N.; Giovannini, T.; Restrepo, A.; Cappelli, C. Ring Vibrations to Sense Anionic Ibuprofen in Aqueous Solution as Revealed by Resonance Raman. *Molecules* **2022**, *27*, 442. [CrossRef]
52. Uribe, L.; Gomez, S.; Egidi, F.; Giovannini, T.; Restrepo, A. Computational hints for the simultaneous spectroscopic detection of common contaminants in water. *J. Mol. Liq.* **2022**, *355*, 118908. [CrossRef]
53. Gómez, S.; Lafiosca, P.; Egidi, F.; Giovannini, T.; Cappelli, C. Uv-resonance Raman spectra of systems in complex environments: A multiscale modeling applied to doxorubicin intercalated into dna. *J. Chem. Inf. Model* **2023**, *63*, 1208–1217. [CrossRef]
54. Reed, A.E.; Curtiss, L.A.; Weinhold, F. Intermolecular interactions from a natural bond orbital, donor-acceptor viewpoint. *Chem. Rev.* **1988**, *88*, 899–926. [CrossRef]
55. Weinhold, F.; Landis, C.; Glendening, E. What is NBO analysis and how is it useful? *Int. Rev. Phys. Chem.* **2016**, *35*, 399–440. [CrossRef]
56. Weinhold, F.; Landis, C.R. *Discovering Chemistry with Natural Bond Orbitals*; Wiley-VCH: Hoboken, NJ, USA, 2012; p. 319.
57. Bader, R. *Atoms in Molecules: A Quantum Theory*; Oxford University Press: Oxford, UK, 1990.
58. Matta, C.; Boyd, R.; Becke, A. *The Quantum Theory of Atoms in Molecules: From Solid State to DNA and Drug Design*; Wiley: Hoboken, NJ, USA, 2007.
59. Alula, M.T.; Mengesha, Z.T.; Mwenesongole, E. Advances in surface-enhanced Raman spectroscopy for analysis of pharmaceuticals: A review. *Vib. Spectrosc.* **2018**, *98*, 50–63. [CrossRef]
60. Balbuena, P.B.; Blocker, W.; Dudek, R.M.; Cabrales-Navarro, F.A.; Hirunsit, P. Vibrational spectra of anhydrous and monohydrated caffeine and theophylline molecules and crystals. *J. Phys. Chem. A* **2008**, *112*, 10210–10219. [CrossRef]
61. Singh, V.B. Spectroscopic signatures and structural motifs in isolated and hydrated caffeine: A computational study. *RSC Adv.* **2014**, *4*, 58116–58126. [CrossRef]
62. Hanes, A.T.; Grieco, C.; Lalisse, R.F.; Hadad, C.M.; Kohler, B. Vibrational relaxation by methylated xanthines in solution: Insights from 2D IR spectroscopy and calculations. *J. Chem. Phys.* **2023**, *158*, 044302. [CrossRef]
63. Gómez, S.; Rojas-Valencia, N.; Restrepo, A. Analysis of conformational preferences in caffeine. *Molecules* **2022**, *27*, 1937. [CrossRef]
64. Zhang, Y.; Chen, J.; Kohler, B. Hydrogen bond donors accelerate vibrational cooling of hot purine derivatives in heavy water. *J. Phys. Chem. A* **2013**, *117*, 6771–6780. [CrossRef]
65. Tavagnacco, L.; Schnupf, U.; Mason, P.E.; Saboungi, M.L.; Cesàro, A.; Brady, J.W. Molecular dynamics simulation studies of caffeine aggregation in aqueous solution. *J. Phys. Chem. B* **2011**, *115*, 10957–10966. [CrossRef]

66. Skarmoutsos, I.; Tzeli, D.; Petsalakis, I.D. Hydration structure and dynamics, ultraviolet–visible and fluorescence spectra of caffeine in ambient liquid water. A combined classical molecular dynamics and quantum chemical study. *J. Mol. Liq.* **2023**, *391*, 123220. [CrossRef]
67. Gómez, S.; Ambrosetti, M.; Giovannini, T.; Cappelli, C. Close-Up Look at Electronic Spectroscopic Signatures of Common Pharmaceuticals in Solution. *J. Phys. Chem. B* **2024**, *128*, 2432–2446. [CrossRef]
68. Edwards, H.G.; Lawson, E.; de Matas, M.; Shields, L.; York, P. Metamorphosis of caffeine hydrate and anhydrous caffeine. *J. Chem. Soc. Perkin Trans.* **1997**, *2*, 1985–1990. [CrossRef]
69. Gómez, S.; Gómez, S.; David, J.; Guerra, D.; Cappelli, C.; Restrepo, A. Dissecting Bonding Interactions in Cysteine Dimers. *Molecules* **2022**, *27*, 8665. [CrossRef]
70. Chen, J.; Kohler, B. Ultrafast nonradiative decay by hypoxanthine and several methylxanthines in aqueous and acetonitrile solution. *Phys. Chem. Chem. Phys.* **2012**, *14*, 10677–10682. [CrossRef]
71. Agilent Technologies, Inc. Available online: https://www.agilent.com/cs/library/certificateofanalysis/8500-6917%20lot%20OC128158.pdf (accessed on 26 May 2024).
72. Yisak, H.; Redi-Abshiro, M.; Chandravanshi, B.S. New fluorescence spectroscopic method for the simultaneous determination of alkaloids in aqueous extract of green coffee beans. *Chem. Cent. J.* **2018**, *12*, 59. [CrossRef]
73. Gunasekaran, S.; Sankari, G.; Ponnusamy, S. Vibrational spectral investigation on xanthine and its derivatives—Theophylline, caffeine and theobromine. *Spectrochim. Acta A Mol. Biomol. Spectrosc.* **2005**, *61*, 117–127. [CrossRef]
74. Rojalin, T.; Kurki, L.; Laaksonen, T.; Viitala, T.; Kostamovaara, J.; Gordon, K.C.; Galvis, L.; Wachsmann-Hogiu, S.; Strachan, C.J.; Yliperttula, M. Fluorescence-suppressed time-resolved Raman spectroscopy of pharmaceuticals using complementary metal-oxide semiconductor (CMOS) single-photon avalanche diode (SPAD) detector. *Anal. Bioanal. Chem.* **2016**, *408*, 761–774. [CrossRef]
75. Xi, W.; Volkert, A.A.; Boller, M.C.; Haes, A.J. Vibrational frequency shifts for monitoring noncovalent interactions between molecular imprinted polymers and analgesics. *J. Phys. Chem. C* **2018**, *122*, 23068–23077. [CrossRef]
76. Nolasco, M.M.; Amado, A.M.; Ribeiro-Claro, P.J. Computationally assisted approach to the vibrational spectra of molecular crystals: Study of hydrogen-bonding and pseudo-polymorphism. *ChemPhysChem* **2006**, *7*, 2150–2161.
77. Rijal, R.; Sah, M.; Lamichhane, H.P.; Mallik, H.S. Quantum chemical calculations of nicotine and caffeine molecule in gas phase and solvent using DFT methods. *Heliyon* **2022**, *8*, e12494. [CrossRef]
78. Lee, S.Y.; Heller, E.J. Time-dependent theory of Raman scattering. *J. Chem. Phys.* **1979**, *71*, 4777–4788. [CrossRef]
79. Santoro, F.; Cappelli, C.; Barone, V. Effective time-independent calculations of vibrational resonance raman spectra of isolated and solvated molecules including duschinsky and herzberg–teller effects. *J. Chem. Theory Comput.* **2011**, *7*, 1824–1839. [CrossRef]
80. Hoehn, R.D.; Carignano, M.A.; Kais, S.; Zhu, C.; Zhong, J.; Zeng, X.C.; Francisco, J.S.; Gladich, I. Hydrogen bonding and orientation effects on the accommodation of methylamine at the air-water interface. *J. Chem. Phys.* **2016**, *144*, 214701. [CrossRef]
81. Macchiagodena, M.; Mancini, G.; Pagliai, M.; Barone, V. Accurate prediction of bulk properties in hydrogen bonded liquids: Amides as case studies. *Phys. Chem. Chem. Phys.* **2016**, *18*, 25342–25354. [CrossRef]
82. Liu, C.; Zhao, D.X.; Yang, Z.Z. Direct evaluation of individual hydrogen bond energy in situ in intra-and intermolecular multiple hydrogen bonds system. *J. Comput. Chem.* **2012**, *33*, 379–390. [CrossRef]
83. Campetella, M.; De Mitri, N.; Prampolini, G. Automated parameterization of quantum-mechanically derived force-fields including explicit sigma holes: A pathway to energetic and structural features of halogen bonds in gas and condensed phase. *J. Chem. Phys.* **2020**, *153*, 044106. [CrossRef]
84. Boys, S.F. Construction of some molecular orbitals to be approximately invariant for changes from one molecule to another. *Rev. Mod. Phys.* **1960**, *32*, 296. [CrossRef]
85. Abraham, M.J.; Murtola, T.; Schulz, R.; Páll, S.; Smith, J.C.; Hess, B.; Lindahl, E. GROMACS: High performance molecular simulations through multi-level parallelism from laptops to supercomputers. *SoftwareX* **2015**, *1*, 19–25. [CrossRef]
86. Wang, J.; Wolf, R.M.; Caldwell, J.W.; Kollman, P.A.; Case, D.A. Development and testing of a general amber force field. *J. Comput. Chem.* **2004**, *25*, 1157–1174. [CrossRef]
87. Jorgensen, W.L.; Chandrasekhar, J.; Madura, J.D.; Impey, R.W.; Klein, M.L. Comparison of simple potential functions for simulating liquid water. *J. Chem. Phys.* **1983**, *79*, 926–935. [CrossRef]
88. Brehm, M.; Kirchner, B. TRAVIS—A Free Analyzer and Visualizer for Monte Carlo and Molecular Dynamics Trajectories. *J. Chem. Inf. Model* **2011**, *51*, 2007–2023. [CrossRef]
89. Brehm, M.; Thomas, M.; Gehrke, S.; Kirchner, B. TRAVIS—A free analyzer for trajectories from molecular simulation. *J. Chem. Phys.* **2020**, *152*, 164105. [CrossRef]
90. Frisch, M.J.; Trucks, G.W.; Schlegel, H.B.; Scuseria, G.E.; Robb, M.A.; Cheeseman, J.R.; Scalmani, G.; Barone, V.; Petersson, G.A.; Nakatsuji, H.; et al. *Gaussian 16 Revision B.01*; Gaussian Inc.: Wallingford, CT, USA, 2016.
91. Rick, S.W.; Stuart, S.J.; Berne, B.J. Dynamical fluctuating charge force fields: Application to liquid water. *J. Chem. Phys.* **1994**, *101*, 6141–6156. [CrossRef]
92. Glendening, E.D.; Badenhoop, J.K.; Reed, A.E.; Carpenter, J.E.; Bohmann, J.A.; Morales, C.M.; Karafiloglou, P.; Landis, C.R.; Weinhold, F. *NBO 7.0*; Theoretical Chemistry Institute, University of Wisconsin: Madison, WI, USA, 2018.

93. Giovannini, T.; Olszowka, M.; Egidi, F.; Cheeseman, J.R.; Scalmani, G.; Cappelli, C. Polarizable embedding approach for the analytical calculation of Raman and Raman optical activity spectra of solvated systems. *J. Chem. Theory Comput.* **2017**, *13*, 4421–4435. [CrossRef] [PubMed]
94. Egidi, F.; Bloino, J.; Cappelli, C.; Barone, V. A robust and effective time-independent route to the calculation of resonance raman spectra of large molecules in condensed phases with the inclusion of Duschinsky, Herzberg–Teller, anharmonic, and environmental effects. *J. Chem. Theory Comput.* **2014**, *10*, 346–363. [CrossRef] [PubMed]

molecules

MDPI

Article

Self-Association and Microhydration of Phenol: Identification of Large-Amplitude Hydrogen Bond Librational Modes

Dmytro Mihrin [1,2,*], Karen Louise Feilberg [2] and René Wugt Larsen [1,*]

1 Department of Chemistry, Technical University of Denmark, Kemitorvet 206, 2800 Kongens Lyngby, Denmark
2 DTU Offshore, Technical University of Denmark, Elektrovej 375, 2800 Kongens Lyngby, Denmark; klfe@dtu.dk
* Correspondence: dmytro@dtu.dk (D.M.); rewl@kemi.dtu.dk (R.W.L.)

Abstract: The self-association mechanisms of phenol have represented long-standing challenges to quantum chemical methodologies owing to the competition between strongly directional intermolecular hydrogen bonding, weaker non-directional London dispersion forces and C–H$\cdots\pi$ interactions between the aromatic rings. The present work explores these subtle self-association mechanisms of relevance for biological molecular recognition processes via spectroscopic observations of large-amplitude hydrogen bond librational modes of phenol cluster molecules embedded in inert neon "quantum" matrices complemented by domain-based local pair natural orbital-coupled cluster DLPNO-CCSD(T) theory. The spectral signatures confirm a primarily intermolecular O-H$\cdots$H hydrogen-bonded structure of the phenol dimer strengthened further by cooperative contributions from inter-ring London dispersion forces as supported by DLPNO-based local energy decomposition (LED) predictions. In the same way, the hydrogen bond librational bands observed for the trimeric cluster molecule confirm a pseudo-C_3 symmetric cyclic cooperative hydrogen-bonded barrel-like potential energy minimum structure. This structure is vastly different from the sterically favored "chair" conformations observed for aliphatic alcohol cluster molecules of the same size owing to the additional stabilizing London dispersion forces and C–H$\cdots\pi$ interactions between the aromatic rings. The hydrogen bond librational transition observed for the phenol monohydrate finally confirms that phenol acts as a hydrogen bond donor to water in contrast to the hydrogen bond acceptor role observed for aliphatic alcohols.

Keywords: phenol cluster molecules; vibrational spectroscopy; neon matrices; large-amplitude librational motion; hydrogen bonding; London dispersion forces; local energy decomposition

1. Introduction

Hydrogen bonding stands as a cornerstone in both the structure and function of supramolecular chemical interactions, governing the majority of biological recognition mechanisms and the properties of solvents, and guiding chemical reactivity. In intricate environments, highly directional hydrogen bonds represent just one among several other classes of non-covalent intermolecular interactions contributing to molecular association mechanisms. Non-directional attractive London dispersion forces constitute an integral component of every non-covalent interaction, contributing a substantial portion of the total association energy within any given molecular aggregate growing with the size of the molecular system. London dispersion forces become particularly important when the primary bond is weaker than typical conventional hydrogen bonds, such as those between O-H$\cdots$N and O-H$\cdots$O motifs [1]. In addition, secondary interactions such as non-conventional O-H$\cdots\pi$ hydrogen bonds and even weaker C-H$\cdots\pi$ contacts often compete effectively with conventional hydrogen bonds in aromatic molecular systems [2,3]. The complex interplay between London dispersion, electrostatic forces, and weak donor-acceptor interactions represents a significant theoretical challenge for predictive ab initio methods targeting realistic supramolecular assemblies. Currently, only high-level post-Hartree-Fock ab initio quantum chemistry approaches such as coupled cluster theory CCSD(T) [4–6] consistently provide accurate predictions of non-covalent interactions, particularly in the presence of extensive van der Waals forces. However, the computational cost of these advanced ab initio methods scales rapidly, rendering them impractical for larger molecular systems, whereas lower-tier methods such as density functional theory (DFT), despite reliance on empirical corrections, often fail to achieve the necessary chemical accuracy [7]. The prototypical homocluster molecules of phenol (PhOH) have been investigated extensively and have earned a place in benchmark databases for biomolecular systems [8], although the computational costs for theoretical CCSD(T) predictions become prohibitive already at the $(PhOH)_2$ level due to the size of the aromatic fragments [9,10]. The introduction of these aromatic structures facilitates the formation of intermolecular contacts involving the π-electron clouds. The self-association mechanisms of phenol thus involve an interplay of both conventional primary O-H$\cdots$O hydrogen bonds, substantial dispersive contributions and secondary C-H$\cdots\pi$ contacts constituting a perfect playground for the validation of lower-tier theoretical methodologies with accurate spectroscopic findings.

The currently available experimental spectroscopic data primarily consist of structural and vibrational observations for phenol dimer $(PhOH)_2$ and phenol trimer $(PhOH)_3$ as well as various phenol microhydrate cluster molecules $PhOH{\cdot}(H_2O)_{1\text{-}3}$ [11–20]. While both microwave spectroscopy guided substitution structures and mid-IR transitions associated with strongly active intramolecular vibrational OH-stretching modes have been reported in the literature for these homo- and microhydrate cluster molecules, there are currently no reported observations of the low-energy transitions associated with the large-amplitude and highly anharmonic OH librational (hindered torsional) modes linked directly to these (cooperative) hydrogen bond networks. This class of large-amplitude hydrogen bond librational motion has been shown to be an accurate probe of the strength and directionality of the intermolecular hydrogen bonds [21–23]. For aliphatic alcohols, the alkyl groups in cyclic cluster molecules have previously been shown experimentally to avoid each other due to steric hindrance resulting in chair-like potential energy minima for alcohol trimers and "up-down-up-down" (relative to the hydrogen bond plane) minima for alcohol tetramers [24]. Furthermore, the spectroscopic detection of these large-amplitude hydrogen bond librational modes reveals the most dominating contributions to the change in vibrational zero-point energy upon complexation (ΔZPE) [25]. These observations help to translate high-level ab initio quantum chemical predictions of electronic equilibrium dissociation energies (D_e) into accurate semi-empirical ground-state dissociation energies (D_0), which is notoriously challenging from first principles quantum chemistry alone due to the highly anharmonic nature of this class of large-amplitude vibrational motion [26–29].

In the present work, we provide for the first time spectroscopic assignments for the experimentally less-accessible vibrational transitions associated with large-amplitude hydrogen bond librational motions for phenol dimer, phenol trimer and isotopic variants of the phenol monohydrate ($PhOH{\cdot}H_2O/PhOH{\cdot}D_2O$) embedded in inert cryogenic "quantum" neon matrices at 4 K. These experimental findings are complemented by systematic theoretical conformation searches and harmonic force field predictions employing both dispersion-corrected DFT and ab initio methodologies of the most stable potential energy minima. In addition, detailed analyses of the total interaction energies in these phenol cluster molecules are carried out by means of a local energy decomposition (LED) scheme to account for the competition between the different classes of intermolecular non-covalent forces.

2. Results

2.1. Phenol Cluster Molecules

Figure 1 shows the mid-infrared absorption spectra of doped neon matrices embedded with phenol at two different mixing ratios (by thermostating the phenol sublimation vessel at 0 °C and at −5 °C, respectively). In the former, most concentrated neon matrix experiment, a so-called pre-annealing spectrum (blue trace) was recorded immediately after the deposition of the sample and a post-annealing spectrum (red trace) was collected after the annealing procedure.

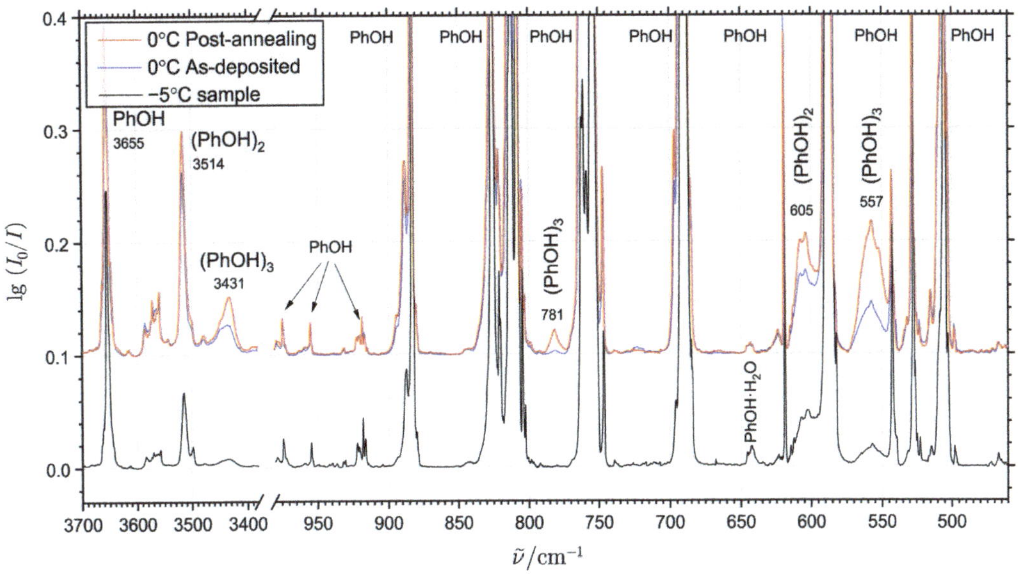

Figure 1. The infrared absorption spectra of phenol embedded in solid neon at 4 K. In the two different experiments shown, the phenol source sublimation vessel was thermostated at 0 °C (**bottom**) and at −5 °C (**top** and **bottom**, respectively) during deposition to achieve different mixing ratios with neon. The spectra of the 0 °C experiment were recorded before (blue trace) and after annealing of the matrix at 9.5 K (red trace). The spectra have been normalized to the phenol monomer transitions in the 650–575 cm^{-1} range. The experimentally observed transitions associated with the intramolecular OH-stretching and the large-amplitude hydrogen bond librational modes of $(PhOH)_2$ and $(PhOH)_3$ are indicated with the respective band origins.

The primary distinct bands observed in the spectra above 3400 cm^{-1} are associated with the strongly IR-active intramolecular OH stretching modes of phenol monomer and the intensity-enhanced donor OH-stretching modes of phenol cluster molecules abundant in sub-% concentrations relative to the monomer. The bands observed at 3655 cm^{-1}, 3515 cm^{-1} and 3431 cm^{-1} belong to the free OH-stretching mode of phenol monomer [11,13], the hydrogen-bonded donor OH-stretching modes of $(PhOH)_2$ [11,13] and $(PhOH)_3$ [11], respectively, in agreement with the observed concentration dependency, the effects of annealing and reported band origins from previous neon [11] and argon matrix isolation [14] and jet spectroscopy investigations [15,16]. The band origins for phenol monomer and phenol dimer embedded in neon have recently been revised with the latest observations at 3655 and 3515 cm^{-1}, respectively [13]. The donor OH-stretching mode of $(PhOH)_2$ has previously been observed at 3530 cm^{-1} in a jet REMPI investigation [15,16], whereas the dangling OH-stretching mode of $(PhOH)_2$ and the free OH-stretching mode of the monomer have previously been assigned at 3654 cm^{-1} and 3657 cm^{-1} in jet expansions, respectively. Three transitions associated with the OH-stretching modes of $(PhOH)_3$ have previously been assigned upon jet-cooling at 3394, 3441 and 3449 cm^{-1} suggesting a cyclic structure of $(PhOH)_3$ [15], which has subsequently been confirmed by jet microwave spectroscopy [17–19].

In the 400–800 cm^{-1} spectral range normally associated with the class of large-amplitude (hindered) torsional motions, several additional broad but still distinct bands, with intensities depending on both sample concentration and matrix annealing are observed. In order to distinguish the size of the cluster molecules, the concentration dependence and growth rate upon annealing of these bands were explored in additional experiments employing the phenol sublimation temperature in the 0–10 °C range during deposition. Figure 2a shows the spectral sampling of the doped neon matrix below 800 cm^{-1} at several different spatial locations limited to a diameter of approximately 3.5 mm of the infrared probe beam. This provides a variation of the local mixing ratio between neon and phenol molecules within the same matrix. The individual spectra have been normalized using the phenol monomer transition intensities in the 650–575 cm^{-1} range, which are not overlapped by the cluster molecule features. Figure 2b shows spectra below 800 cm^{-1} recorded before (black trace) and during annealing of the matrix at different steps (colored traces), with the latter calculated by subtracting the cold absorbance spectrum from the annealing spectra.

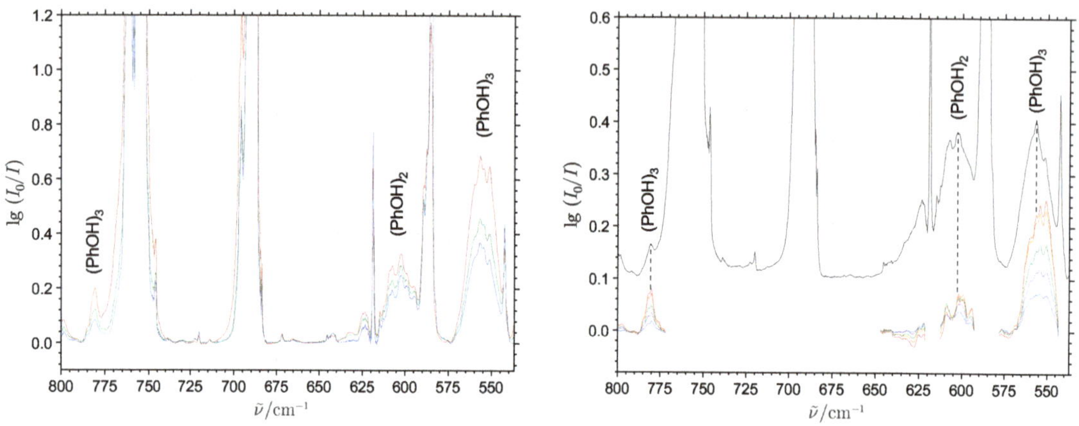

(**a**) Spatial matrix spot probing (**b**) Matrix annealing

Figure 2. (**a**) The spectral dependence of the phenol/neon mixing ratio by spatial spot probing. The three shown spectra collected for spots with increasing phenol/ratio (the blue and red traces collected for the lowest and highest phenol/neon mixing ratios, respectively) have been normalized to the monomer transitions. The proposed dimer and trimer absorption features are differentiated based on their different growth rates relative to the monomer bands. (**b**) The evolution of the neon matrix spectra over time during annealing at 9.5 K. A series of difference spectra (annealing spectrum subtracted the pre-annealing spectrum) collected at 3 min intervals during the annealing procedure is shown. The colored difference spectra (the blue trace collected after 3 min and the red trace collected after 15 min, respectively) show excess absorption relative to the pre-annealing spectrum (black trace) demonstrating the progressive formation of phenol cluster molecules.

The experimental findings from the combined spatial spot probing and annealing of the doped neon matrices allow the identification of two different sets of bands belonging to phenol cluster molecules. The most distinct absorption feature below 800 cm^{-1}, behaving as a cluster molecule band even for very low phenol/neon mixing ratios and always appearing in the spectra prior to annealing, is located at 605 cm^{-1}. Its band intensity correlates well with the OH-stretching band at 3515 cm^{-1} and this transition is therefore assigned to the only large-amplitude hydrogen bond librational (strongly hindered OH torsional) fundamental expected for (PhOH)$_2$. The two other distinct bands observed at 557 cm^{-1} and 781 cm^{-1} show higher stoichiometry compared to the assigned (PhOH)$_2$ features. The most intense far-infrared band located at 557 cm^{-1} reveals the same concentration and annealing dependencies as the mid-infrared OH-stretching band at 3431 cm^{-1} assigned for

$(PhOH)_3$. The intensity of the band at 781 cm^{-1} similarly shows strong correlation with the 557 cm^{-1} and 3431 cm^{-1} bands with respect to the phenol mixing ratio and annealing procedures; however, in the high concentration experiments the spectral overlap with some broad satellite bands around 762 cm^{-1} impedes the exact correlations, tentatively assigned to CH out-of-plane bending motions of $(PhOH)_2/(PhOH)_3$. These exclusively experimental considerations suggest that the linked 557 cm^{-1} and 781 cm^{-1} bands should be assigned to transitions associated with large-amplitude hydrogen bond librational motion of the most stable conformation of $(PhOH)_3$, which will be supported by complementary theoretical predictions in the following.

Figure 3 shows the optimized molecular geometries of phenol and its hydrogen-bonded cluster molecules $(PhOH)_2$ and $(PhOH)_3$ using various theoretical methodologies, and provides the relative zero-point energy corrected ground-state dissociation energies for the four most stable conformations predicted for the $(PhOH)_3$ system. The computational predictions of the molecular geometry for the $(PhOH)_2$ system remain challenging even for modern computer resources. The MP2 quantum chemical method, which in general tends to provide consistent ab initio predictions for conventional hydrogen-bonded cluster molecules, runs into an issue in the treatment of the intermolecular interactions between the aromatic rings of the phenol fragments. Previously, a qualitatively correct structure has only been obtained with this methodology in conjunction with counterpoise corrections to account for the small basis set [10]. Considering the large basis set required to achieve accurate vibrational frequency predictions, this theoretical approach becomes prohibitively expensive. The optimized geometry of $(PhOH)_2$ has been calculated using the MP2/AVQZ methodology (Figure 3a), which has shown significant underestimation of the hydrogen bonding in the complex, compared to what is observed in the experiment. Even with this relatively large basis set, the conventional MP2 method suffers from a strong over-binding effect between the aromatic rings, which in this case directly competes with the hydrogen bond. In order to achieve the desired accuracy for all cluster molecules investigated, the empirically scaled SCS-MP2 method and the dispersion-corrected hybrid PW6B95-D4 DFT functional have been selected, both known to perform comparatively well in the description of intermolecular non-covalent interactions [7]. These methods have reproduced experimental data with reasonable consistency; however, there is variance in the predicted dissociation energies and harmonic vibrational frequencies between the two different approaches.

The predicted harmonic frequencies of vibrational modes for the conformations of $(PhOH)_2$, $(PhOH)_3$ and $PhOH{\cdot}H_2O$ are given in Table S1 in the Electronic Supplementary Information (ESI). For the $(PhOH)_2$ system, for which only one conformation has been optimized (Figure 3b), both methods predict one highly IR-active transition where the band associated with the hydrogen bond librational mode is experimentally observed in the present work. For the even more challenging $(PhOH)_3$ system, four different conformations have been optimized. Both methodologies show a large energy gap between the most stable cyclic pseudo-C_3 symmetric barrel-like potential energy minimum structure involving three cooperative $OH{\cdots}O$ hydrogen bonds and the other three predicted energy minima (Figure 3c). The less stable conformations 2–4 differ in the arrangements of the aromatic rings, or substitute the conventional intermolecular $OH{\cdots}O$ hydrogen bonds for the non-conventional $OH{\cdots}\pi$ bond or the stacking of aromatic rings. The differences in the optimized geometries result in different rankings of the dissociation energy calculated for the less stable conformations (Figure 3d,f).

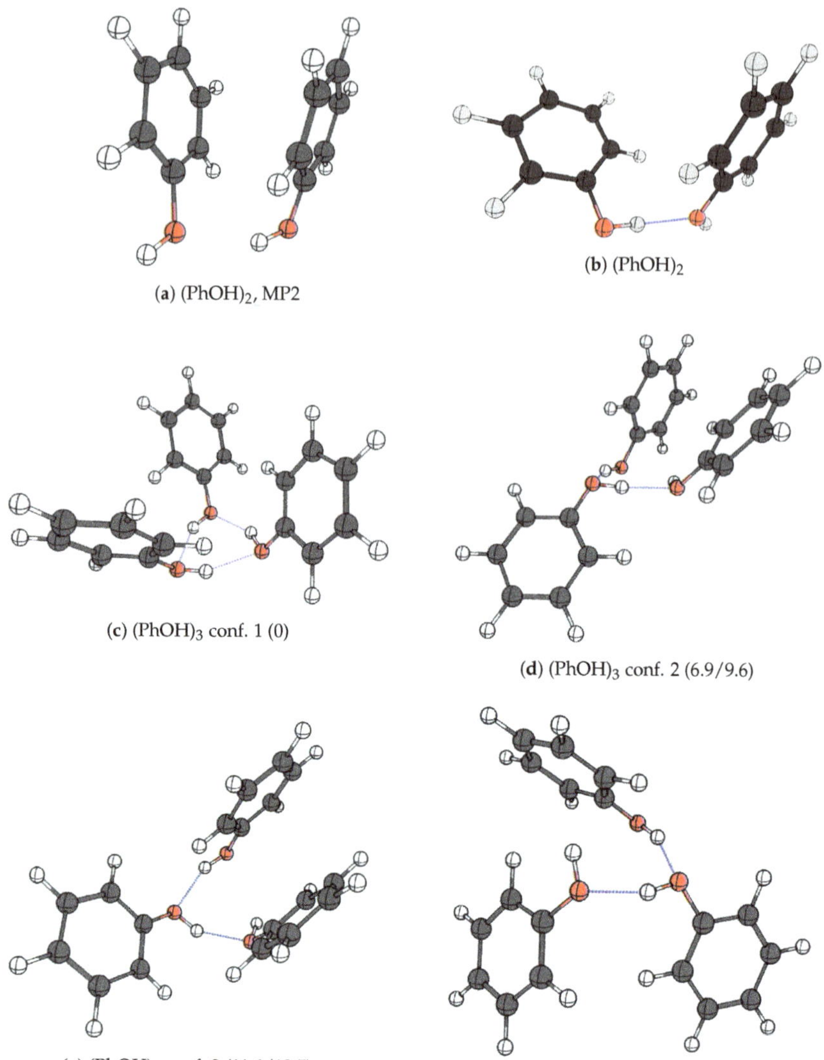

(**a**) $(PhOH)_2$, MP2

(**b**) $(PhOH)_2$

(**c**) $(PhOH)_3$ conf. 1 (0)

(**d**) $(PhOH)_3$ conf. 2 (6.9/9.6)

(**e**) $(PhOH)_3$ conf. 3 (11.6/13.5)

(**f**) $(PhOH)_3$ conf. 4 (7.7/8.5)

Figure 3. The optimized potential energy minima structures of the pure phenol cluster molecules (**a**) $(PhOH)_2$ (MP2/AVQZ level), (**b**) $(PhOH)_2$ (PW6B95-D4/ma-def2-QZVP level) and (**c**–**f**) the four different conformations of $(PhOH)_3$. The relative zero-point energy corrected dissociation energies D_0 (the change of zero-point energy (ΔZPE) calculated at PW6B95-D4/ma-def2-QZVP//SCS-MP2/AVQZ levels with electronic energies obtained at the DLPNO-CCSD(T)/AVQZ level, both values in $kJ \cdot mol^{-1}$) are given in brackets for each conformation.

The simulated vibrational spectra for the four different conformations of $(PhOH)_3$ are shown in Figure 4. In this simulation, the calculated harmonic frequencies of the normal modes have been scaled using a separate factor for the modes involving large-amplitude hindered torsional motion of the OH-groups, which are estimated from the experimental spectra. The predicted pseudo-C_3-symmetric global potential energy minimum (conf. 1) has two near-degenerate highly IR-active librational modes (Figure 5b) estimated around 550 cm^{-1}. Owing to the pseudo-C_3 symmetry, there are no other predicted intense transitions associated with hydrogen bond librational modes in the 500–700 cm^{-1} range, unlike the asymmetric conformations 2–4, which all have at least two distinct IR-active hydrogen bond librational transitions in this region (indicated by filled colored band areas). In the case of conformation 2, one of the predicted hydrogen bond librational transitions has severe spectral overlaps with other less perturbed intramolecular transitions (DFT simulation); however, the two remaining highly IR-active hydrogen bond librational transitions should still be observed clearly between 500 and 600 cm^{-1} for this conformation, which is not supported in the present experiments. The second distinct transition assigned experimentally for $(PhOH)_3$ in the present experimental work at 781 cm^{-1} is also supported by the DFT and SCS-MP2 quantum chemical predictions. While the positions of the intramolecular transitions seem better predicted by the DFT approach, both methods agree on the position of the third IR-active transition associated with the concerted hydrogen bond librational motion of the most stable C_3 conformation of $(PhOH)_3$ (Figure 5c), involving all three O-H$\cdots$O motifs, which is close to the experimental value. The relative transition intensity ratio between the two hydrogen bond librational transitions at 557 and 781 cm^{-1} of $(PhOH)_3$ is also qualitatively reproduced, although the overlap with the transitions attributed with perturbed intramolecular out-of-plane CH bending modes of $(PhOH)_2$ and $(PhOH)_3$ at higher concentrations makes more accurate analyses impossible. Together with the near-degenerate donor OH-stretching transitions observed at 3431 cm^{-1}, the two transitions associated with hydrogen bond librational motion allow us to unambiguously conclude that only the most stable pseudo-C_3 barrel-like conformation is observed in the present neon matrix experiments, both in the lower concentration regime and after being formed via annealing under cryogenic conditions. This is in strong contrast to previous findings for cyclic trimers of aliphatic alcohols, which have been shown to form "chair" structures due to steric hindrance [24,30].

2.2. The Phenol Monohydrate

Figure 6 shows the spectra collected for neon matrices doped with pure phenol, pure H_2O and pure D_2O together with spectra of phenol/H_2O and phenol/D_2O mixtures in the OH-stretching range (3400–3700 cm^{-1}) and the relevant range for the transitions associated with large-amplitude hydrogen bond librational motion (500–670 cm^{-1}). In between the hydrogen-bonded OH-stretching transitions for $(PhOH)_2$, $(H_2O)_2$ and $(H_2O)_3$, it is evident that the simultaneous deposition of phenol and H_2O reveal a new strongly IR-active distinct band at 3499 cm^{-1}, which has recently been assigned to the phenol monohydrate PhOH$\cdot H_2O$ [13]. The corresponding neon matrix isolation experiments for phenol/D_2O show that the assigned signal is indeed due to most stable conformation of this mixed cluster molecule, as the band shifts only slightly (and broadens somewhat due to spectral overlap with traces of the phenol-HDO complex), whereas the less stable conformation with water as hydrogen bond donor to phenol would absorb in the entirely different OD stretching region. This is supported by the quantum chemical results showing that the predicted zero-point energy corrected dissociation energy D_0 of the global potential energy minimum (Figure 7a) is 9 kJ$\cdot$mol^{-1} higher than predicted for the local minimum (Figure 7b). The donor OH stretching transitions of the phenol hydrates PhOH$\cdot(H_2O)_{1\text{-}3}$ have previously been observed in jet expansions at 3524 cm^{-1}, 3388 cm^{-1} and 3345 cm^{-1} [20], respectively, and the OH-stretching transition of the phenol monohydrate embedded in argon matrices has been assigned at 3461 cm^{-1} [14].

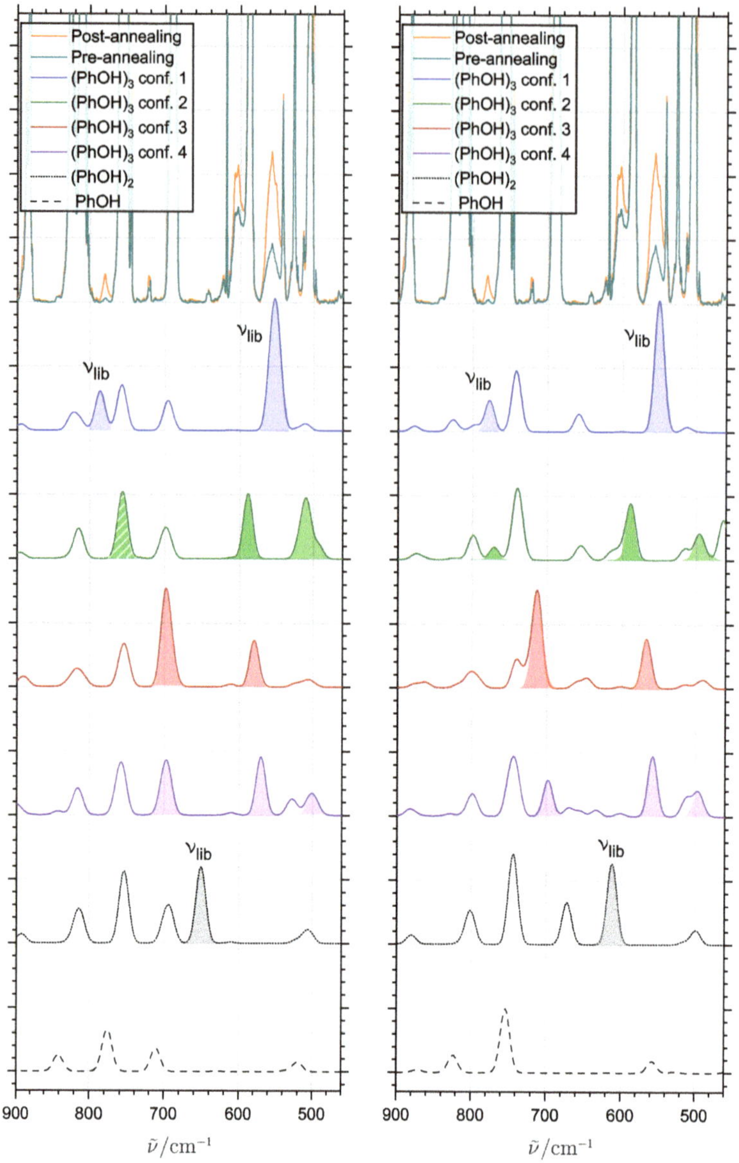

(**a**) DFT Simulation (**b**) SCS-MP2 Simulation

Figure 4. The simulated vibrational spectra of PhOH, $(PhOH)_2$ and the predicted conformations of $(PhOH)_3$ using the DFT (PW6B95-D4) and SCS-MP2 methodologies. Harmonic vibrational mode frequencies have been scaled separately using a scaling factor of 0.95 for the large-amplitude hydrogen bond librational modes and 0.97 for less perturbed intramolecular transitions. The predicted large-amplitude hydrogen bond librational transitions are marked with filled areas. The striped area indicates spectral overlap between a librational transition and a less perturbed intramolecular transition. The band positions for the experimentally assigned librational transitions of $(PhOH)_2$ and $(PhOH)_3$ are indicated on the trace for the identified conformation. The experimental pre- and post-annealing spectra are provided above the simulations.

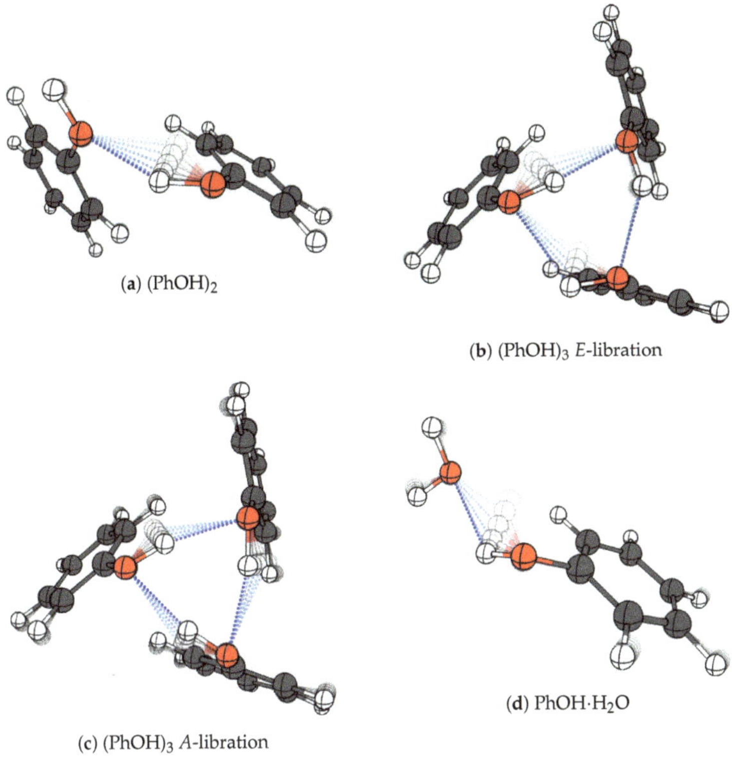

Figure 5. Animations of the large-amplitude hydrogen bond librational modes of (**a**) $(PhOH)_2$, (**b**,**c**) $(PhOH)_3$ and (**d**) $PhOH{\cdot}H_2O$ associated with the experimentally assigned absorption bands.

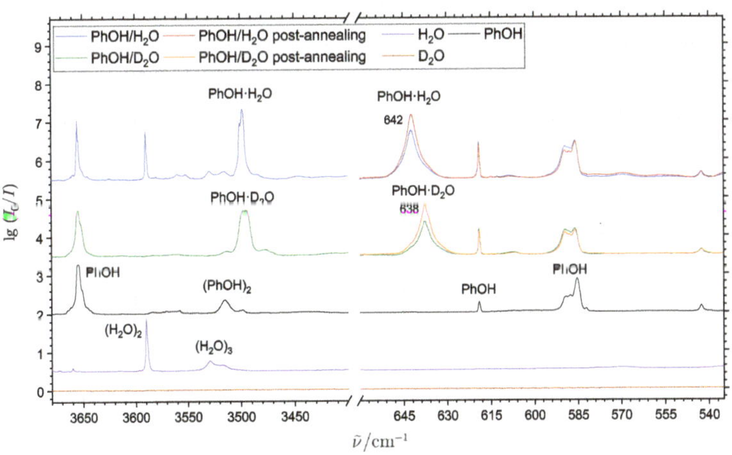

Figure 6. The infrared absorption spectra of phenol (black trace, small traces of H_2O), H_2O (purple trace) and D_2O (brown trace) together with spectra of phenol/H_2O (blue trace) and phenol/D_2O (green trace) mixtures. For the two mixtures the spectra after annealing of the matrix to 9.5 K (red and orange traces, respectively) are shown below 650 cm^{-1}. The new assigned transitions associated with the large-amplitude hydrogen bond librational modes of the $PhOH{\cdot}H_2O$ and $PhOH{\cdot}D_2O$ monohydrates are indicated in the spectra with their respective band positions.

In the spectral region below 700 cm^{-1}, a highly IR-active distinct band is observed at 642 cm^{-1} in the neon matrices doped simultaneously with phenol and H_2O. Furthermore, the band intensity responds significantly to annealing, indicating complex formation events in the neon matrix. The band shifts slightly down to 638 cm^{-1} in the experiment where phenol is deposited together with D_2O. The small isotopic shift of this transition confirms the proposed structure of the phenol monohydrate PhOH·H_2O where the phenol molecule acts as the hydrogen bond donor and the observed transition is straightforwardly assigned to the large-amplitude donor librational motion of PhOH·H_2O as visualized in Figure 5d. We do not observe any features in the low-energy part of the spectrum which could be clearly attributed to the less stable conformation.

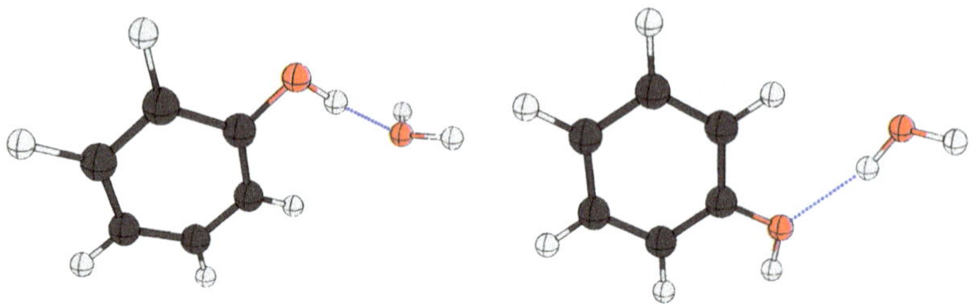

(**a**) Conf. 1 (**b**) Conf. 2

Figure 7. The optimized global (**a**) and local (**b**) potential energy minima structures of the phenol monohydrate (PhOH·H_2O) employing the SCS-MP2/AVQZ level of theory.

2.3. *Supporting Theoretical Analysis*

The results of the LED analyses for the phenol cluster molecules PhOH·H_2O, $(PhOH)_2$ and $(PhOH)_3$ are summarized in Table 1. The optimized potential energy minima structures obtained from the PW6B95-D4/ma-def2-QZVP and SCS-MP2/AVTZ methodologies were used for the subsequent LED analyses at the DLPNO-CCSD(T)/AVQZ level of theory. The electronic equilibrium dissociation energies D_e and the zero-point corrected dissociation energies D_0 obtained from the higher DLPNO-CCSD(T) level on the optimized geometries are denoted D_e(CC) and D_0(CC), respectively. In the case of the smaller $(PhOH)_2$ and PhOH·H_2O systems, geometry optimizations were also performed using the highest feasible SCS-MP2/AVQZ level to access the potential influence of the basis set size. The basis set does not have a significant influence on the dissociation energies for $(PhOH)_2$ as seen from the minor 0.3 kJ·mol^{-1} (0.1 %) increase in D_e(CC), when calculated on the optimized geometry employing the larger AVQZ basis set. However, the molecular properties obtained directly from the method, specifically the value of change in zero-point energy upon complexation ΔZPE, is much more dependent on the basis set. In that respect, the cheaper DFT approach performs better for $(PhOH)_2$, likely due to lower basis set requirements. In general, when approaching the basis set limit, the SCS-MP2 method tends to underestimate non-covalent interactions, as can be seen from the results for the PhOH·H_2O system. Still, the SCS-MP2 wave function approach provides more consistent results when dealing with the larger $(PhOH)_3$ conformations, when evaluated against the high-level DLPNO-CCSD(T) energies based on the same respective optimized molecular geometries probably due the larger contributions from London dispersion forces.

Table 1. The electronic equilibrium dissociation energies (D_e) for PhOH·H_2O, $(PhOH)_2$ and $(PhOH)_3$ calculated both directly using the theoretical approaches PW6B95-D4/ma-def2-QZVP (denoted as DFT) and SCS-MP2 and the higher DLPNO-CCSD(T)/AVQZ level (D_e (CC)) on the respective optimized geometries, the change of zero-point energy (ΔZPE), the zero-point energy corrected dissociation energies (D_0, D_0(CC)), the interfragment dispersion energies ($\Delta E_{disp,\,frag}$) and the sums of the dispersive ($\sum E_{disp}$) and non-dispersive parts of D_e ($\sum E_{N/D}$, given as D_e(CC)− $\sum E_{disp}$ [31,32]). The SCS-MP2 results for PhOH·H_2O are obtained using the AVQZ basis set. All values are given in units of kJ·mol^{-1}.

	DFT							SCS-MP2						
	PhOH·H_2O		$(PhOH)_2$	$(PhOH)_3$				PhOH·H_2O		$(PhOH)_2$	$(PhOH)_3$			
	Conf. 1	Conf. 2		Conf. 1	Conf. 2	Conf. 3	Conf. 4	Conf. 1	Conf. 2		Conf. 1	Conf. 2	Conf. 3	Conf. 4
D_e	27.9	18.3	27.8	76.3	74.4	71.7	73.1	26.6	17.8	31.7	95.5	85.4	85.2	87.0
D_e(CC)	28.9	19.8	30.0	89.2	80.8	78.0	81.6	28.8	19.7	30.1 * 30.4	90.1	81.7	78.5	82.4
D_0	20.9	11.7	22.5	66.6	63.2	61.0	62.4	19.7	11.6	24.5	79.6	70.6	71.2	71.9
D_0(CC)	21.8	13.2	24.8	79.5	69.6	67.3	70.9	22.0	13.6	23.0	74.3	66.9	64.5	67.3
ΔZPE	7.1	6.5	5.2	9.7	11.2	10.7	10.7	6.9	6.2	7.2 * 4.2	15.8	14.7	14.0	15.1
$\Delta E_{disp,\,frag}$	8.5	9.1	17.8	15.7 15.9 15.9	25.6 12.6 12.2	27.0 17.5 15.5	18.6 14.3 13.1	8.3	8.5	21.8	20.5 20.1 20.1	26.3 12.9 12.5	31.8 18.5 16.9	20.8 17.9 14.1
$\sum E_{disp}$	8.5	9.1	17.8	47.6	50.4	59.9	46.0	8.3	8.5	21.8 * 19.7	60.7	51.6	67.2	52.8
$\sum E_{N/D}$	20.4	10.7	12.3	41.7	30.5	18.1	35.7	20.6	11.3	8.3 * 10.6	29.4	30.0	11.3	29.6

* Optimized geometries and harmonic vibrational frequencies predicted at the SCS-MP2/aug-cc-pVQZ level.

The interfragment dispersion energies, obtained from the LED analysis, highlight the primary differences between the intermolecular forces dominating in the PhOH·H_2O and the $(PhOH)_2$ systems. Both for the DFT and the SCS-MP2 approach, the resulting electronic equilibrium and zero-point energy corrected dissociation energies are almost identical for the global potential energy minima conformations. However, whereas the interfragment dispersion energy of $(PhOH)_2$ constitutes as much as 72% of the D_e-value, this contribution constitutes only 29% for the less electron-rich PhOH·H_2O system.

While the increase in the basis set size in the calculated SCS-MP2 model of $(PhOH)_2$ leads to only a minor change in the cumulative value of D_e, it causes over 2 kJ·mol^{-1} difference in the dispersion energy, suggesting a noticeable change in the predicted balance of intermolecular forces and the strong competition with the hydrogen bond. When comparing the predictions for the $(PhOH)_2$ to $(PhOH)_3$ systems with the AVTZ basis, the interfragment dispersion energy of the $(PhOH)_3$ system is lower by 1.3 kJ·mol^{-1}, and the proportion of E_{disp} in the total dissociation energy is also lowered to 67%. The distance between the aromatic rings increases, e.g., the calculated interatomic distance between C-atoms in the *para*-positions increases from 6 to 6.25 Å. Despite the cooperative effect, the hydrogen bond length increases from 1.93 to 1.98 Å, while at the same time due to mutual rotation of the rings, the hydrogen in the *meta*-position becomes oriented towards the π-cloud of the adjacent fragment. The same can be observed for the DFT geometries. The DFT structures overall show significantly lower weighting of the dispersive interactions for the $(PhOH)_2$ and $(PhOH)_3$ structures.

The effect of hydrogen bond cooperativity is common for cyclic cluster molecules and can be probed experimentally by the relative average hydrogen bond librational energy compared to the single hydrogen bond librational transition energy of the respective dimer [24,30,33]. In the case of $(PhOH)_3$, we observe experimental evidence of cooperativity in the larger average observed hydrogen bond librational energy of 631 cm^{-1} (assuming the assigned 557 cm^{-1} transition is degenerate) relative to the observed librational band at 605 cm^{-1} for $(PhOH)_2$. It is worth noting that only the high-level wavefunction-based methods properly capture this effect. The DFT approach predicts an almost exactly three-times larger dissociation energy for the $(PhOH)_3$ system with three cooperative intermolecular hydrogen bonds relative to the dissociation energy of $(PhOH)_2$. The high-level DLPNO-CCSD(T) predictions of the DFT optimized molecular structures capture the experimental findings that $D_0((PhOH)_3) > 3\times D_0((PhOH)_2)$ and so does the SCS-MP2 approach even before the DLPNO-CCSD(T) refinement.

While we cannot quantitatively isolate the CH$\cdots\pi$ interactions using the LED approach, we can roughly estimate the relative roles of this contact between the phenol cluster molecules via selected structural parameters. In the $(PhOH)_2$ system, the O-H$\cdots$O angle is 127°, and the planes of the aromatic rings are oriented near perpendicular around the hydrogen bond axis. The bond angle in the $(PhOH)_3$ system is 90°. The structure of $(PhOH)_4$ was calculated using the DFT approach to extend this series (see ESI), and the bond angle is 103° for this system. The CH$\cdots\pi$ interaction is estimated as a function of the distance from the hydrogen in the *ortho*-position to the center of the adjacent phenol fragment. This distance is 3.75 Å, 3.30 Å and 3.37 Å for $(PhOH)_{2-4}$, respectively. The relatively unobstructed rotation of the rings in the dimer results in optimal orientation that does not benefit a strong CH$\cdots\pi$ contact. In the $(PhOH)_3$ and $(PhOH)_4$ systems, the cooperative hydrogen bond networks constrict the mutual orientation of the rings, and therefore the hydrogen bond angles become smaller, but this orientation favors the CH$\cdots\pi$ interaction. The CH$\cdots\pi$ bond distance is shorter in $(PhOH)_3$, but then increases again when transitioning to $(PhOH)_4$. This appears to be due to strengthening of the cooperative hydrogen bond network resulting in an opening of the phenyl ring "barrel"-like structure.

3. Materials and Methods

3.1. Experimental Details

Phenol powder (Sigma-Aldrich, Søborg, Denmark, ≥99.5%) was purified by several cycles of vacuum sublimation over pre-baked molecular sieves (4Å) to eliminate residual air and moisture. Milli-Q grade regular water and isotopically enriched D_2O (Sigma-Aldrich, Søborg, Denmark, 99% D-atoms) samples were degassed under vacuum.

The experimental setup consists of a Bruker Vertex 80V Fourier-transform spectrometer used with a DE-204 4 K (Advanced Research Systems, Inc., Macungie, PA, USA) closed-cycle cryocooler. A transmission sample holder with a wedged diamond cold window (Diamond Materials, GmbH, Freiburg im Breisgau, Germany) which can be cooled to 4.5–5.0 K is mounted onto the cold plate of the cooler. The cryocooler is surrounded by a rotatable vacuum shroud equipped with two wedged diamond windows and a gate valve allowing the insertion of the inlet system into the vacuum space of the cryostat. The wedged diamond windows minimize spectral interference fringes from internal reflections in the windows. The optical setup of the spectrometer consisted of a globar radiation source, a germanium-coated KBr beam splitter together with a broad-band HgCdTe detector for the complete infrared spectral range (4000–450 cm^{-1}). A spectral resolution of 0.6 cm^{-1} was used throughout the measurements.

The inert gas matrix is obtained by simultaneous deposition of neon gas (99.999%, Air Liquide Danmark A/S, Taastrup, Denmark) with vapors of the samples through separate inlet tubes, which are brought to within 5 mm of the cold sample window using a motorized stage. The neon is supplied from an MKS G-series mass flow controller, passing a 1 m-long LN_2-cooled coiled trap for impurities, at a flow rate of 6–8 sccm depending on the desired mixing ratio. Water vapor was supplied from a variable-pressure (30–80 Pa) supply volume through a small-flow Swagelok metering valve, and phenol vapor was deposited directly from the sublimation vessel thermostated in the temperature range −5–10 ± 0.2 °C throughout the deposition process.

Following acquisition of the initial ("pre-annealing") matrix spectra, the samples were annealed by raising the temperature to 9 K for 60 min using a LakeShore controller operating a resistive heater and a Si-diode temperature sensor attached to the cold head. Subsequently, the heater was deactivated, allowing the matrix to cool again to 4–5 K before recording the "post-annealing" spectra. This annealing procedure softens the neon matrix and allows the diffusion of phenol molecules within the matrix and triggers the self-association of phenol into hydrogen-bonded cluster molecules.

The deposition of the samples via separate inlet tubes creates intentional spatial inhomogeneity in the phenol/water/neon mixing ratio, which is exploited to enable spatial "spot-to-spot" matrix probing and differentiate between cluster sizes in the same experiment. The effective diameter of the probed region of the matrix is comparable to the employed 3.5 mm aperture size. The background spectra were all collected of the evacuated cryostat at room temperature.

3.2. Computational Details

The molecular geometries of the phenol cluster molecules were predicted using the CREST (2.12, Bonn, Germany) [34] conformational search utility employing the GFN2-xTB methodology [35]. The 20 most stable structures suggested by the utility for each cluster molecule were then further optimized using several quantum chemical methods in ORCA (5.0.4, Max-Planck-Institut für Kohlenforschung, Germany) [36]: the MP2 [37] and SCS-MP2 [38] approaches with aug-cc-pVnZ [39] (n = T, Q) basis sets (denoted AVTZ, AVQZ), and the hybrid dispersion-corrected PW6B95-D4 [40–42] functional with the ma-def2-QZVP basis set [43,44] (later denoted as the DFT method). The optimized geometries of the most stable structures were then used for the respective harmonic vibrational frequencies and the dissociation energy calculations.

The electronic equilibrium dissociation energies (D_e) of the cluster molecules were obtained by performing domain-based local pair natural orbital coupled cluster (DLPNO-CCSD(T)) [45,46] single-point calculations on the respective DFT and SCS-MP2 geometries. The cut-offs for the DLPNO methodology were set by the TightPNO keyword [47]. The zero-point corrected ground-state dissociation energies (D_0) of the cluster molecules were calculated as the sum of the DLPNO-CCSD(T) electronic dissociation energies (D_e) and the ΔZPE values from the respective base calculations.

The intermolecular London dispersion energies of the cluster molecules were extracted using the DLPNO-based local energy decomposition (LED) scheme [31]. The procedure is described in the literature [31,32].

The resolution of the identity approximation was used in all of the calculations with the appropriate fitting basis sets [48–51]. The numerical "chain-of-spheres" approximation (COSX [52,53]) was used with the MP2 level calculations. The calculations were all carried out using the DTU Computing Center (DCC) cluster [54].

4. Conclusions

The self-association and microhydration molecular recognition mechanisms of phenol have been explored experimentally by means of the less accessible vibrational transitions associated with large-amplitude hydrogen bond librational motion. The present work provides for the first time unambiguous spectroscopic assignments for the large-amplitude and highly anharmonic hydrogen bond librational modes of the phenol dimer (605 cm^{-1}), the phenol trimer (557 and 781 cm^{-1}) and two isotopic variants of the phenol monohydrate (642 cm^{-1} for PhOH·H_2O and 638 for PhOH·D_2O) embedded in inert cryogenic "quantum" neon matrices at 4 K. While accurate microwave spectroscopic information concerned with the molecular structures of these systems has been reported in the literature previously, these new experimental findings provide important constraints for future high-level quantum chemical models of the corresponding intermolecular potential energy surfaces.

The experimental findings have been complemented by extensive conformation search sampling and further computational optimizations employing both dispersion-corrected DFT and wavefunction ab initio methodologies of the most stable potential energy minima. In addition, the local energy decompositions of the total interaction energies for these phenol cluster molecules have been extracted. The local energy decompositions reveal an interplay of several binding mechanisms and active sites in the phenol cluster molecules resulting in a complex internal competition of non-covalent forces. The pure cluster molecules of phenol were found to exhibit OH$\cdots$O cooperative hydrogen bonds, CH$\cdots\pi$ contacts and strong competing London dispersion forces between the aromatic rings. The larger phenol cluster molecules were calculated to have larger relative non-dispersive energy components compared to phenol dimer.

The accurate determination of the molecular geometries and intermolecular potential energy surfaces of these complexes is still challenging for the currently feasible levels of theory, which for systems of this size are limited to *meta*-GGA DFT or MP2 approaches with incomplete basis sets. The inconsistency between some of the energy parameters predicted by the methodologies tested here highlights the necessity of the inclusion of the *state-of-the-art* electronic structure methods in order to obtain reliable ab initio descriptions of these prototypical phenol cluster molecules.

Supplementary Materials: The following supporting information can be downloaded at: https://www.mdpi.com/article/10.3390/molecules29133012/s1. The material contains the optimized potential energy minima geometries of the $(PhOH)_{1\text{-}4}$ and $PhOH{\cdot}H_2O$ cluster molecules in XYZ format and the predicted harmonic vibrational mode frequencies for the respective structures.

Author Contributions: Conceptualization D.M., K.L.F. and R.W.L.; methodology—D.M., K.L.F. and R.W.L.; software—D.M.; validation—D.M., K.L.F. and R.W.L.; formal analysis—D.M., K.L.F. and R.W.L.; investigation—D.M., K.L.F. and R.W.L.; resources—D.M., K.L.F. and R.W.L.; data curation—D.M., K.L.F. and R.W.L.; writing—original draft preparation,—D.M.; writing—review and editing—D.M., K.L.F. and R.W.L.; visualization—D.M.; supervision—K.L.F. and R.W.L.; project administration—K.L.F. and R.W.L.; funding acquisition—K.L.F. and R.W.L. All authors have read and agreed to the published version of the manuscript.

Funding: This research was funded by the Danish Offshore Technology Centre (DTU Offshore).

Institutional Review Board Statement: Not applicable.

Informed Consent Statement: Not applicable.

Data Availability Statement: Data are contained within the article and Supplementary Materials.

Acknowledgments: The authors acknowledge the DTU Computing Center for access to the high-performance computing (HPC) services and the technical support provided. Finally, Alexandre Voute and the mechanical workshop at DTU Chemistry is acknowledged for help and advice in the designs of the experimental setup.

Conflicts of Interest: The authors declare no conflicts of interest.

References

1. Chand, A.; Sahoo, D.K.; Rana, A.; Jena, S.; Biswal, H.S. The Prodigious Hydrogen Bonds with Sulfur and Selenium in Molecular Assemblies, Structural Biology, and Functional Materials. *Accounts Chem. Res.* **2020**, *53*, 1580–1592. [CrossRef] [PubMed]
2. Mata, R.A.; Zhanabekova, T.; Obenchain, D.A.; Suhm, M.A. Dispersion Control over Molecule Cohesion: Exploiting and Dissecting the Tipping Power of Aromatic Rings. *Accounts Chem. Res.* **2024**, *57*, 1077–1086. [CrossRef] [PubMed]
3. Andersen, J.; Heimdal, J.; Nelander, B.; Larsen, R.W. Competition between weak OH$\cdots\pi$ and CH$\cdots$O hydrogen bonds: THz spectroscopy of the C_2H_2—H_2O and C_2H_4—H_2O complexes. *J. Chem. Phys.* **2017**, *146*, 194302. [CrossRef] [PubMed]
4. Řezáč, J.; Hobza, P. Describing Noncovalent Interactions beyond the Common Approximations: How Accurate Is the "Gold Standard", CCSD(T) at the Complete Basis Set Limit? *J. Chem. Theory Comput.* **2013**, *9*, 2151–2155. [CrossRef] [PubMed]
5. Mehta, N.; Martin, J.M.L. Explicitly Correlated Double-Hybrid DFT: A Comprehensive Analysis of the Basis Set Convergence on the GMTKN55 Database. *J. Chem. Theory Comput.* **2022**, *18*, 5978–5991. [CrossRef] [PubMed]
6. Becke, A.D.; Santra, G.; Martin, J.M.L. A double-hybrid density functional based on good local physics with outstanding performance on the GMTKN55 database. *J. Chem. Phys.* **2023**, *158*, 151103. [CrossRef] [PubMed]
7. Goerigk, L.; Hansen, A.; Bauer, C.; Ehrlich, S.; Najibi, A.; Grimme, S. A look at the density functional theory zoo with the advanced GMTKN55 database for general main group thermochemistry, kinetics and noncovalent interactions. *Phys. Chem. Chem. Phys.* **2017**, *19*, 32184–32215. [CrossRef]

8. Jurečka, P.; Šponer, J.; Černý, J.; Hobza, P. Benchmark database of accurate (MP2 and CCSD(T) complete basis set limit) interaction energies of small model complexes, DNA base pairs, and amino acid pairs. *Phys. Chem. Chem. Phys.* **2006**, *8*, 1985–1993. [CrossRef] [PubMed]
9. León, I.; Lesarri, A.; Fernández, J.A. Noncovalent interactions in isolated molecular aggregates: From single molecules to nanostructures. In *Intra- and Intermolecular Interactions Between Non-Covalently Bonded Species*; Elsevier: Amsterdam, The Netherlands, 2021; pp. 143–188. [CrossRef]
10. Kolář, M.; Hobza, P. Accurate Theoretical Determination of the Structure of Aromatic Complexes Is Complicated: The Phenol Dimer and Phenol·Methanol Cases. *J. Phys. Chem. A* **2007**, *111*, 5851–5854. [CrossRef]
11. Plokhotnichenko, A.M.; Radchenko, E.D.; Blagoi, Y.P.; Karachevtsev, V.A. Dimers of phenol in argon and neon matrices. *Low Temp. Phys.* **2001**, *27*, 666–675. [CrossRef]
12. Parthasarathi, R.; Subramanian, V.; Sathyamurthy, N. Hydrogen Bonding in Phenol, Water, and Phenol–Water Clusters. *J. Phys. Chem. A* **2005**, *109*, 843–850. [CrossRef] [PubMed]
13. Bödecker, M.; Mihrin, D.; Suhm, M.A.; Wugt Larsen, R. Regularities and anomalies in neon matrix shifts of hydrogen-bonded O-H stretching fundamentals. *J. Phys. Chem. A* **2024**.
14. Gor, G.Y.; Tapio, S.; Domanskaya, A.V.; Räsänen, M.; Nemukhin, A.V.; Khriachtchev, L. Matrix-isolation study of the phenol–water complex and phenol dimer. *Chem. Phys. Lett.* **2011**, *517*, 9–15. [CrossRef]
15. Ebata, T.; Watanabe, T.; Mikami, N. Evidence for the Cyclic Form of Phenol Trimer: Vibrational Spectroscopy of the OH Stretching Vibrations of Jet-Cooled Phenol Dimer and Trimer. *J. Phys. Chem.* **1995**, *99*, 5761–5764. [CrossRef]
16. Ebata, T.; Kayano, M.; Sato, S.; Mikami, N. Picosecond IR−UV Pump−Probe Spectroscopy. IVR of OH Stretching Vibration of Phenol and Phenol Dimer. *J. Phys. Chem. A* **2001**, *105*, 8623–8628. [CrossRef]
17. Weichert, A.; Riehn, C.; Brutschy, B. High-Resolution Rotational Coherence Spectroscopy of the Phenol Dimer. *J. Phys. Chem. A* **2001**, *105*, 5679–5691. [CrossRef]
18. Schmitt, M.; Böhm, M.; Ratzer, C.; Krügler, D.; Kleinermanns, K.; Kalkman, I.; Berden, G.; Meerts, W.L. Determining the Intermolecular Structure in the S_0 and S_1 States of the Phenol Dimer by Rotationally Resolved Electronic Spectroscopy. *ChemPhysChem* **2006**, *7*, 1241–1249. [CrossRef] [PubMed]
19. Seifert, N.A.; Steber, A.L.; Neill, J.L.; Pérez, C.; Zaleski, D.P.; Pate, B.H.; Lesarri, A. The interplay of hydrogen bonding and dispersion in phenol dimer and trimer: Structures from broadband rotational spectroscopy. *Phys. Chem. Chem. Phys.* **2013**, *15*, 11468. [CrossRef]
20. Tanabe, S.; Ebata, T.; Fujii, M.; Mikami, N. OH stretching vibrations of phenol—$(H_2O)_n$ ($n = 1 - 3$) complexes observed by IR-UV double-resonance spectroscopy. *Chem. Phys. Lett.* **1993**, *215*, 347–352. [CrossRef]
21. Yu, Z.; Klemperer, W. Asymmetry in angular rigidity of hydrogen-bonded complexes. *Proc. Natl. Acad. Sci. USA* **2005**, *102*, 12667–12669. [CrossRef]
22. Andersen, J.; Heimdal, J.; Wugt Larsen, R. The influence of large-amplitude librational motion on the hydrogen bond energy for alcohol–water complexes. *Phys. Chem. Chem. Phys.* **2015**, *17*, 23761–23769. [CrossRef] [PubMed]
23. Mihrin, D.; Andersen, J.; Jakobsen, P.W.; Larsen, R.W. Highly localized H_2O librational motion as a far-infrared spectroscopic probe for microsolvation of organic molecules. *Phys. Chem. Chem. Phys.* **2019**, *21*, 1717–1723. [CrossRef] [PubMed]
24. Wugt Larsen, R.; Suhm, M.A. The benefits of alternation and alkylation: Large amplitude hydrogen bond librational modes of alcohol trimers and tetramers. *Phys. Chem. Chem. Phys.* **2010**, *12*, 8152–8157. [CrossRef] [PubMed]
25. Andersen, J.; Heimdal, J.; Wugt Larsen, R. Spectroscopic identification of ethanol-water conformers by large-amplitude hydrogen bond librational modes. *J. Chem. Phys.* **2015**, *143*, 224315. [CrossRef] [PubMed]
26. Mihrin, D.; Voute, A.; Jakobsen, P.W.; Feilberg, K.L.; Wugt Larsen, R. The effect of alkylation on the micro-solvation of ethers revealed by highly localized water librational motion. *J. Chem. Phys.* **2022**, *156*, 084305. [CrossRef] [PubMed]
27. Andersen, J.; Voute, A.; Mihrin, D.; Heimdal, J.; Berg, R.W.; Torsson, M.; Larsen, R.W. Probing the global potential energy minimum of $(CH_2O)_2$: THz absorption spectrum of $(CH_2O)_2$ in solid neon and *para*-hydrogen. *J. Chem. Phys.* **2017**, *146*, 244311. [CrossRef] [PubMed]
28. Andersen, J.; Heimdal, J.; Mahler, D.W.; Nelander, B.; Larsen, R.W. Communication: THz absorption spectrum of the CO_2–H_2O complex: Observation and assignment of intermolecular van der Waals vibrations. *J. Chem. Phys.* **2014**, *140*, 091103. [CrossRef] [PubMed]
29. Mihrin, D.; Jakobsen, P.W.; Voute, A.; Manceron, L.; Wugt Larsen, R. High-resolution synchrotron terahertz investigation of the large-amplitude hydrogen bond librational band of $(HCN)_2$. *Phys. Chem. Chem. Phys.* **2018**, *20*, 8241–8246. [CrossRef] [PubMed]
30. Wugt Larsen, R.; Suhm, M.A. Cooperative organic hydrogen bonds: The librational modes of cyclic methanol clusters. *J. Chem. Phys.* **2006**, *125*, 154314. [CrossRef]
31. Schneider, W.B.; Bistoni, G.; Sparta, M.; Saitow, M.; Riplinger, C.; Auer, A.A.; Neese, F. Decomposition of Intermolecular Interaction Energies within the Local Pair Natural Orbital Coupled Cluster Framework. *J. Chem. Theory Comput.* **2016**, *12*, 4778–4792. [CrossRef]
32. Altun, A.; Neese, F.; Bistoni, G. Effect of Electron Correlation on Intermolecular Interactions: A Pair Natural Orbitals Coupled Cluster Based Local Energy Decomposition Study. *J. Chem. Theory Comput.* **2018**, *15*, 215–228. [CrossRef] [PubMed]

33. Kollipost, F.; Andersen, J.; Mahler, D.W.; Heimdal, J.; Heger, M.; Suhm, M.A.; Wugt Larsen, R. The effect of hydrogen bonding on torsional dynamics: A combined far-infrared jet and matrix isolation study of methanol dimer. *J. Chem. Phys.* **2014**, *141*, 174314. [CrossRef] [PubMed]
34. Pracht, P.; Bohle, F.; Grimme, S. Automated exploration of the low-energy chemical space with fast quantum chemical methods. *Phys. Chem. Chem. Phys.* **2020**, *22*, 7169–7192. [CrossRef] [PubMed]
35. Bannwarth, C.; Ehlert, S.; Grimme, S. GFN2-xTB—An Accurate and Broadly Parametrized Self-Consistent Tight-Binding Quantum Chemical Method with Multipole Electrostatics and Density-Dependent Dispersion Contributions. *J. Chem. Theory Comput.* **2019**, *15*, 1652–1671. [CrossRef] [PubMed]
36. Neese, F. Software update: The ORCA program system—Version 5.0. *Wires Comput. Mol. Sci.* **2022**, *12*, e1606. [CrossRef]
37. Bernholdt, D.E.; Harrison, R.J. Large-scale correlated electronic structure calculations: The RI-MP2 method on parallel computers. *Chem. Phys. Lett.* **1996**, *250*, 477–484. [CrossRef]
38. Grimme, S. Improved second-order Møller–Plesset perturbation theory by separate scaling of parallel- and antiparallel-spin pair correlation energies. *J. Chem. Phys.* **2003**, *118*, 9095–9102. [CrossRef]
39. Kendall, R.A.; Dunning, T.H.; Harrison, R.J. Electron affinities of the first-row atoms revisited. Systematic basis sets and wave functions. *J. Chem. Phys.* **1992**, *96*, 6796–6806. [CrossRef]
40. Zhao, Y.; Truhlar, D.G. Design of Density Functionals That Are Broadly Accurate for Thermochemistry, Thermochemical Kinetics, and Nonbonded Interactions. *J. Phys. Chem. A* **2005**, *109*, 5656–5667. [CrossRef]
41. Caldeweyher, E.; Ehlert, S.; Hansen, A.; Neugebauer, H.; Spicher, S.; Bannwarth, C.; Grimme, S. A generally applicable atomic-charge dependent London dispersion correction. *J. Chem. Phys.* **2019**, *150*, 154122 . [CrossRef]
42. Caldeweyher, E.; Bannwarth, C.; Grimme, S. Extension of the D3 dispersion coefficient model. *J. Chem. Phys.* **2017**, *147*, 034112. [CrossRef] [PubMed]
43. Zheng, J.; Xu, X.; Truhlar, D.G. Minimally augmented Karlsruhe basis sets. *Theor. Chem. Accounts* **2010**, *128*, 295–305. [CrossRef]
44. Weigend, F.; Ahlrichs, R. Balanced basis sets of split valence, triple zeta valence and quadruple zeta valence quality for H to Rn: Design and assessment of accuracy. *Phys. Chem. Chem. Phys.* **2005**, *7*, 3297. [CrossRef] [PubMed]
45. Riplinger, C.; Pinski, P.; Becker, U.; Valeev, E.F.; Neese, F. Sparse maps—A systematic infrastructure for reduced-scaling electronic structure methods. II. Linear scaling domain based pair natural orbital coupled cluster theory. *J. Chem. Phys.* **2016**, *144*, 024109. [CrossRef] [PubMed]
46. Riplinger, C.; Sandhoefer, B.; Hansen, A.; Neese, F. Natural triple excitations in local coupled cluster calculations with pair natural orbitals. *J. Chem. Phys.* **2013**, *139*, 134101. [CrossRef] [PubMed]
47. Liakos, D.G.; Sparta, M.; Kesharwani, M.K.; Martin, J.M.L.; Neese, F. Exploring the Accuracy Limits of Local Pair Natural Orbital Coupled-Cluster Theory. *J. Chem. Theory Comput.* **2015**, *11*, 1525–1539. [CrossRef] [PubMed]
48. Weigend, F. Hartree–Fock exchange fitting basis sets for H to Rn. *J. Comput. Chem.* **2007**, *29*, 167–175. [CrossRef] [PubMed]
49. Weigend, F. A fully direct RI-HF algorithm: Implementation, optimised auxiliary basis sets, demonstration of accuracy and efficiency. *Phys. Chem. Chem. Phys.* **2002**, *4*, 4285–4291. [CrossRef]
50. Weigend, F. Accurate Coulomb-fitting basis sets for H to Rn. *Phys. Chem. Chem. Phys.* **2006**, *8*, 1057. [CrossRef]
51. Weigend, F.; Köhn, A.; Hättig, C. Efficient use of the correlation consistent basis sets in resolution of the identity MP2 calculations. *J. Chem. Phys.* **2002**, *116*, 3175–3183. [CrossRef]
52. Neese, F.; Wennmohs, F.; Hansen, A.; Becker, U. Efficient, approximate and parallel Hartree–Fock and hybrid DFT calculations. A 'chain-of-spheres' algorithm for the Hartree–Fock exchange. *Chem. Phys.* **2009**, *356*, 98–109. [CrossRef]
53. Kossmann, S.; Neese, F. Efficient Structure Optimization with Second-Order Many-Body Perturbation Theory: The RIJCOSX-MP2 Method. *J. Chem. Theory Comput.* **2010**, *6*, 2325–2338. [CrossRef] [PubMed]
54. DTU Computing Center. *DTU Computing Center Resources;* Technical University of Denmark: Kongens Lyngby, Denmark, 2021. [CrossRef]

molecules

MDPI

Article

Chalcogen-Bond-Assisted Formation of the N→C Dative Bonds in the Complexes between Chalcogenadiazoles/Chalcogenatriazoles and Fullerene C_{60}

Yu Zhang and Weizhou Wang *

College of Chemistry and Chemical Engineering, and Henan Key Laboratory of Function-Oriented Porous Materials, Luoyang Normal University, Luoyang 471934, China; yzhpaper@yahoo.com

* Correspondence: wzw@lynu.edu.cn; Tel.: +86-379-686-18320

Abstract: The existence of the N→C dative bonds in the complexes between N-containing molecules and fullerenes have been verified both theoretically and experimentally. However, finding stable N→C dative bonds is still a highly challenging task. In this work, we investigated computationally the N→C dative bonds in the complexes formed by fullerene C_{60} with 1,2,5-chalcogenadiazoles, 2,1,3-benzochalcogenadiazoles, and 1,2,4,5-chalcogenatriazoles, respectively. It was found that the N→C dative bonds are formed along with the formation of the N–Ch···C (Ch = S, Se, Te) chalcogen bonds. In the gas phase, from S-containing complexes through Se-containing complexes to Te-containing complexes, the intrinsic interaction energies become more and more negative, which indicates that the N–Ch···C chalcogen bonds can facilitate the formation of the N→C dative bonds. The intrinsic interaction energies are compensated by the large deformation energy of fullerene C_{60}. The total interaction energies of Te-containing complexes are negative, while both total interaction energies of the S-containing complexes and Se-containing complexes are positive. This means that the N→C dative bonds in the Te-containing complexes are more easily observed in experiments in comparison with those in the S-containing complexes and Se-containing complexes. This study provides a new theoretical perspective on the experimental observation of the N→C dative bonds in complexes involving fullerenes. Further, the formation of stable N→C dative bonds in the complexes involving fullerenes can significantly change the properties of fullerenes, which will greatly simulate and expand the application range of fullerenes.

Keywords: N→C dative bond; chalcogen bond; chalcogenadiazole; fullerene C_{60}

1. Introduction

The "dative bond" is not a new term. In 1999, the International Union of Pure and Applied Chemistry (IUPAC) defined the dative bond as "the coordination bond formed upon interaction between molecular species, one of which serves as a donor and the other as an acceptor of the electron pair to be shared in the complex formed" [1]. Unlike covalent bonds, dative bonds have more significant polarity, weaker strength, and longer bond lengths. The distinctive features and nature of dative bonds have been reviewed extensively [2–5]. In a relatively recent review article, Nandi and Kozuch detailed the historical development of dative bonds and provided an outlook on their future advancement [5]. Currently, controversies related to dative bonds have become increasingly rare.

The well-known dative bond is the central bond in ammonia borane ($H_3N{\to}BH_3$). Like the case in $H_3N{\to}BH_3$, the dative bond is always represented by an arrow from the electron donor to the electron acceptor. There are many different kinds of dative bonds [2–5]. Fullerenes are of considerable interest due to their unusual structures and peculiar properties [6]. Through combined experimental and theoretical studies, Hobza and colleagues have discovered the existence of N→C dative bonds in the complexes between piperidine and fullerene molecules such as C_{60} and C_{70} [7–9]. The charge transfer from piperidine to fullerenes significantly changes the properties of fullerenes. Subsequently, it was found by quantum chemical calculations that the N→C dative bonds also exist in the complexes between piperidine and cyclo[*n*]carbons [10,11]. Further, the P→C dative bonds were computationally found in the complexes involving molecular carbon materials [12,13]. The N→C/P→C dative bonds in these complexes are unstable due to the large deformation energies of molecular carbon materials. This greatly limits the experimental observation and further applications of N→C/P→C dative bonds. It is still a highly challenging task to find stable N→C/P→C dative bonds in the complexes involving important molecular carbon materials.

The N→C dative bond in the 1:1 complex between piperidine and fullerene C_{60} is not very stable, whereas it is stable in the 2:1 complex between piperidine and fullerene C_{60} [7]. This is mainly caused by two factors. First, the strong N–H···N hydrogen bond between two piperidine molecules further stabilizes the 2:1 complex. Second, the stability of the N→C dative bond is greatly enhanced by the σ-bond cooperativity with the N–H···N hydrogen bond [7,14]. Note that the σ-bond cooperativity is a very common phenomenon between noncovalent bonds of the same or different types [15]. As is well known, besides hydrogen bonds, there are many other types of noncovalent bonds [16–20]. By similar principles, other noncovalent bonds should also enhance the N→C dative bonds in the complexes involving important molecular carbon materials. In this study, we focused on the chalcogen bonds [21–23] and explored how and why they can enhance the N→C dative bonds. Fullerene C_{60} was selected as a model molecule for the ball-shaped molecular carbon materials. The small molecules 1,2,5-chalcogenadiazoles, 2,1,3-benzochalcogenadiazoles, and 1,2,4,5-chalcogenatriazoles contain both nitrogen atoms and chalcogen atoms. They are ideal model molecules for studying the chalcogen-bond-enhanced N→C dative bonds. This is also one of the reasons why we are studying the influence of chalcogen bonds on the N→C dative bonds first.

2. Results and Discussion

2.1. Complexes between Chalcogenadiazoles and Fullerene C_{60}

At the outset of the research, we designed a series of complexes containing the N→C dative bonds (Figure 1). After the initial structures of the complexes O-1, O-2, S-1, S-2, Se-1, Se-2, Te-1, and Te-2 were fully optimized at the PBE0-D3/def2-TZVPP level of theory, we found that only the complexes O-1 and O-2 did not exist and instead transformed into corresponding complexes bound by the π···π stacking interactions. It can be clearly seen from Figure 1 that the fullerene C_{60} molecules undergo significant deformation in the PBE0-D3/def2-TZVPP optimized structures of the complexes S-1, S-2, Se-1, Se-2, Te-1, and Te-2. Table 1 lists the N···C_1 and Ch···C_2 interatomic distances, intrinsic interaction energies, total interaction energies, total dipole moments and Mulliken charges on N, C_1, Ch and C_2 for the complexes S-1, S-2, Se-1, Se-2, Te-1, and Te-2.

Table 1. The N···C_1 and Ch···C_2 interatomic distances (d, Å), intrinsic interaction energies (ΔE^{INTR}, kcal/mol), total interaction energies (ΔE, kcal/mol), total dipole moments (μ, Debye) and Mulliken charges (q, e) on N, C_1, Ch and C_2 of the complexes studied.

Complex	d(N···C_1)	d(Ch···C_2)	ΔE^{INTR}	Δ	μ	q(N)	q(C_1)	q(Ch)	q(C_2)
S-1	1.504	2.569	−4.52	20.85	8.14	0.081	1.121	0.351	−0.507
S-2	1.500	2.557	−8.75	18.13	10.18	0.090	1.153	0.381	−0.545
Se-1	1.480	2.467	−20.05	11.40	5.60	0.007	0.923	0.341	−0.221
Se-2	1.478	2.468	−24.04	8.35	7.59	−0.004	0.952	0.383	−0.241
Te-1	1.467	2.471	−42.54	−4.11	3.39	−0.142	1.019	0.443	−0.245
Te-2	1.466	2.464	−46.88	−7.32	4.92	−0.150	1.043	0.497	−0.245

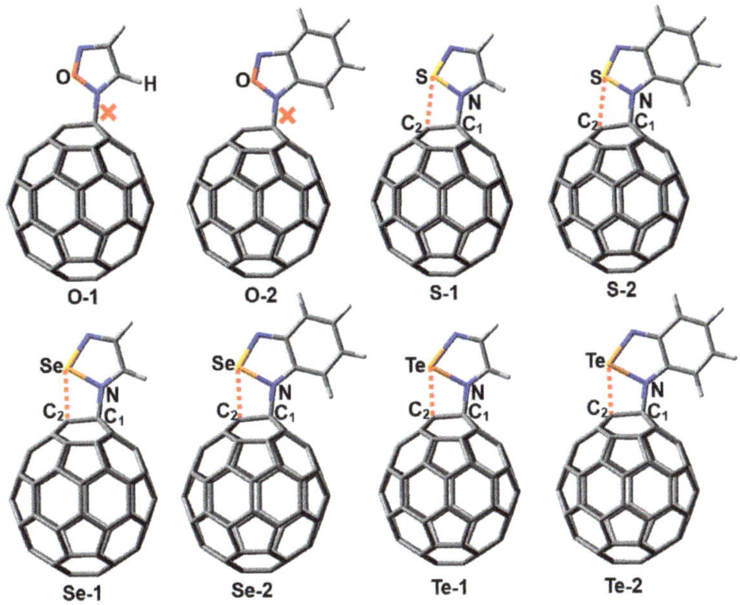

Figure 1. PBE0-D3/def2-TZVPP optimized structures of the complexes S-1, S-2, Se-1, Se-2, Te-1 and Te-2. The red dashed line indicates the possibility of a chalcogen bond.

The Van der Waals radii of S, Se, Te and C are 1.80, 1.90, 2.06 and 1.70 Å, respectively [24]. The $Ch\cdots C_2$ interatomic distances in Table 1 are all smaller than the sum of the radii of two interacting atoms, which provides preliminary support for classifying the $Ch\cdots C_2$ interactions as chalcogen bonds. The $N\cdots C_1$ interatomic distances are in the range of 1.466–1.507 Å, which is clearly less than the interatomic distance of 1.606 Å for the N→C dative bond in the complex between piperidine and fullerene C_{60} [7]. This means that the covalent character of the N→C dative bonds in this study is significantly enhanced. In Table 1, the intrinsic interaction energies and total interaction energies contain the contributions of all the noncovalent interactions because it is difficult to completely separate N→C dative bonds from other noncovalent interactions. From the S-containing complexes through Se-containing complexes to Te-containing complexes, the intrinsic interaction energies become more and more negative. It is well known that the strength of a chalcogen bond increases in the order of S < Se < Te. Hence, such a result indicates that the N–Ch···C chalcogen bonds facilitate the formation of the N→C dative bonds. The intrinsic interaction energies are compensated by the large deformation energies of the monomers. Unlike the intrinsic interaction energies, the total interaction energies do not neglect the deformation energies of the monomers. As shown in Table 1, the total interaction energies of Te-containing complexes are negative, while both total interaction energies of the S-containing complexes and Se-containing complexes are positive. These results indicate that the N→C dative bonds in the Te-containing complexes are more easily observed in experiments in comparison with those in the S-containing complexes and Se-containing complexes.

At the PBE0-D3/def2-TZVPP theory level, the calculated total dipole moments of the monomers are in the range of 0–1.78 Debye. Compared to the monomers, the total dipole moments of the complexes significantly increase. This also conforms to the characteristics of dative bond formation. It is very convenient to employ the Mulliken charge for qualitative estimation of possible changes in electron distribution, although it is often criticized that Mulliken charge is arbitrary and heavily basis set dependent. Table 1 summarizes the Mulliken charges on N, C_1, Ch and C_2. The Mulliken charges on Ch atoms are all positive and the Mulliken charges on C_2 atoms are all negative, which once again demonstrates the formation of chalcogen bonds.

In fact, we also calculated the complexes involving the 6:5 bonds of fullerene C_{60} at the PBE0-D3/def2-TZVPP theory level. Figure S1 in the Supplementary Materials illustrates the optimized structures and corresponding total interaction energies of the complexes S-2(6:5), Se-1(6:5), Se-2(6:5), Te-1(6:5) and Te-2(6:5). Unlike the results in Figure 1 and Table 1, the expected complex S-1(6:5) in Figure S1 does not exist, and the total interaction energies of the Te-containing complexes in Figure S1 are not negative. In line with our predictions based on the molecular electrostatic potentials of fullerene C_{60}, the interactions between 6:5 bonds of fullerene C_{60} and N–Ch bonds of chalcogenazoles are much weaker than the interactions between 6:6 bonds of fullerene C_{60} and N–Ch bonds of chalcogenazoles. On the other hand, Table S1 in the Supplementary Materials also lists the calculated results for the complexes S-1, S-2, Se-1, Se-2, Te-1 and Te-2 at the low-cost PBE0-D3/6-31G(d) or PBE0-D3/SDD level of theory. Comparing the calculated results of Tables 1 and S1, it can be observed that the low-cost PBE0-D3/6-31G(d) or PBE0-D3/SDD calculations yield qualitatively consistent results with those from high-cost PBE0-D3/def2-TZVPP calculations. Therefore, if the sole purpose is to conduct large-scale computational searches for stable N→C dative bonds in large complexes involving large molecular carbon materials, we recommend utilizing low-cost PBE0-D3/6-31G(d) or PBE0-D3/SDD calculations.

2.2. Molecular Electrostatic Potential Maps of Chalcogenadiazoles

So far, there is still one question left unanswered: Why do the complexes O-1 and O-2 in Figure 1 not exist? To address this question, we plotted the electrostatic potential mapped electron density surfaces for 1,2,5-chalcogenadiazoles and 2,1,3-benzochalcogenadiazoles. Figure 2 shows that, except for 1,2,5-oxadiazole and 2,1,3-benzooxadiazole, other chalcogenadiazoles all have two σ-holes on the chalcogen atoms. In 1,2,5-oxadiazole and 2,1,3-benzooxadiazole, the electrostatic potentials on the extensions of the N–O bonds are all negative, which means that the N–O···C chalcogen bonds cannot be formed in the two complexes. When N–O···C chalcogen bonds are absent, N→C dative bonds are also absent; when N–Ch···C (Ch = S, Se, Te) chalcogen bonds are present, N→C dative bonds are also present. Clearly, the formation of chalcogen bonds determines whether N→C dative bonds form or not.

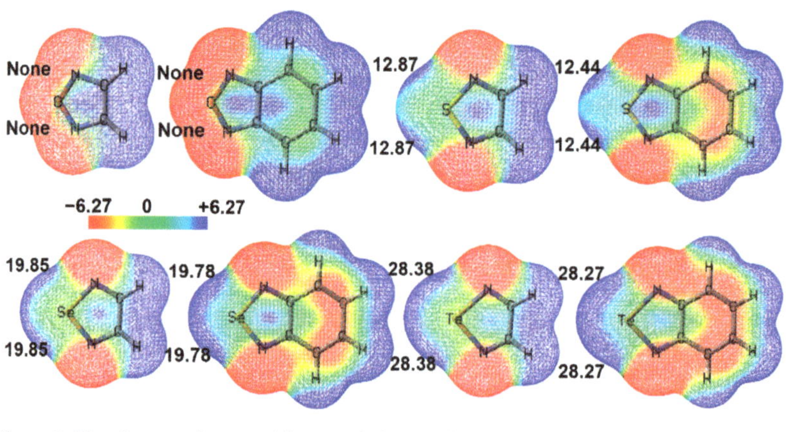

Figure 2. The electrostatic potential mapped electron density surfaces (isoval = 0.001 au) of 1,2,5-chalcogenadiazoles and 2,1,3-benzochalcogenadiazoles. The most positive electrostatic potential of each σ-hole ($V_{S,max}$) is also shown. The electrostatic potentials are given in kcal/mol.

Figure 3 shows the correlation between the most positive electrostatic potentials of σ-holes of 1,2,5-chalcogenadiazoles and the intrinsic interaction energies of S-1, Se-1 and Te-1.

Figure 4 shows the correlation between the most positive electrostatic potentials of σ-holes of 2,1,3-benzochalcogenadiazoles and the intrinsic interaction energies of S-2, Se-2 and Te-2. The correlation coefficients (R-Square) in Figures 3 and 4 are both close to 1, indicating a very strong correlation between $V_{S,max}$ and ΔE^{INTR}. The value of $V_{S,max}$ determines the strength of a chalcogen bond. At the same time, the more stable the complex, the more stable the N→C dative bond. Once again, it proves that the formation of chalcogen bonds determines the formation of N→C dative bonds.

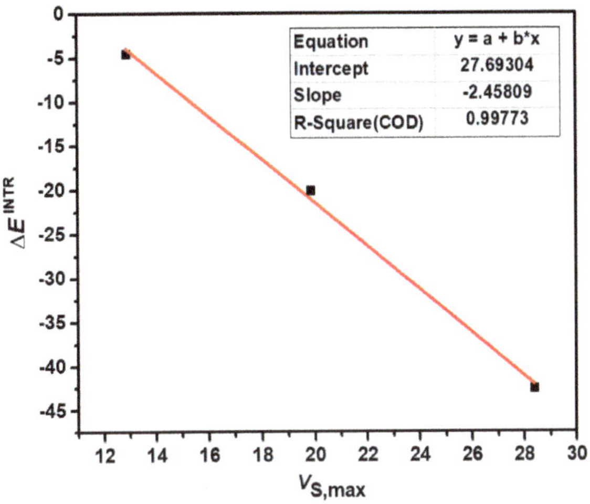

Figure 3. The correlation between $V_{S,max}$ (kcal/mol) of 1,2,5-chalcogenadiazoles and ΔE^{INTR} (kcal/mol) of S-1, Se-1 and Te-1.

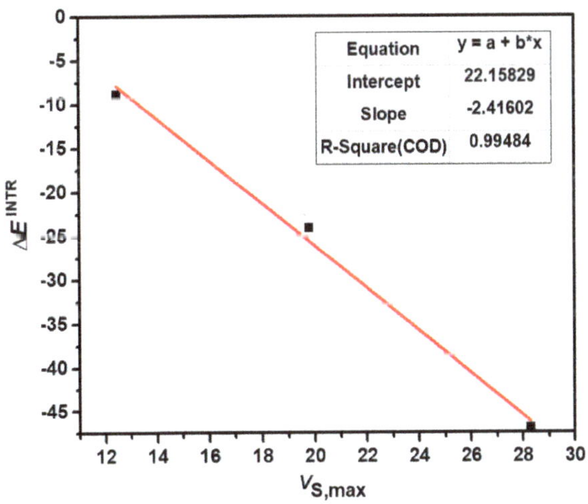

Figure 4. The correlation between $V_{S,max}$ (kcal/mol) of 2,1,3-benzochalcogenadiazoles and ΔE^{INTR} (kcal/mol) of S-2, Se-2 and Te-2.

2.3. AIM Analyses

As can be seen in Figure 1, according to the bonding situation of the nitrogen atom, it can be determined that a N→C dative bond is formed between the nitrogen atom and the carbon atom. In previous studies, the dative bonds with significant covalent bond components were also referred to as "dative/covalent bonds", "covalent/dative bond" or "covalent dative bonds" [8,10,11,25]. Therefore, we will no longer discuss the nature of N→C dative bonds in this section and instead focus on the nature of chalcogen bonds.

The AIM theory was often used to study the nature of the noncovalent bonds [26–30]. Figure 5 shows the bond critical points and bond paths for the complexes Se-1 and Se-2. The molecular graphs of S-1 and Te-1 are almost the same as the one of Se-1, and the molecular graphs of S-2 and Te-2 are almost the same as the one of Se-2 (Figure S2). For the sake of clarity, only the bond critical points and bond paths are shown in Figures 5 and S2. The electron density, Laplacian of electron density, eigenvalues of the Hessian of electron density and ellipticity at the Ch···C_2 bond critical point of each complex are summarized in Table 2.

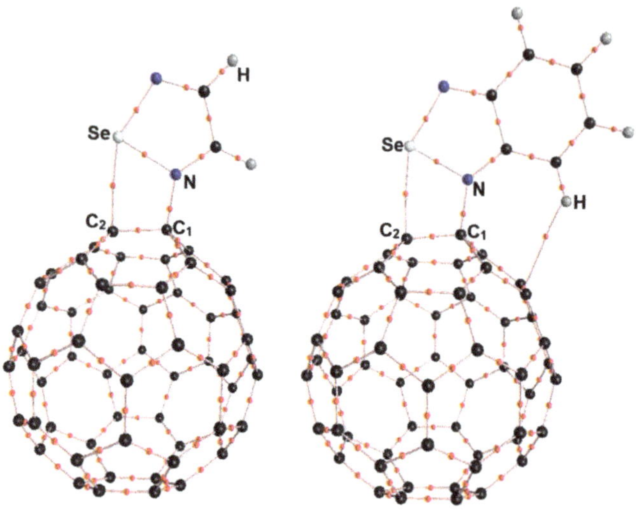

Figure 5. The bond critical points (small red dots) and bond paths of the complexes Se-1 and Se-2.

Table 2. The electron density (ρ_b), Laplacian of electron density ($\nabla^2\rho_b$), eigenvalues of the Hessian of electron density (λ_1, λ_2, λ_3) and ellipticity (ε) at the Ch···C_2 bond critical point of each complex. All the values are given in atomic units.

Complex	ρ_b	$\nabla^2\rho_b$	λ_1	λ_2	λ_3	ε
S-1	0.0393	0.0659	−0.0336	−0.0257	0.1253	0.3091
S-2	0.0398	0.0683	−0.0345	−0.0259	0.1287	0.3305
Se-1	0.0550	0.0493	−0.0509	−0.0491	0.1493	0.0370
Se-2	0.0544	0.0517	−0.0495	−0.0489	0.1502	0.0136
Te-1	0.0645	0.0302	−0.0618	−0.0544	0.1464	0.1355
Te-2	0.0651	0.0313	−0.0623	−0.0559	0.1496	0.1138

The bond paths between Ch and C_2 and the corresponding bond critical points can be clearly seen in the representative Figure 5. This is a necessary condition for the formation of the N–Ch···C_2 chalcogen bond. The electron densities at the bond critical points of the N–Ch···C_2 chalcogen bonds range from 0.0393 to 0.0651 au. In contrast, the electron densities at the bond critical points of hydrogen bonds range from 0.002 to 0.035 au [26,27]. The larger electron density means a much stronger noncovalent bond. Here, the N–Ch···C_2 chalcogen bonds should belong to strong chalcogen bonds. The Laplacian of electron density is the sum of λ_1, λ_2 and λ_3. According to AIM theory, the Laplacian of electron density is positive at the bond critical point of a noncovalent bond [31]. The values of $\nabla^2\rho_b$ in Table 2 are all positive at the Ch···C_2 bond critical point of each complex, which proves once again that the Ch···C_2 contacts are of chalcogen bonds. The ellipticity ε can be calculated using the formula $\lambda_1/(\lambda_2 - 1)$, and it can be used to assess the π character and structural stability of a bond. In Table 2, the values of ε are all very small, which indicates that the N–Ch···C_2 chalcogen bonds are all very stable. This is consistent with the results of electron densities at the Ch···C_2 bond critical points.

2.4. *Complexes between Chalcogenatriazoles and Fullerene C_{60}*

It is meaningful to expand the complexes studied above to include more complexes. First, the fullerene C_{60} in the complexes can be replaced by other fullerene molecules or other molecular carbon materials. The results should be very similar. Second, the chalcogenadiazole can be replaced by its derivatives or other similar molecules. Here, we studied the complexes between 1,2,4,5-chalcogenatriazoles and fullerene C_{60}.

Figure 6 shows the PBE0-D3/def2-TZVPP optimized structures of the complexes N–S, N–Se and N–Te along with the N···C_1 and Ch···C_2 interatomic distances. Similarly, the Ch···C_2 interatomic distances in Figure 6 are all smaller than the sum of the radii of two interacting atoms, which indicates the formation of the N–Ch···C_2 chalcogen bonds. At the same time, the bonding situation in the three complexes clearly shows the formation of the N→C dative bonds. The intrinsic interaction energies of the complexes N–S, N–Se and N–Te are −0.11, −18.55 and −43.62 kcal/mol, respectively, whereas the total interaction energies of the complexes N–, N–Se and N–Te are +27.22, +17.44 and +1.21 kcal/mol, respectively. The positive total interaction energies indicate that these three complexes are difficult to form in the gas phase. In fact, the calculated total interaction energy of the complex between piperidine and fullerene C_{60} is also positive, but it has been detected by both FTIR spectra and NMR spectroscopy [7]. As pointed out above, the influence of the chemical environment around the complexes is also very important and must be considered during the experimental process. Aside from the 1,2,4,5-chalcogenatriazoles, we believe that there should be many similar nitrogen-containing organic compounds that can form stable chalcogen-bond-assisted N→C dative bonds with molecular carbon materials.

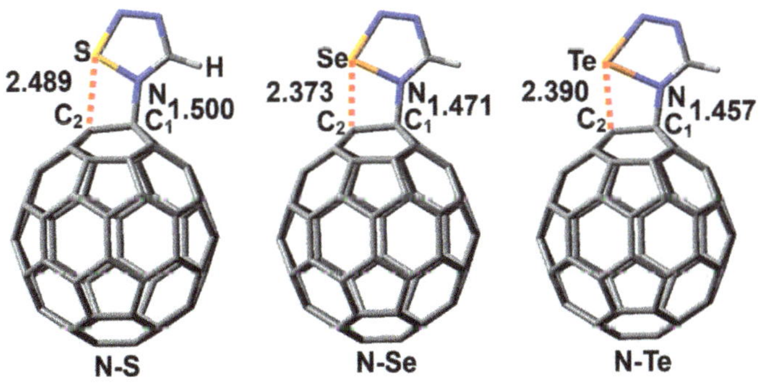

Figure 6. PBE0-D3/def2-TZVPP optimized structures of the complexes N–S, N–Se and N–Te. The red dashed lines represent the chalcogen bonds. The numbers shown are the interatomic distances (Å). The red dashed lines represent chalcogen bonds.

3. Materials and Methods

Unless otherwise stated, the structures and properties of all the monomers and complexes were calculated using the PBE0 density functional with Grimme's D3 dispersion correction in conjunction with the def2-TZVPP basis set [32–34]. The Becke–Johnson damping function and ultrafine integration grid were used in all PBE0-D3/def2-TZVPP calculations [35]. The basis set superposition error was corrected with the conventional counterpoise method [36].

The calculations of noncovalent systems have always required extreme caution [37,38]. Bickelhaupt and colleagues carried out benchmark calculations for the chalcogen bonds with the type of $D_2Ch\cdots A^-$ (Ch = S, Se; D, A = F, Cl) [38]. They found that some density functionals with specific dispersion corrections could give varying results for the chalcogen bonds. In this study, we calculated the complexes at the PBE0-D3/def2-TZVPP theory level. Our previous publications have shown that the PBE0-D3/def2-TZVPP calculations are reliable for the study of noncovalent systems [39,40]. Further, benchmark calculations on the dative bonds have been performed by employing a relatively small complex between fullerene C_{20} and piperidine [7]. The results again proved the reliability of the PBE0-D3/def2-TZVPP calculations for the study of the dative bonds.

It is also highly meaningful to study the reliability of results at low computational costs. Due to the generally large size of molecular carbon material systems, low-cost computations make studying these systems feasible. Using a small basis set 6-31G(d) for the S/Se-containing complexes and the small pseudopotential basis set SDD for the Te-containing complexes, the structures, harmonic frequencies, and energies of the complexes considered in this study were calculated at the PBE0-D3, PBE0, B3LYP-D3, B3LYP and M06-2X levels of theory, respectively. Let us add here that calculations involving heavy atom systems require the use of pseudopotential basis sets. The frequency calculations show that the structures of the complexes are true minima on their respective potential energy surfaces. For the geometries and energies of the complexes studied, quantitative differences exist between low-cost and high-cost computational results, but qualitative consistency can be achieved.

The fullerene C_{60} has two different types of carbon–carbon bonds as follows: the 6:6 bonds between two hexagons [1.401(10) Å] and the 6:5 bonds between a hexagon and a pentagon [1.458(6) Å] [41]. The 6:6 bonds are shorter than the 6:5 bonds and are more akin to the nature of double bonds. Figure 7 shows the electrostatic potential mapped electron density surface of fullerene C_{60}. The electrostatic potentials above the 6:6 bonds are more negative than those above 6:5 bonds. The C atoms of fullerene C_{60} act as the electron donors upon the formation of the N–Ch···C chalcogen bonds. Therefore, only the interactions between 6:6 bonds of fullerene C_{60} and N–Ch bonds of chalcogenazoles were considered in this study.

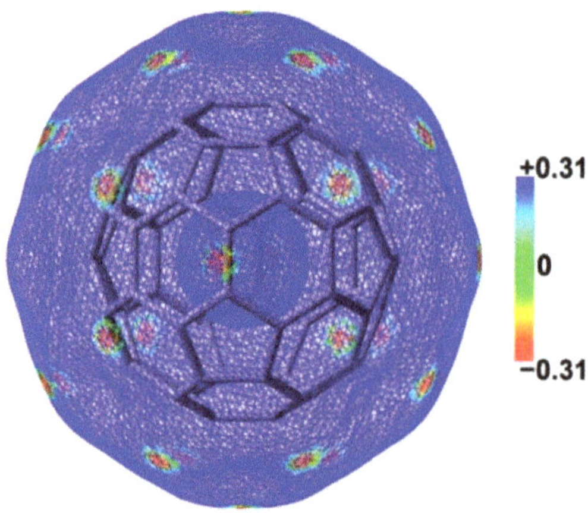

Figure 7. The electrostatic potential mapped electron density surface (isoval = 0.001 au) of fullerene C_{60}. The color bar is in kcal/mol.

The formation of the N–Ch···C chalcogen bonds was analyzed employing Bader's "atoms in molecules" (AIM) theory [31]. Despite some controversies, the AIM theory remains a very useful tool for analyzing noncovalent interactions. The AIM analyses were performed with the AIM2000 software, version 2.0 [42]. The most positive electrostatic potential of each σ-hole of chalcogen atoms in 1,2,5-chalcogenadiazoles and 2,1,3-benzochalcogenadiazoles was calculated using the Multiwfn software, version 3.7 [43,44]. Other calculations were carried out with the GAUSSIAN 09 program package [45].

4. Conclusions

The N→C dative bonds in the complexes between chalcogenadiazoles and fullerene C_{60} have been investigated by employing quantum chemical calculations. The research scope has been further expanded to the complexes between 1,2,4,5-chalcogenatriazoles and fullerene C_{60}. The results clearly show that the N–Ch···C chalcogen bonds facilitate the formation of the N→C dative bonds in these complexes. From then S-containing complexes through Se-containing complexes to Te-containing complexes, the complexes become increasingly stable as the N–Ch···C chalcogen bonds become stronger. Attractive N→C dative bonds and N–Ch···C chalcogen bonds together can overcome the significant repulsive deformation energy of fullerene C_{60}. We therefore predict that the N→C dative bonds in the complexes between Te-containing molecules and fullerenes should be more easily observed experimentally and further applied. Indeed, following this line of thought, more stable complexes bound by the N→C dative bonds can be designed and synthesized. The formation of stable N→C dative bonds in the complexes containing fullerenes can significantly change the properties of fullerenes. Such a molecular modification strategy will greatly simulate and expand the application range of fullerenes.

In this study, we only focused on the chalcogen-bond-assisted N→C dative bonds. Naturally, the other noncovalent bonds such as halogen bonds, pnictogen bonds, tetrel bonds, etc., can also facilitate the N→C/P→C dative bonds in the complexes involving important molecular carbon materials. The relevant research is currently ongoing in our laboratory. On the other hand, all the calculations in this study were performed in the gas phase. It must be pointed out that the chalcogen-bond-assisted N→C dative bonds may be more stable in certain solvents. The solvation effect is also one of our future research focuses. We did not consider the case for the planar molecular carbon materials in this study. This issue also awaits further investigation in the future.

Supplementary Materials: The following supporting information can be downloaded at: https://www.mdpi.com/article/10.3390/molecules29112685/s1, Figure S1: The PBE0-D3/def2-TZVPP optimized structures and corresponding total interaction energies of the complexes involving the 6:5 bonds of fullerene C_{60}; Figure S2: The molecular graphs of the complexes S-1, S-2, Te-1 and Te-2; Table S1: The calculated results at the PBE0-D3/6-31G(d) or PBE0-D3/SDD level of theory; Cartesian atomic coordinates for the complexes and monomers optimized at the PBE0-D3/def2-TZVPP theory level; Cartesian atomic coordinates for the complexes optimized at a variety of low-cost levels of theory.

Author Contributions: Y.Z. performed all the calculations; W.W. designed and supervised this project; Y.Z. and W.W. wrote and revised the paper. All authors have read and agreed to the published version of the manuscript.

Funding: This research was funded by the Natural Science Foundation of Henan Province of China, grant number 232300421147.

Institutional Review Board Statement: Not applicable.

Informed Consent Statement: Not applicable.

Data Availability Statement: Data are contained within the article.

Acknowledgments: We thank the Natural Science Foundation of Henan Province of China for the financial support. W.W. thanks the National Supercomputing Center in Shenzhen for the computational support.

Conflicts of Interest: The authors declare no conflicts of interest.

References

1. Minkin, V.I. Glossary of Terms Used in Theoretical Organic Chemistry (IUPAC Recommendations 1999). *Pure Appl. Chem.* **1999**, *71*, 1919–1981. [CrossRef]
2. Haaland, A. Covalent versus Dative Bonds to Main Group Metals, a Useful Distinction. *Angew. Chem. Int. Ed. Engl.* **1989**, *28*, 992–1007. [CrossRef]
3. Lepetit, C.; Maraval, V.; Canac, Y.; Chauvin, R. On the Nature of the Dative Bond: Coordination to Metals and Beyond. The Carbon Case. *Coord. Chem. Rev.* **2016**, *308*, 59–75. [CrossRef]
4. Zhao, L.; Hermann, M.; Holzmann, N.; Frenking, G. Dative Bonding in Main Group Compounds. *Coord. Chem. Rev.* **2017**, *344*, 163–204. [CrossRef]
5. Nandi, A.; Kozuch, S. History and Future of Dative Bonds. *Chem. Eur. J.* **2020**, *26*, 759–772. [CrossRef] [PubMed]
6. Hirsch, A. *The Chemistry of the Fullerenes*; WILEY-VCH Verlag GmbH & Co. KGaA: Weinheim, Germany, 2008.
7. Lamanec, M.; Lo, R.; Nachtigallová, D.; Bakandritsos, A.; Mohammadi, E.; Dračínský, M.; Zbořil, R.; Hobza, P.; Wang, W. The Existence of a N→C Dative Bond in the C_{60}–Piperidine Complex. *Angew. Chem. Int. Ed.* **2021**, *60*, 1942–1950. [CrossRef] [PubMed]
8. Lo, R.; Lamanec, M.; Wang, W.; Manna, D.; Bakandritsos, A.; Dračínský, M.; Zbořil, R.; Nachtigallová, D.; Hobza, P. Structure-Directed Formation of the Dative/Covalent Bonds in Complexes with $C_{70}\cdots$Piperidine. *Phys. Chem. Chem. Phys.* **2021**, *23*, 4365–4375. [CrossRef]
9. Lo, R.; Manna, D.; Lamanec, M.; Wang, W.; Bakandritsos, A.; Dračínský, M.; Zbořil, R.; Nachtigallová, D.; Hobza, P. Addition Reaction between Piperidine and C_{60} to Form 1,4-Disubstituted C_{60} Proceeds through van der Waals and Dative Bond Complexes: Theoretical and Experimental Study. *J. Am. Chem. Soc.* **2021**, *143*, 10930–10939. [CrossRef] [PubMed]
10. Lo, R.; Manna, D.; Hobza, P. Cyclo[*n*]carbons Form Strong N→C Dative/Covalent Bonds with Piperidine. *J. Phys. Chem. A* **2021**, *125*, 2923–2931. [CrossRef]
11. Nandi, A.; Martin, J.M.L. Heavy-Atom Tunneling in the Covalent/Dative Bond Complexation of Cyclo [18]carbon–Piperidine. *J. Phys. Chem. B* **2022**, *126*, 1799–1804. [CrossRef]
12. Lo, R.; Manna, D.; Hobza, P. Tuning the P–C Dative/Covalent Bond Formation in R_3P–C_{60} Complexes by Changing the R Group. *Chem. Commun.* **2021**, *57*, 3363–3366. [CrossRef] [PubMed]
13. Lo, R.; Manna, D.; Hobza, P. P-Doped Graphene–C_{60} Nanocomposite: A Donor–Acceptor Complex with a P–C Dative Bond. *Chem. Commun.* **2022**, *58*, 1045–1048. [CrossRef]
14. Steiner, T. The Hydrogen Bond in the Solid State. *Angew. Chem. Int. Ed.* **2002**, *41*, 48–76. [CrossRef]
15. Riel, A.M.S.; Rowe, R.K.; Ho, E.N.; Carlsson, A.-C.C.; Rappé, A.K.; Berryman, O.B.; Ho, S.P. Hydrogen Bond Enhanced Halogen Bonds: A Synergistic Interaction in Chemistry and Biochemistry. *Acc. Chem. Res.* **2019**, *52*, 2870–2880. [CrossRef] [PubMed]
16. Scheiner, S. The Pnicogen Bond: Its Relation to Hydrogen, Halogen, and Other Noncovalent Bonds. *Acc. Chem. Res.* **2013**, *46*, 280–288. [CrossRef] [PubMed]
17. Fanfrlík, J.; Holub, J.; Růžičková, Z.; Řezáč, J.; Lane, P.D.; Wann, D.A.; Hnyk, D.; Růžička, A.; Hobza, P. Competition between Halogen, Hydrogen and Dihydrogen Bonding in Brominated Carboranes. *ChemPhysChem* **2016**, *17*, 3373–3376. [CrossRef]
18. Brammer, L. Halogen Bonding, Chalcogen Bonding, Pnictogen Bonding, Tetrel Bonding: Origins, Current Status and Discussion. *Faraday Discuss.* **2017**, *203*, 485–507. [CrossRef]
19. Alkorta, I.; Elguero, J.; Frontera, A. Not Only Hydrogen Bonds: Other Noncovalent Interactions. *Crystals* **2020**, *10*, 180. [CrossRef]
20. Zierkiewicz, W.; Michalczyk, M.; Scheiner, S. Noncovalent Bonds through Sigma and Pi-Hole Located on the Same Molecule. Guiding Principles and Comparisons. *Molecules* **2021**, *26*, 1740. [CrossRef]
21. Wang, W.; Ji, B.; Zhang, Y. Chalcogen Bond: A Sister Noncovalent Bond to Halogen Bond. *J. Phys. Chem. A* **2009**, *113*, 8132–8135. [CrossRef]
22. Aakeroy, C.B.; Bryce, D.L.; Desiraju, G.R.; Frontera, A.; Legon, A.C.; Nicotra, F.; Rissanen, K.; Scheiner, S.; Terraneo, G.; Metrangolo, P.; et al. Definition of the Chalcogen Bond. *Pure Appl. Chem.* **2019**, *91*, 1889–1892. [CrossRef]
23. Kolb, S.; Oliver, G.A.; Werz, D.B. Chemistry Evolves, Terms Evolve, but Phenomena Do Not Evolve: From Chalcogen–Chalcogen Interactions to Chalcogen Bonding. *Angew. Chem. Int. Ed.* **2020**, *59*, 22306–22310. [CrossRef]
24. Bondi, A. van der Waals Volumes and Radii. *J. Phys. Chem. A* **1964**, *68*, 441–451. [CrossRef]
25. Lo, R.; Manna, D.; Miriyala, V.M.; Nachtigallová, D.; Hobza, P. Trends in the Stability of Covalent Dative Bonds with Variable Solvent Polarity Depend on the Charge Transfer in the Lewis Electron-Pair System. *Phys. Chem. Chem. Phys.* **2023**, *25*, 25961–25964. [CrossRef]
26. Koch, U.; Popelier, P.L.A. Characterization of C–H–O Hydrogen Bonds on the Basis of the Charge Density. *J. Phys. Chem.* **1995**, *99*, 9747–9754. [CrossRef]
27. Popelier, P.L.A. Characterization of a Dihydrogen Bond on the Basis of the Electron Density. *J. Phys. Chem. A* **1998**, *102*, 1873–1878. [CrossRef]
28. Wang, W.; Wong, N.-B.; Zheng, W.; Tian, A. Theoretical Study on the Blueshifting Halogen Bond. *J. Phys. Chem. A* **2004**, *108*, 1799–1805. [CrossRef]
29. Zeng, Y.; Zhang, X.; Li, X.; Zheng, S.; Meng, L. Ab Initio and AIM Studies on Typical π-Type and Pseudo-π-Type Halogen Bonds: Comparison with Hydrogen Bonds. *Int. J. Quantum Chem.* **2011**, *111*, 3725–3740. [CrossRef]
30. Grabowski, S.J. QTAIM Characteristics of Halogen Bond and Related Interactions. *J. Phys. Chem. A* **2012**, *116*, 1838–1845. [CrossRef]

31. Bader, R.F.W. *Atoms in Molecules—A Quantum Theory*; Oxford University Press: Oxford, UK, 1990.
32. Adamo, C.; Barone, V. Toward Reliable Density Functional Methods without Adjustable Parameters: The PBE0 Model. *J. Chem. Phys.* **1999**, *110*, 6158–6169. [CrossRef]
33. Grimme, S.; Antony, J.; Ehrlich, S.; Krieg, H. A Consistent and Accurate Ab Initio Parametrization of Density Functional Dispersion Correction (DFT-D) for the 94 Elements H-Pu. *J. Chem. Phys.* **2010**, *132*, 154104. [CrossRef]
34. Weigend, F.; Ahlrichs, R. Balanced Basis Sets of Split Valence, Triple Zeta Valence and Quadruple Zeta Valence Quality for H to Rn: Design and Assessment of Accuracy. *Phys. Chem. Chem. Phys.* **2005**, *7*, 3297–3305. [CrossRef]
35. Grimme, S.; Ehrlich, S.; Goerigk, L. Effect of the Damping Function in Dispersion Corrected Density Functional Theory. *J. Comput. Chem.* **2011**, *32*, 1456–1465. [CrossRef] [PubMed]
36. Boys, S.F.; Bernardi, F. The Calculation of Small Molecular Interactions by the Difference of Separate Total Energies. Some Procedures with Reduced Errors. *Mol. Phys.* **1970**, *19*, 553–566. [CrossRef]
37. Anderson, L.N.; Aquino, F.W.; Raeber, A.E.; Chen, X.; Wong, B.M. Halogen Bonding Interactions: Revised Benchmarks and a New Assessment of Exchange vs Dispersion. *J. Chem. Theory Comput.* **2018**, *14*, 180–190. [CrossRef]
38. de Azevedo Santos, L.; Ramalho, T.C.; Hamlin, T.A.; Bickelhaupt, F.M. Chalcogen Bonds: Hierarchical Ab Initio Benchmark and Density Functional Theory Performance Study. *J. Comput. Chem.* **2021**, *42*, 688–698. [CrossRef]
39. Wang, W.; Sun, T.; Zhang, Y.; Wang, Y.B. The Benzene$\cdots$Naphthalene Complex: A more Challenging System than the Benzene Dimer for newly Developed Computational Methods. *J. Chem. Phys.* **2015**, *143*, 114312. [CrossRef] [PubMed]
40. Wang, W.; Zhang, Y.; Wang, Y.B. Highly Accurate Benchmark Calculations of the Interaction Energies in the Complexes $C_6H_6\cdots C_6X_6$ (X = F, Cl, Br, and I). *Int. J. Quantum Chem.* **2017**, *117*, e25345. [CrossRef]
41. Hedberg, K.; Hedberg, L.; Bethune, D.S.; Brown, C.A.; Dorn, H.C.; Johnson, R.D.; De Vries, M. Bond Lengths in Free Molecules of Buckminsterfullerene, C_{60}, from Gas-Phase Electron Diffraction. *Science* **1991**, *254*, 410–412. [CrossRef]
42. Biegler-König, F.; Schönbohm, J.; Bayles, D. AIM2000-A Program to Analyze and Visualize Atoms in Molecules. *J. Comput. Chem.* **2001**, *22*, 545–559.
43. Lu, T.; Chen, F. Multiwfn: A Multifunctional Wavefunction Analyzer. *J. Comput. Chem.* **2012**, *33*, 580–592. [CrossRef] [PubMed]
44. Lu, T.; Chen, F. Quantitative Analysis of Molecular Surface Based on Improved Marching Tetrahedra Algorithm. *J. Mol. Graph. Model.* **2012**, *38*, 314–323. [CrossRef] [PubMed]
45. Frisch, M.J.; Trucks, G.W.; Schlegel, H.B.; Scuseria, G.E.; Robb, M.A.; Cheeseman, J.R.; Scalmani, G.; Barone, V.; Mennucci, B.; Petersson, G.A.; et al. *Gaussian 09, Revision D.01*; Gaussian, Inc.: Wallingford, CT, USA, 2013.

molecules

MDPI

Article

Characteristics of Intermolecular Interactions between Encapsulated Molecules and the Lantern-Like Carcerand Superphanes

Mirosław Jabłoński

Faculty of Chemistry, Nicolaus Copernicus University in Toruń, Gagarina 7, 87-100 Torun, Poland; teojab@chem.umk.pl; Tel.: +48-056-611-4695

Abstract: The main topic of the article is to provide the characteristics of individual intermolecular interactions present between three lantern-like superphanes and the H_2O, NH_3, HF, HCN, and MeOH molecules trapped inside them. Despite the large cavity, the freedom of the trapped molecules is significantly limited by the presence of numerous interaction sites on the side chains of the superphane molecule. It is shown that the molecule trapped inside the superphane is stabilized mainly by only one or, less often, two strong hydrogen bonds involving the imino nitrogen atom, but QTAIM calculations also suggest the presence of many other intermolecular interactions, mainly hydrogen bonds involving imino or central hydrogen atoms from the side chains of the superphane molecule. Moreover, it is also shown that the structural simplification of the side chains does not significantly affect both the size of the superphane molecule and the obtained encapsulation energies, which is important in modeling this type of carceplexes. Noticeably, the parent superphane considered here was previously synthesized by the group of Qing He, so the results obtained will help in understanding this type and similar systems.

Keywords: superphane; cyclophane; carcerand; carceplex; endohedral complex; encapsulation; intermolecular interaction; hydrogen bond; inclusion complex; guest–host interaction

1. Introduction

Undoubtedly, superphanes constitute the most unusual subgroup of cyclophanes [1–3]. This uniqueness is primarily related to the aesthetically beautiful symmetry of their structure, most often comparable with a six-bladed pinwheel, a lantern, or a barrel. Namely, superphanes consist of two parallel or almost parallel benzene rings (or other aromatic rings [2]) joined together by as many as six bridges [4]. In the simplest representative of superphanes, $[2_6]$superphane (i.e., $[2_6](1,2,3,4,5,6)$cyclophane), these are ethylene bridges [5–10], while the somewhat larger $[3_6]$superphane (i.e., $[3_6](1,2,3,4,5,6)$cyclophane) has six trimethylene bridges [11–19]. The structures of both of these simple superphanes are shown in Figure 1.

Although the trapping abilities of these superphanes are severely limited or even impossible due to too small size of the internal cavity, one may expect that these abilities increase significantly in larger superphanes with longer side chains [19]. Experimental verification of this possibility may, however, be greatly hampered due to the challenging complexity of the organic synthesis of superphanes. On the other hand, trapping can be greatly facilitated by using multiple binding sites or even charged groups on the side chains of the carcerand superphane molecule. Indeed, very recently, the group of Qing He for the first time synthesized a large-sized lantern-like superphane **1** (Figure 2) and showed that it was able to trap a water dimer [20]. This dimer was stabilized by numerous hydrogen bonds involving disubstituted benzene rings (position 2) and imino protons.

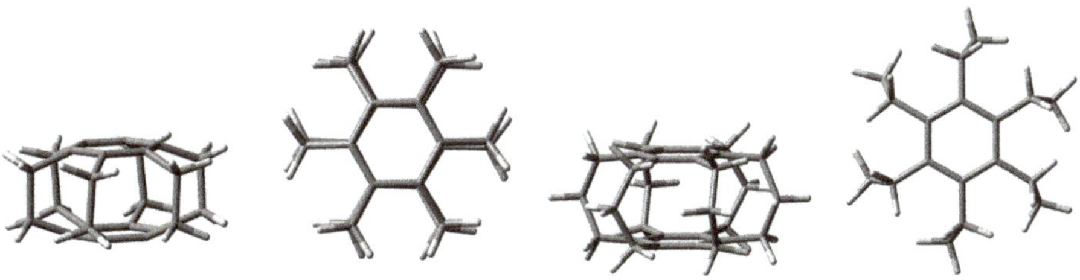

Figure 1. Side and top views of the $[2_6]$ (**two from the left**) and $[3_6]$ (**two from the right**) superphanes.

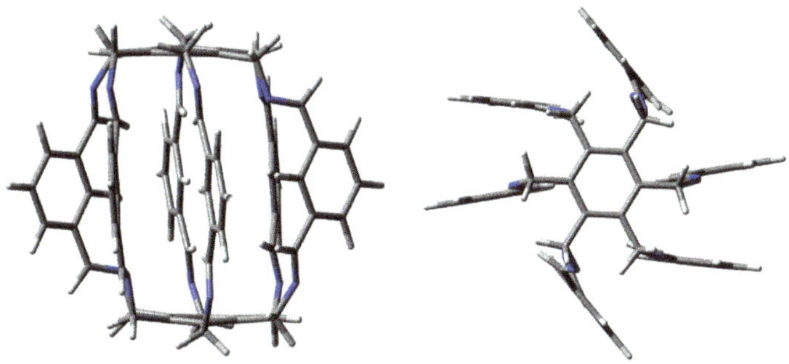

Figure 2. Side and top views of the lantern-like superphane (**1**) synthesized by He et al. [20].

Shortly thereafter, the same group in a series of papers showed the possibility of trapping small molecules and anions by various similar superphanes [21–24]. The trapped species included ($2Cl^-{\cdot}H_2O$) and MeOH [21], ReO_4^-, DMSO, and ($H_2O{\cdot}MeOH$) [22], $H_2PO_4^-$ and AsO_4^{3-} [23], or I_2 and I_3^- [24]. The trapped species interact with many binding sites of a given superphane through hydrogen bonds. As has often been emphasized [20–24], the obtained superphanes are characterized by extremely high selectivity towards trapped species over many other types of competing ions, and the obtained carceplexes show high thermal stability in a wide range of pH. Therefore, the obtained superphanes may be extremely important, e.g., in the removal of toxic ions (such as AsO_4^{3-}) from wastewater. Similar superphanes were also obtained by the group of Jovica D. Badjić [25,26]. Similarly, this group explored the possibility of selectively trapping a wide variety of ions [26], showing that the barrel-shaped hexapodal superphane they obtained (see Figure 3) easily binds tetrahedral oxyanions, such as SO_4^{2-} or HPO_4^- [25]. High selectivity and different ion trapping times result from the specific slotted structure of the side surface of the superphane. It is also noted that the accommodation of the trapped ion is facilitated by the adaptation of the appropriate conformation of the trapping superphane host molecule.

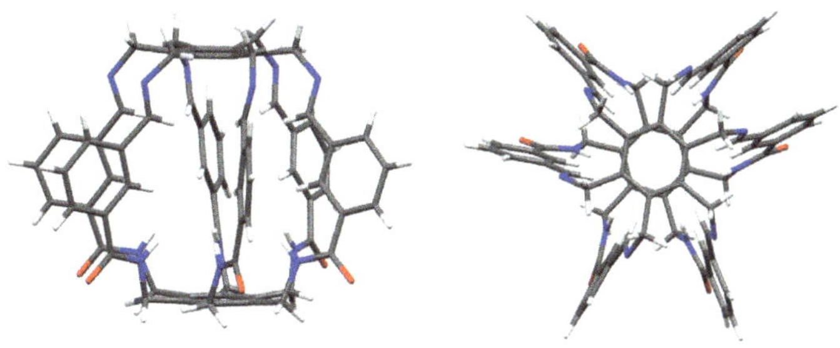

Figure 3. Side and top views of the superphane synthesized by Badjić et al. [25].

It is also worth mentioning the recent articles by Oh et al. [27] and by Zhao et al. [28]. The obtained carcerands contain only three rather than six side chains, but they also show high recognition towards tetrahedral oxyanions, such as $H_2PO_4^-$ and SO_4^{2-}. Thus, these examples are evidence that it is not necessarily the number of side chains closing the internal cavity that is crucial in trapping anions, but rather the number of binding sites. In turn, importantly, these binding sites act actively by forming numerous hydrogen bonds to the trapped species. Thus, these are clear examples of the key role of hydrogen bonds in the recognition and binding of various chemical species.

The research results show that the trapped species are stabilized inside the superphane cage thanks to many hydrogen bonds [20–26]. The article aims to characterize the hydrogen bonds formed between five fundamental molecules (H_2O, NH_3, HF, HCN, and MeOH) trapped in the cavity of superphane **1** and the donor and acceptor centers of this superphane molecule. In addition to the geometric parameters of the found hydrogen bonds, an important issue will be the total guest$\cdots$host interaction energy, and especially the estimated interaction energies of the most important individual hydrogen bonds and other intermolecular interactions found. In contrast to the unavailability of this parameter experimentally, the energy of an individual interaction can be relatively easily estimated based on theoretical studies. Another important goal is to investigate the influence of the type of side chain on the structure of the obtained carceplexes, especially on the pattern of the hydrogen bonds formed and their energy. For this reason, the side chains in the parent superphane **1** were gradually reduced to those present in superphanes **2** and **3**. The **1** $\rightarrow$ **2** step involves replacing the benzene ring in the side chains with the -CH=CH-CH_2- fragment having a double bond, while the **2** $\rightarrow$ **3** replacement involves saturating this bond, i.e., introducing the -CH_2-CH_2-CH_2- fragment (see Figure 4). As a result of these changes, the obtained superphanes **2** and **3** have a significantly simplified structure with the lack of as many as six side benzene rings (Figure 5), but the binding sites are retained (Figure 4).

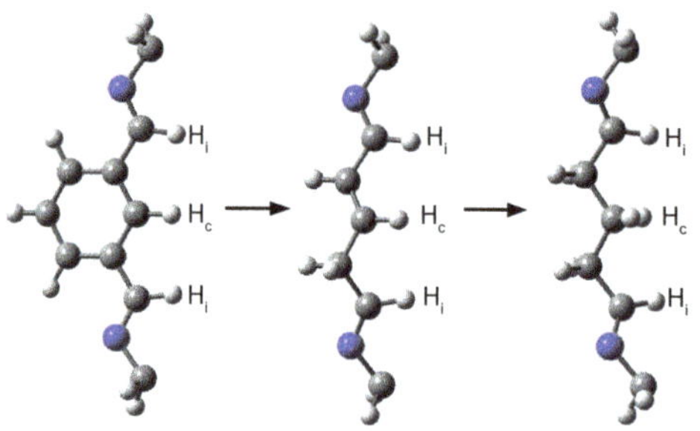

1 2 3

Figure 4. Side chains in superphanes **1**, **2**, and **3**. Hydrogen atoms participating in hydrogen bonds with the guest molecule are labeled as follows: H_i—imino H atom, H_c—central H atom.

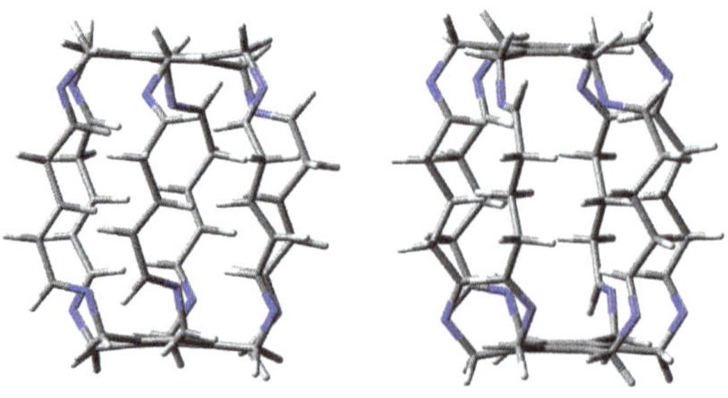

2 3

Figure 5. Superphanes **2** and **3**.

2. Results and Discussion

2.1. Characteristics of Intermolecular Interactions in Carceplexes Guest@1

As seen from Figures 2 and 5, the considered superphanes **1**, **2**, and **3** are large enough that the guest molecules (H_2O, NH_3, HF, HCN, or MeOH) trapped inside them have quite a lot of freedom. However, this freedom should be somewhat suppressed by the presence of distinct binding sites, such as the imino nitrogen and the proton from either the central or the imino C-H bond (Figure 4). The specific slotted structure of superphanes and their limited internal spaces result in relatively small distances between the atoms of the superphane and the guest molecule. Consequently, the molecular graphs obtained for the considered carceplexes quite often feature numerous bond paths between the superphane atoms and the molecules trapped in them, thus suggesting the occurrence of many intermolecular interactions. The multitude and diversity of intermolecular interactions are visible in the representative molecular graphs of **14a** and **15a**, i.e., the most stable forms of the carceplexes HCN@**1** and MeOH@**1**, respectively (Figure 6). The former of these systems contains 11 bond paths between the trapped guest molecule (i.e., HCN) and the superphane, while the latter one features as many as 17 such bond paths.

The general characteristics (type, values of the electron density at the bond critical point (ρ_{bcp}), interaction energies according to Equation (2) (BE), and their sum ($\sum$BE)) of the various individual interactions between the guest molecule and the superphane atoms indicated by the presence of corresponding bond paths are given in Table 1. Additionally, the encapsulation energy (E_b) (Equation (1)) is also shown.

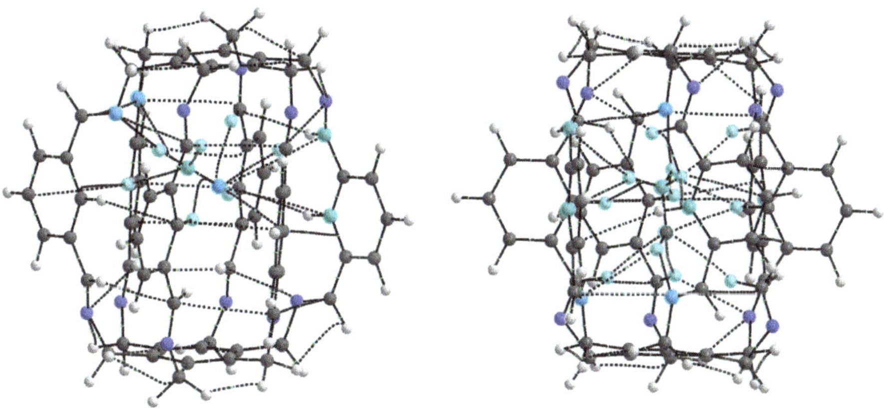

Figure 6. Molecular graphs (ring critical points and cage critical points have been removed for better clarity) of carceplexes **14a** (i.e., the form **a** of HCN@**1**) and **15a** (i.e., the form **a** of MeOH@**1**). Atoms of the guest molecule and all host atoms connected to them through bond paths are marked in light blue.

Table 1. Characteristics of individual intermolecular interactions in carceplexes guest@**1**. The dominant interactions and their energies (BE) are marked in boldface.

Carceplex	SP	Guest	E_b [a]	$\sum$BE [b]	Intermolecular Interaction			
					n [c]	Type [d]	ρ_{bcp} [e]	BE [f]
11a	1	H_2O	−18.8	−14.2	2	**N···H-O**	0.024 and 0.030	**−4.7** and **−5.9**
					1	C-H_i···O	0.006	−0.7
					3	C-H_c···O	0.005–0.010	−0.4 to −1.4
11b	1	H_2O	−16.8	−11.6	2	**N···H-O**	0.006 and 0.022	−0.6 and **−4.1**
					2	C-H_i···O	0.010 and 0.013	−1.5 and −2.2
					2	C-H_c···H-O	0.008 and 0.010	−1.0 and −1.6
					1	C(π)···O	0.006	−0.6
					1	C-H_i···H-O	0.004	0.0

Table 1. *Cont.*

Carceplex	SP	Guest	E_b [a]	$\sum$BE [b]	Intermolecular Interaction			
					n [c]	Type [d]	ρ_{bcp} [e]	BE [f]
12	1	NH_3	−12.9	−9.9	1	C-H$_i$···N	0.012	−1.8
					5	**C-H$_c$···N**	0.002–0.019	0.3 to **−3.4**
					2	C-H$_i$···H-N	0.004 and 0.008	−0.2 and −1.1
					2	N···H-N	0.006 and 0.010	−0.6 and −1.4
13a	1	HF	−19.7	−16.4	1	**N···H-F**	0.052	**−10.8**
					2	C-H$_i$···F	0.006 and 0.010	−0.5 and −1.5
					3	C-H$_c$···F	0.004–0.010	−0.2 to −1.6
					1	C(π)···F	0.003	0.2
					1	N···F	0.006	−0.7
13b	1	HF	−18.5	−14.7	1	**N···H-F**	0.046	**−9.5**
					2	C-H$_i$···F	0.004 and 0.012	−0.2 and −2.0
					3	C-H$_c$···F	0.003–0.011	0.1 to −1.7
					1	N···F	0.007	−0.8
14a	1	HCN	−16.2	−13.9	1	**N···H-C**	0.027	**−5.3**
					1	N···(H-C)	0.009	−1.2
					1	C-H$_c$···C	0.005	−0.4
					1	C-H$_i$···C	0.010	−1.5
					3	C-H$_c$···N	0.004–0.015	−0.2 to −2.6
					2	C-H$_i$···N	0.006 and 0.008	−0.5 and −1.2
					1	C(π)···N	0.003	0.1
					1	C$_i$···N	0.002	0.2
14b	1	HCN	−16.2	−14.6	2	**N···H-C**	0.008 and 0.029	−1.1 and **−5.8**
					1	C-H$_c$···C	0.007	−0.8
					1	C-H$_i$···C	0.009	−1.2
					1	C(π)···C	0.004	0.0
					3	C-H$_c$···N	0.005–0.014	−0.4 to −2.3
					2	C-H$_i$···N	0.006–0.008	−0.6 and −0.9
					1	C(π)···N	0.002	0.3
15a	1	MeOH	−19.2	−13.0	2	**N···H-O**	0.004 and 0.012	−0.2 and **−1.9**
					1	**C-H$_c$···O**	0.011	**−1.7**
					3	C-H$_i$···O	0.003–0.008	0.2 to −1.1
					3	C(π)···H-C	0.001–0.009	0.5 to −1.2
					2	**C-H$_c$···H-C**	0.002 and 0.011	0.3 and **−1.8**
					3	C-H$_i$···H-C	0.007–0.009	−0.8 to −1.2
					1	**C-H$_c$···(H-C)**	0.010	**−1.6**
					1	C$_i$···H-C	0.005	−0.5
					1	N···H-C	0.003	0.1
15b	1	MeOH	−18.3	−13.3	2	N···H-O	0.006 and 0.010	−0.6 and −1.5
					3	**C-H$_c$···O**	0.010–0.013	−1.4 to **−2.1**
					1	C(π)···O	0.004	−0.1
					2	C-H$_i$···H-C	0.010 and 0.010	−1.5 and −1.5
					2	C$_i$···H-C	0.007 and 0.008	−0.8 and −1.0
					2	C(π)···H-C	0.005 and 0.006	−0.3 and −0.7

[a] Encapsulation energy (in kcal/mol) given by Equation (1). [b] Total binding energy (in kcal/mol) obtained by summing over individual interactions, i.e., $\sum BE = \sum_i BE_i$. [c] Number of intermolecular interactions of a given type. [d] Type of a given interaction in the format atom/bond(superphane)···atom/bond(guest). The subscripts i and c refer to the imino group and the central position, respectively (see Figure 4), while C(π) denotes the carbon atom from the central benzene ring. [e] The value of the electron density (in au) at the bond critical point (bcp) of a given interaction. [f] Binding energy (in kcal/mol) according to Equation (2) for a single interaction.

2.1.1. Intermolecular Interactions in H_2O@1

Geometry optimizations of several starting H_2O@**1** structures led to two forms, **11a** and **11b** (Figure 7), with clearly different characteristics of the intermolecular interactions between the trapped water molecule and the superphane **1** (Table 1). The binding energy of the more stable form **11a** (−18.8 kcal/mol) is dominated by two strong (−5.9 and −4.7 kcal/mol) hydrogen bonds between the O-H bonds of the water molecule and the nitrogen atoms of the imino groups of the superphane **1**. The N· · · H distances are 1.969 and 2.068 Å, respectively, and the N-H-O angles are 164° and 148°. So both of these hydrogen bonds are far from linear (Figure 7). According to QTAIM (i.e., Quantum Theory of Atoms in Molecules [29–31]), the water molecule in **11a** is additionally weakly (BE is from −0.4 to −1.4 kcal/mol) stabilized by as many as four hydrogen bonds between the oxygen atom of water and the C-H bonds of **1**, with one hydrogen atom coming from the imino group, while as many as three from the benzene rings in the side chains. The total binding energy of all the interactions demonstrated by the presence of bond paths is −14.2 kcal/mol.

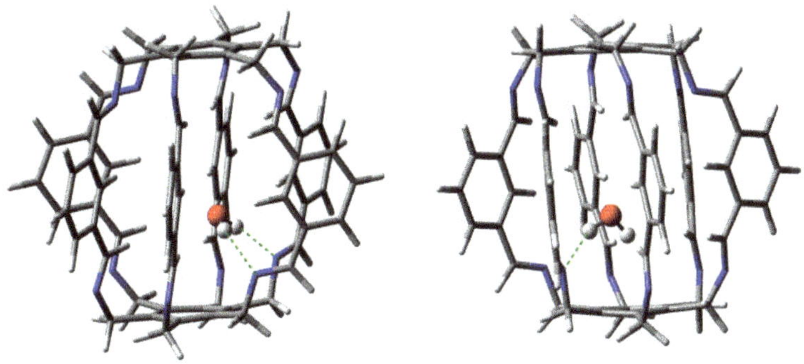

Figure 7. Structures of carceplexes **11a** (**left**) and **11b** (**right**). The dominant N· · · H-O hydrogen bonds are marked with a green dashed line.

The slightly less stable (E_b = −16.8 kcal/mol) form **11b** is dominated by only one N· · · H-O hydrogen bond with an energy of −4.1 kcal/mol, while the second N· · · H-O hydrogen bond is, according to QTAIM, very weak (only −0.6 kcal/mol). This is also visible in the values of the lengths and angles of both hydrogen bonds, which are 2.108 and 2.724 Å and 154° and 120°, respectively. Additionally, according to QTAIM, the water molecule is also stabilized by two C-H_i · · · O hydrogen bonds and two C-H_c · · · H-O interactions, for which BE is from −1.0 to −2.2 kcal/mol (Table 1). Comparison of the type of bond paths in carceplexes **11a** and **11b** shows that changing the position of the water molecule causes the C-H_c · · · O bond paths to 'switch' to C-H_c · · · H-O. Moreover, the molecular graph for **11b** also shows energetically irrelevant C(π)· · · O and C-H_i · · · H-O bond paths. The total binding energy of all the intermolecular interactions traced by bond paths is −11.6 kcal/mol.

Taking into account the results obtained for both forms of carceplex H_2O@**1**, it can be concluded that the trapped water molecule is stabilized mainly by two (in **11a**) or one (in **11b**) strong N· · · H-O hydrogen bonds, but these bonds are accompanied by numerous interactions that involve imino or central C-H bonds.

2.1.2. Intermolecular Interactions in NH_3@**1**

Carceplex NH_3@**1** is the only one for which only one form was obtained (Figure 8). The interaction between NH_3 and **1** is weaker (−12.9 kcal/mol) than the previously discussed H_2O· · · **1** in H_2O@**1**. The NH_3 · · · **1** interaction is mainly dominated by C-H_c · · · N hydrogen bonds (BE of the strongest of them is −3.4 kcal/mol) and, to a much lesser extent, C-H_i · · · N (−1.8 kcal/mol). According to QTAIM, the trapped ammonia molecule interacts with the superphane **1** also through its N-H bonds, forming two N· · · H-N hydrogen bonds (BE are −0.6 and −1.4 kcal/mol) and two weak C-H_i · · · H-N contacts (−0.2 and −1.1 kcal/mol). Despite the relative weakness of the interactions (perhaps except one C-H_c · · · N hydrogen bond), their large number (10) makes the $\sum$BE value relatively large at −9.9 kcal/mol.

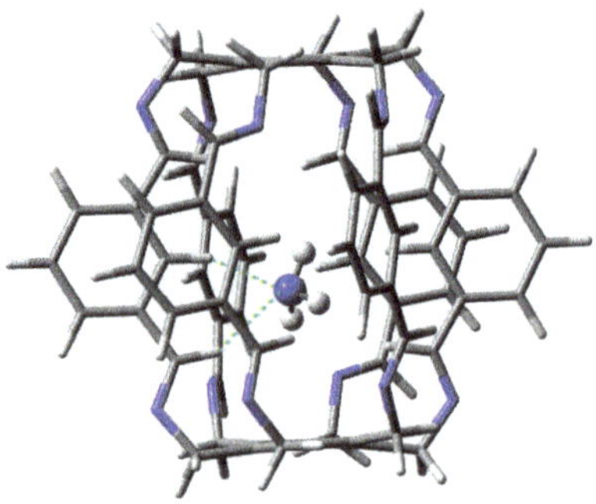

Figure 8. Structure of the **12** (i.e., NH_3@**1**) carceplex. The green dashed lines indicate the two strongest C-H$\cdots$N hydrogen bonds, with BEs of −3.4 and −1.8 kcal/mol.

Both C-H$\cdots$N hydrogen bonds marked with the green dashed line in Figure 8 are quite long, 2.272 and 2.489 Å. These bonds are also clearly far from linear, the C-H-N angle being 158° and 152°, respectively.

2.1.3. Intermolecular Interactions in HF@**1**

Carceplexes HF@**1** deserve special attention. Although the total binding energy (−19.7 kcal/mol for form **13a** and −18.5 kcal/mol for form **13b**) is only slightly higher than for the H_2O@**1** carceplexes, the HF$\cdots$**1** binding is mainly due to the formation of the very strong (−10.8 kcal/mol in **13a** and −9.5 kcal/mol in **13b**) hydrogen bond N$\cdots$H-F. This is also visible in the very short N$\cdots$H distance, which is 1.701 and 1.747 Å in **13a** and **13b**, respectively. This is a clear confirmation of the fact that the HF molecule is an excellent proton donor. This has also been recently demonstrated with examples of hydrogen bonds to the carbene electron lone pair [32]. The N$\cdots$H-F hydrogen bonds in **13a** and **13b** are closer to being linear, the N-H-F angle amounts to 175° and 166°, respectively. The structures of both obtained forms of carceplex HF@**1** (i.e., **13a** and **13b**) along with N$\cdots$H-F hydrogen bonds are shown in Figure 9.

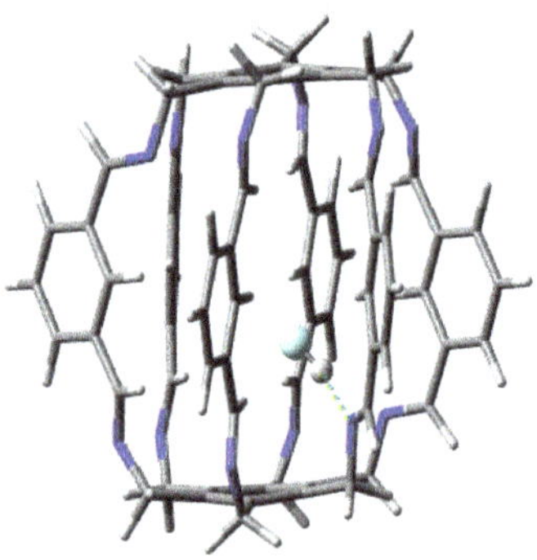

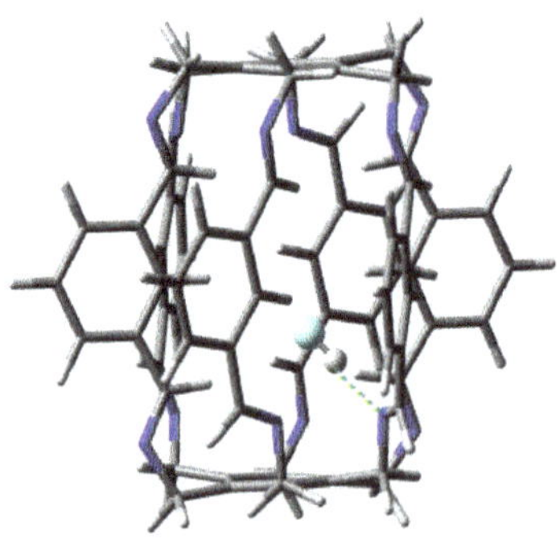

Figure 9. Structures of carceplexes **13a** (**left**) and **13b** (**right**). The dominant N$\cdots$H-F hydrogen bonds are marked with a green dashed line.

The HF molecule contains a strongly electronegative fluorine atom with electron lone pairs. It is therefore not surprising that the obtained molecular graphs suggest the presence of many hydrogen bonds to this atom (see Table 1). Namely, in the case of both forms, the molecular graphs show two C-$H_i\cdots$F bond paths and three C-$H_c\cdots$F bond paths. Although, of course, these hydrogen bonds are much weaker (up to −2.0 kcal/mol in **13b**) than N$\cdots$H-F, their multitude suggests a not necessarily negligible contribution to the total binding of the trapped HF molecule. The ∑BE values are −16.4 and −14.7 kcal/mol in **13a** and **13b**, respectively.

2.1.4. Intermolecular Interactions in HCN@1

The HCN molecule can be both a proton donor and a proton acceptor, so it was interesting to see its main sites of interaction after trapping. Geometry optimizations led to two forms of the HCN@1 carceplex with equal binding energies (−16.2 kcal/mol) and very similar arrangements of the HCN molecule (Figure 10).

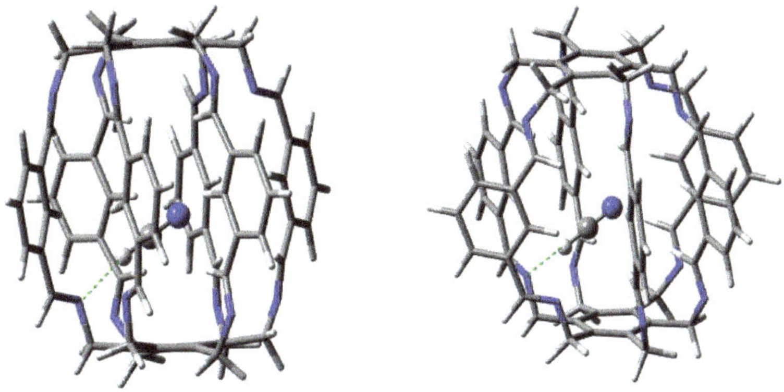

Figure 10. Structures of carceplexes **14a** (**left**) and **14b** (**right**). The dominant N···H-C hydrogen bonds are marked with a green dashed line.

QTAIM calculations indicate a clear dominance of one of the two N···H-C hydrogen bonds (in **14a**, one of the bond paths leads to the critical point of the H-C bond of the HCN molecule). Its energy is −5.3 and −5.8 kcal/mol in **14a** and **14b**, respectively. Its length is 2.007 and 1.984 Å, respectively, and the N-H-C angle is 166° and 155°, respectively, so the slightly shorter and stronger bond in **14b** is more bent. The second N···H-C hydrogen bond is much weaker, with an energy of only ca. −1 kcal/mol. However, this leading N···H-C type hydrogen bond is accompanied by many different interactions with the C atom, and especially N (Table 1), which results, of course, from the presence of an electron lone pair on this atom. This situation is well illustrated by the molecular graph of carceplex **14a** shown earlier in Figure 6. Among these interactions, the greatest stabilization comes from one C-H_c···N hydrogen bond (−2.6 and −2.3 kcal/mol in **14a** and **14b**, respectively), again showing that the central C-H bond (Figure 4) is a good proton donor. The ∑BE values are ca. −14 kcal/mol and are only slightly lower than the E_b value (−16.2 kcal/mol).

2.1.5. Intermolecular Interactions in MeOH@1

The MeOH molecule is the largest of all guest molecules considered here. Additionally, it contains an extensive methyl group, so the presence of many different bond paths between this molecule and the host superphane **1** should be expected. Indeed, this is confirmed by the results presented in Table 1 and the molecular graph shown for **15a** in Figure 6. This form features as many as four different interactions stronger than −1.5 kcal/mol. These are: N···H-O (−1.9 kcal/mol), C-H_c···H-C (−1.8 kcal/mol), C-H_c···O (−1.7 kcal/mol), and C-H_c···(H-C) (−1.6 kcal/mol). However, the molecular graph (Figure 6) also shows the presence of three bond paths for the C-H_i···O interaction, three for C(π)···H-C, three for C-H_i···H-C, and one for C_i···H-C and N···H-C. Although these interactions are much weaker (and some are even slightly destabilizing according to Emamian's formula 2), the energy of some of them is ca. −1 kcal/mol. The binding energy of all interactions traced by the presence of bond paths is −13.0 kcal/mol, much lower (by ca. 6 kcal/mol) than the encapsulation energy (−19.2 kcal/mol), showing that other long-range interactions are also important.

The molecular graph of the second found form of MeOH@**1** carceplex, i.e., **15b**, is considerably poorer in the variety of bond path types (Table 1). Somewhat surprisingly, according to QTAIM, the strongest are two C-$H_c \cdots$O hydrogen bonds with energies of −2.1 and −2.0 kcal/mol (the third C-$H_c \cdots$O is weaker with BE of −1.4 kcal/mol), while the standard N$\cdots$H-O hydrogen bonds have energies of only −1.5 and −0.6 kcal/mol. Interactions with the methyl group in MeOH are also relatively important, especially with the participation of (two) protons of the imino groups. The binding energies of both these C-$H_i \cdots$H-C contacts are −1.5 kcal/mol. Although the total number of interactions indicated by the presence of bond paths is much smaller (12) than in **15a** (17), the $\sum$BE value is similar (−13.3 kcal/mol).

A large number of diverse (weak) interactions results in the fact that the encapsulation energies of these carceplexes (−19.2 and −18.3 kcal/mol for **15a** and **15b**, respectively) belong to the highest among all obtained and are similar to those for HF@**1**. It is quite significant that the standard N$\cdots$H-O hydrogen bond, i.e., to the highly polarized O-H bond, is, according to QTAIM, not the dominant interaction (especially in **15b**). Instead, QTAIM calculations suggest a large energetic contribution from interactions involving the oxygen atom and the methyl group of the MeOH molecule. It is worth adding that the distances between the hydroxyl hydrogen atom of methanol and the nearest nitrogen atoms in the superphane are quite large; in the case of carceplex **15a**, the distances are 2.400 and as much as 2.844 Å, while in the case of **15b**, 2.468 and as much as 2.702 Å. Moreover, the N$\cdots$H-O hydrogen bonds are significantly bent. It can therefore be assumed that the large nonlinearity of the N$\cdots$H-O hydrogen bonds together with their considerable length is the reason for their weakness. The structures of the obtained MeOH@**1** carceplex forms (i.e., **15a** and **15b**) are shown in Figure 11.

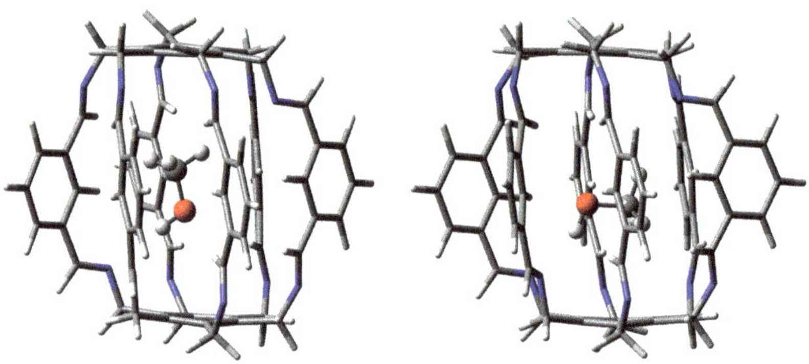

Figure 11. Structures of carceplexes **15a** (**left**) and **15b** (**right**).

2.2. *Influence of the Type of Side Chain*

Another aim of the research on the carceplexes considered here was to investigate the influence of the type of side chain on the height of the superphane molecule and the strength of the guest$\cdots$superphane interaction. These issues will be discussed in the next two subsections.

2.2.1. Influence of the Type of Side Chain on the Height of the Cage

Table 2 shows the values of the distance between the centers of opposite benzene rings, which can be taken as a parameter describing the height of the cage created by the superphane molecule.

It is worth noting first that the structural reduction of side chains in the superphane molecule (see Figure 4) leads to a slight decrease in the cage's height, most significantly in the case of **1** → **2** (from 9.15 to 8.96 Å). Encapsulation does not cause significant changes in the height of the superphane molecule cage. This result confirms the relatively large cavity space and stiffness of the side chains.

Table 2. The distance (in Å) between the centers of the two confining benzene rings of a superphane.

Superphane		H_2O			NH_3		HF		HCN		MeOH	
		a	b	c	a	b	a	b	a	b	a	b
1	9.15	9.18	9.16	n/a	9.11	n/a	9.21	9.18	9.10	9.10	9.08	9.05
2	8.96	8.97	9.01	n/a	8.91	8.93	9.03	n/a	8.94	n/a	8.81	n/a
3	9.05	9.09	9.07	9.08	9.02	9.04	9.12	9.10	9.02	n/a	8.99	n/a

2.2.2. Influence of the Type of Side Chain on the Encapsulation Energy

The obtained encapsulation energy values of all considered carceplexes are given in Table 3.

Table 3. Encapsulation energies (in kcal/mol) for various forms of the guest@superphane carceplexes.

Superphane	H_2O			NH_3		HF		HCN		MeOH	
	a	b	c	a	b	a	b	a	b	a	b
1	−18.8	−16.8	n/a	−12.9	n/a	−19.7	−18.5	−16.2	−16.2	−19.2	−18.3
2	−19.6	−16.5	n/a	−13.1	−13.0	−23.1	n/a	−18.4	n/a	−17.5	n/a
3	−16.6	−15.6	−14.2	−12.9	−12.5	−18.3	−16.4	−15.8	n/a	−18.0	n/a

The obtained results show that the structural reduction of the side chains of superphane **1** does not significantly affect the encapsulation energies. This result is important as it shows that the guest@**1** carceplexes can be successfully modeled by the less computationally demanding guest@**2** and guest@**3** carceplexes. Importantly, except for MeOH, the **1** → **2** exchange leads to an increase in the encapsulation energies (i.e., they become more negative) of the most stable forms, while the **1** → **3** exchange leads to a decrease in these energies (i.e., they become less negative). Thus, the guest molecule is most stable when trapped by superphane **2** and least stable in a carceplex formed by superphane **3**. It would be interesting to see whether this relationship would hold for the species considered by Qing He's [20–24] and Jovica D. Badjić's [25,26] groups, but so far superphanes **2** and **3** have not been synthesized yet.

The greatest increase in stabilization occurred for the HF molecule (from −19.7 kcal/mol in **13a** to −23.1 kcal/mol in **23**, i.e., by 3.4 kcal/mol), so it is particularly worth taking a closer look at this case. The structure and molecular graph of the most stable carceplex **23** are shown in Figure 12.

Similarly to carceplexes **13a** and **13b**, the stabilization of the system is dominated by the N···H-F hydrogen bond, but in the case of carceplex 23, this bond is much stronger. According to Equation (2), its energy is −12.1 kcal/mol, while −10.8 and −9.5 kcal/mol in **13a** and **13b**, respectively. The exceptional strength of this bond is also reflected in its extremely short length, which is only 1.657 Å. QTAIM calculations suggest, however, that such a high stabilization of carceplex **23** is also due to the significant participation of three relatively strong C-H···F hydrogen bonds, two involving the central H atoms (−2.7 and −2.0 kcal/mol) and one involving hydrogen from the imino group (−1.6 kcal/mol). The remaining interactions visible in Figure 12 are not significant (−0.8 kcal/mol for the third hydrogen bond C-H_c···F, −0.2 kcal/mol for the second hydrogen bond C-H_i···F, and −0.6 kcal/mol for the contact N···F). The total binding energy for all HF···**2** interactions visible in Figure 12 is −20.1 kcal/mol, which is greater than −16.4 and −14.7 kcal/mol for **13a** and **13b**, respectively (Table 1).

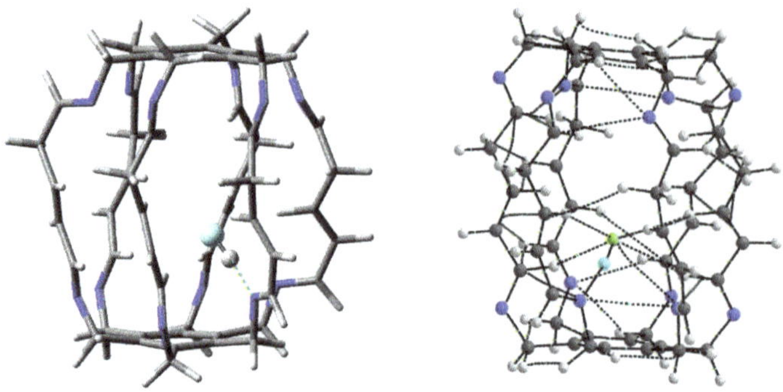

Figure 12. Structure (**left**) and molecular graph (**right**) of the most stable HF@**2** (i.e., **23**) carceplex. In the subfigure on the left, the most important N$\cdots$H-F hydrogen bond is marked with a green dashed line. In turn, in the molecular graph on the right, ring critical points and cage critical points have been removed for greater clarity. Additionally, the hydrogen atom of the HF molecule is highlighted in light blue.

On the contrary, the strongest weakening via the **1** $\rightarrow$ **3** structural modification has occurred for water. Namely, the encapsulation energy decreased from -18.8 kcal/mol in **11a** to -16.6 kcal/mol in **31a**, i.e., by 2.2 kcal/mol. QTAIM calculations suggest that this weakening of carceplex stabilization is mainly due to a slight weakening of both N$\cdots$H-O hydrogen bonds, from -5.9 and -4.7 kcal/mol in **11a** to -5.1 and -4.3 kcal/mol in **31a**, however, this weakening is partially compensated by slightly stronger C-H$\cdots$O bonds (their total binding energy is -5.3 kcal/mol in **31a** while only -3.6 kcal/mol in **11a**). The weakening of the N$\cdots$H-O bonds is also suggested by their lengthening, from 1.969 and 2.068 Å in **11a** to 2.033 and 2.094 Å in carceplex **31a**. Moreover, both of these bonds are strongly non-linear (158° and 154°, respectively).

3. Methodology

Geometry optimizations of the superphanes **1** (see Figure 2), **2** and **3** (see Figure 5) and their carceplexes with H_2O, NH_3, HF, HCN, or MeOH were performed at the ωB97X-D/6-31G(d) level of theory, i.e., within the ωB97X-D exchange-correlation functional [33] of Density Functional Theory [34,35] and the 6-31G(d) basis set [36,37]. As shown [38], within the group of the 200 tested exchange-correlation functionals, ωB97X-D proved to be one of the best for general purposes, including the reliable description of intermolecular interactions. Moreover, as shown earlier [9], this functional well reproduced the crystallographic structure of [2_6]superphane. Frequency analysis were performed to confirm true minima on the potential energy surface. In each case, all frequencies were positive, confirming that stable forms were always obtained. It should be noted that the multiplicity of binding sites in the side chains of the tested superphanes may lead to many minima on the potential energy surface, i.e., to various forms of the obtained carceplexes. Therefore, geometry optimizations have started with several different initial geometries, which in some cases have led to different final carceplex geometries, although in some cases only one such complex has been obtained. For convenience, the obtained guest@superphane carceplexes have been marked as **ijk**, where **i** denotes the superphane molecule (i.e., **1**, **2**, or **3**; see Figures 2 and 5), **j** numbers the guest molecule, **1** for H_2O, **2** for NH_3, **3** for HF, **4** for HCN, and **5** for MeOH, while **k**, being **a**, **b**, or possibly **c**, refers to the corresponding forms of carceplex **ij** in order from the most stable to the least stable.

Despite the relatively small size of the basis set used for geometry optimization (due to the large size of the considered superphanes, many degrees of freedom, and therefore high computational cost), the total energies were then calculated (single point calculations) using larger 6-31++G(d) basis set, containing diffuse functions on all atoms. Geometry optimizations, frequency analysis, and single point calculations were performed using the

Gaussian 16 package [39]. Graphical representations of the systems were obtained with the GaussView 6 program [40].

The encapsulation energy was calculated as the difference between the total energy of the complex (carceplex) and the sum of the total energies of the superphane and the guest molecule trapped in it:

$$E_b = E(\text{carceplex}) - E(\text{superphane}) - E(\text{guest}) \quad (1)$$

One of the goals of this work is to assess the strength of individual hydrogen bonds and other intermolecular interactions involved in the overall superphane$\cdots$guest binding. Determining the energies of individual hydrogen bonds is not easy using experimental methods, but it is relatively simple using tools of theoretical chemistry. For this purpose, Espinosa's formula $E = \frac{1}{2}V_{bcp}$ [41] is most often used, based on the value of the potential energy density determined at the so-called bond critical point (bcp) [29–31] of a given bond (interaction). However, due to numerous objections to this formula (see [42] and especially point 2.5.1 in the review article [43]), it was not used here. Firstly, due to the negativity of V_{bcp}, the obtained value of E is always negative, thus describing each interaction as stabilizing. Secondly, it should be recalled that Espinosa's formula was actually derived only for X-H$\cdots$O hydrogen bonds (X = C, N, O) and its application to other types of interactions is not necessarily correct. For this reason, improved versions of this formula can be found in the literature [42,44,45]. It can also be objected that the values of fitting bond energies were obtained using a mixture of various theoretical methods. On the other hand, Emamian et al. [46] showed that among various wave function–based descriptors of hydrogen bonds, the electron density determined at the bond critical point correlates best with the binding energy. Therefore, their Equation (2) was used in this work:

$$\text{BE} = -223.08 \cdot \rho_{bcp} + 0.7423 \quad (2)$$

where ρ_{bcp} is the value (in au) of the electron density at bcp of a given hydrogen bond. As Emamian et al. emphasized in the abstract of their article [46], this formula can be used to estimate the binding energy of individual hydrogen bonds, e.g., in biomolecules. It should be noted, however, that, as Equation (2) shows, the value of the electron density at the bond critical point itself could in principle be used as a measure of the strength of a given interaction. Molecular graphs and electron density values at the bond critical points of individual interactions were obtained using the AIMAll program [47].

4. Conclusions

The main topic of the article was to provide the characteristics of intermolecular interactions present between trapped H_2O, NH_3, HF, HCN, or MeOH molecules and the lantern-like superphane (**1**) acting as a host, previously synthesized by the group of Qing He [20]. It has been shown that despite the large internal cavity of the superphane molecule, the freedom of the trapped molecules is significantly limited by the presence of numerous interaction sites. In addition to imino nitrogen atoms, which most often form only one or, rarely, two strong hydrogen bonds to the trapped guest molecule, the presence of numerous weaker hydrogen bonds involving central or imino C-H bonds deserves special attention. QTAIM calculations suggest that other interactions indicated by the presence of bond paths are less important.

It has also been shown that a slight structural simplification of the side chains of the encapsulating superphane **1** (consequently leading to superphanes **2** and then **3**) has no significant effect on the cage height and has only a minor effect on the encapsulation energy of a given guest molecule. Therefore, it is concluded that the carceplexes obtained by the Qing He group as well as other similar systems can be successfully modeled by using these slightly simplified superphanes **2** and **3**. The use of these superphanes also leads to a lower computational cost while, as shown, maintaining the main conclusions regarding the characteristics of the interactions in the guest@**1** system.

Funding: This research received no external funding.

Data Availability Statement: Data available from the author on reasonable request.

Conflicts of Interest: The author declares no conflict of interest.

References

1. Gleiter, R.; Hopf, H. (Eds.) *Modern Cyclophane Chemistry*; Wiley-VCH: Weinheim, Germany, 2004.
2. Ghasemabadi, P.G.; Yao, T.; Bodwell, G.J. Cyclophanes containing large polycyclic aromatic hydrocarbons. *Chem. Soc. Rev.* **2015**, *44*, 6494–6518. [CrossRef] [PubMed]
3. Zhang, X.-X.; Li, J.; Niu, Y.-Y. A Review of Crystalline Multibridged Cyclophane Cages: Synthesis, Their Conformational Behavior, and Properties. *Molecules* **2022**, *27*, 7083. [CrossRef]
4. Gleiter, R.; Kratz, D. "Super" Phanes. *Acc. Chem. Res.* **1993**, *26*, 311–318. [CrossRef]
5. Sekine, Y.; Brown, M.; Boekelheide, V. [2.2.2.2.2.2](1,2,3,4,5,6)Cyclophane: Superphane. *J. Am. Chem. Soc.* **1979**, *101*, *11*, 3126–3127. [CrossRef]
6. Iwamura, H.; Katoh, M.; Kihara, H. How Strained Is the "Flat" Benzene Ring in Superphane ? *Tetrahedron Lett.* **1980**, *21*, 1757–1760. [CrossRef]
7. Sekine, Y.; Boekelheide, V. A Study of the Synthesis and Properties of $[2_6]$(1,2,3,4,5,6)Cyclophane (Superphane). *J. Am. Chem. Soc.* **1981**, *103*, 1777–1785. [CrossRef]
8. Dodziuk, H.; Vetokhina, V.; Hopf, H.; Luboradzki, R.; Gaweł, P.; Waluk, J. Electronic states of cyclophanes with small bridges. *J. Chem. Soc.* **2012**, *136*, 074201. [CrossRef]
9. Jabłoński, M. Does the Presence of a Bond Path Really Mean Interatomic Stabilization? The Case of the Ng@Superphane (Ng = He, Ne, Ar, and Kr) Endohedral Complexes. *Symmetry* **2021**, *13*, 2241. [CrossRef]
10. Jabłoński, M. Determining Repulsion in Cyclophane Cages. *Molecules* **2022**, *27*, 3969. [CrossRef] [PubMed]
11. Sakamoto, Y.; Miyoshi, N.; Shinmyozu, T. Synthesis of a "Molecular Pinwheel": [3.3.3.3.3.3](1,2,3,4,5,6)Cyclophane. *Angew. Chem. Int. Ed. Engl.* **1996**, *35*, 549–550. [CrossRef]
12. Sakamoto, Y.; Miyoshi, N.; Hirakida, M.; Kusumoto, S.; Kawase, H.; Rudzinski, J.M.; Shinmyozu, T. Syntheses, Structures, and Transannular π–π Interactions of Multibridged $[3_n]$Cyclophanes. *J. Am. Chem. Soc.* **1996**, *118*, 12267–12275. [CrossRef]
13. Hori, K.; Sentou, W.; Shinmyozu, T. *Ab Initio* Molecular Orbital Study on Inversion Mechanism of Trimethylene Bridges of $[3_3]$(1,3,5)- and $[3_6]$(1,2,3,4,5,6) Cyclophanes. *Tetrahedron Lett.* **1997**, *38*, 8955–8958. [CrossRef]
14. Bettinger, H.F.; Schleyer, P.v.R.; Schaefer III, H.F. $[3_6]$(1,2,3,4,5,6)Cyclophane–A Molecular Pinwheel and Its Correlated Inversion: NMR and Energetic Considerations. *J. Am. Chem. Soc.* **1998**, *120*, 1074–1075. [CrossRef]
15. Yasutake, M.; Sakamoto, Y.; Onaka, S.; Sako, K.; Tatemitsu, H.; Shinmyozu, T. Crystal structural properties of a pinwheel compound: $[3_6]$(1,2,3,4,5,6)cyclophane. *Tetrahedron Lett.* **2000**, *41*, 7933–7938. [CrossRef]
16. Yasutake, M.; Koga, T.; Sakamoto, Y.; Komatsu, S.; Zhou, M.; Sako, K.; Tatemitsu, H.; Onaka, S.; Aso, Y.; Inoue, S.; et al. An Alternative Synthetic Route of $[3_5]$(1,2,3,4,5)Cyclophane, and Structural Properties of Multibridged $[3_n]$Cyclophanes and Their Charge-Transfer Complexes in the Solid State. *J. Am. Chem. Soc.* **2002**, *124*, 10136–10145. [CrossRef]
17. Nogita, R.; Matohara, K.; Yamaji, M.; Oda, T.; Sakamoto, Y.; Kumagai, T.; Lim, C.; Yasutake, M.; Shimo, T.; Jefford, C.W.; et al. Photochemical Study of $[3_3]$(1,3,5)Cyclophane and Emission Spectral Properties of $[3_n]$Cyclophanes (n = 2–6). *J. Am. Chem. Soc.* **2004**, *126*, 13732–13741. [CrossRef] [PubMed]
18. Fujitsuka, M.; Tojo, S.; Shinmyozu, T.; Majima, T. Intramolecular dimer radical anions of $[3_n]$ cyclophanes: Transannular distance dependent stabilization energy. *Chem. Commun.* **2009**, 1553–1555. [CrossRef] [PubMed]
19. Jabłoński, M. Bader's Topological Bond Path Does Not Necessarily Indicate Stabilizing Interaction–Proof Studies Based on the Ng@$[3_n]$cyclophane Endohedral Complexes. *Molecules* **2023**, *28*, 6353. [CrossRef] [PubMed]
20. Li, A.; Xiong, S.; Zhou, W.; Zhai, H.; Liu, Y.; He, Q. Superphane: A new lantern-like receptor for encapsulation of a water dimer. *Chem. Commun.* **2021**, *57*, 4496–4499. [CrossRef] [PubMed]
21. Li, A.; Liu, Y.; Zhou, W.; Jiang, Y.; He, Q. Superphanes: Facile and efficient preparation, functionalization and unique properties. *Tetrahedron Chem.* **2022**, *1*, 100006. [CrossRef]
22. Zhou, W.; Li, A.; Gale, P.A.; He, Q. A highly selective superphane for ReO_4^- recognition and extraction. *Cell Rep. Phys. Sci.* **2022**, *3*, 100875. [CrossRef]
23. Zhou, W.; Wang, F.; Li, A.; Bai, S.; Feng, X.; He, Q. A Superphane-based carcerand for arsenic detoxification via imprisoning arsenate. *Cell Rep. Phys. Sci.* **2023**, *4*, 101295. [CrossRef]
24. Zhou, W.; Li, A.; Zhou, M.; Xu, Y.; Zhang, Y.; He, Q. Nonporous amorphous superadsorbents for highly effective and selective adsorption of iodine in water. *Nat. Commun.* **2023**, *14*, 5388. [CrossRef]
25. Xie, H.; Finnegan, T.J.; Gunawardana, V.W.L.; Pavlović, R.Z.; Moore, C.E.; Badjić, J.D. A Hexapodal Capsule for the Recognition of Anions. *J. Am. Chem. Soc.* **2021**, *143*, 3874–3880. [CrossRef]
26. Xie, H.; Gunawardana, V.W.L.; Finnegan, T.J.; Xie, W.; Badjić, J.D. Picking on Carbonate: Kinetic Selectivity in the Encapsulation of Anions. *Angew. Chem. Int. Ed.* **2022**, *61*, e202116518. [CrossRef]
27. Oh, J.H.; Kim, J.H.; Kim, D.S.; Han, H.J.; Lynch, V.M.; Sessler, J.L.; Kim, S.K. Synthesis and Anion Recognition Features of a Molecular Cage Containing Both Hydrogen Bond Donors and Acceptors. *Org. Lett.* **2019**, *21*, 4336–4339. [CrossRef] [PubMed]

28. Zhao, Z.; Xiong, S.; Zhang, J.; P, J.; Ding, W.; Chen, X.; He, Q.; Zhang, Z. A hexapyrrolic molecular cage and the anion-binding studies in chloroform. *J. Mol. Struct.* **2023**, *1293*, 136232. [CrossRef]
29. Bader, R.F.W. *Atoms in Molecules: A Quantum Theory*; Oxford University Press: New York, NY, USA, 1990.
30. Popelier, P.L.A. *Atoms in Molecules. An Introduction*; Longman: Singapore, 2000.
31. Matta, C.F.; Boyd, R.J. *The Quantum Theory of Atoms in Molecules*; Wiley-VCH: Weinheim, Germany, 2007.
32. Jabłoński, M. On the Coexistence of the Carbene···H-D Hydrogen Bond and Other Accompanying Interactions in Forty Dimers of N-Heterocyclic-Carbenes (I, IMe_2, I^iPr_2, I^tBu_2, $IMes_2$, $IDipp_2$, IAd_2; I = imidazol-2-ylidene) and Some Fundamental Proton Donors (HF, HCN, H_2O, MeOH, NH_3). *Molecules* **2022**, *27*, 5712.
33. Chai, J.-D.; Head-Gordon, M. Long-range corrected hybrid density functionals with damped atom–atom dispersion corrections. *Phys. Chem. Chem. Phys.* **2008**, *10*, 6615–6620. [CrossRef] [PubMed]
34. Hohenberg, P.; Kohn, W. Inhomogeneous Electron Gas. *Phys. Rev.* **1964**, *136*, B864–B871. [CrossRef]
35. Parr, R.G.; Yang, W. *Density-Functional Theory of Atoms and Molecules*; Oxford University Press: New York, NY, USA, 1989.
36. Jensen, F. *Introduction to Computational Chemistry*; John Wiley & Sons Ltd.: Chichester, UK, 2007.
37. Pritchard, B.P.; Altarawy, D.; Didier, B.; Gibson, T.D.; Windus, T.L. New Basis Set Exchange: An Open, Up-to-Date Resource for the Molecular Sciences Community. *J. Chem. Inf. Model.* **2019**, *59*, 4814–4820. [CrossRef] [PubMed]
38. Mardirossian, N.; Head-Gordon, M. Thirty years of density functional theory in computational chemistry: An overview and extensive assessment of 200 density functionals. *Mol. Phys.* **2017**, *115*, 2315–2372. [CrossRef]
39. Frisch, M.J.; Trucks, G.W.; Schlegel, H.B.; Scuseria, G.E.; Robb, M.A.; Cheeseman, J.R.; Scalmani, G.; Barone, V.; Petersson, G.A.; Nakatsuji, H.; et al. *Gaussian 16, Revision C.01*; Gaussian, Inc.: Wallingford, CT, USA, 2019.
40. Dennington, R.; Keith, T.; Millam, J. (Eds.) *GaussView, Version 6*; Semichem Inc.: Shawnee Mission, KS, USA, 2019.
41. Espinosa, E.; Molins, E.; Lecomte, C. Hydrogen bond strength revealed by topological analyses of experimentally observed electron densities. *Chem. Phys. Lett.* **1998**, *285*, 170–173. [CrossRef]
42. Jabłoński, M.; Monaco, G. Different Zeroes of Interaction Energies As the Cause of Opposite Results on the Stabilizing Nature of C–H···O Intramolecular Interactions. *J. Chem. Inf. Model.* **2013**, *53*, 1661–1675. [CrossRef]
43. Jabłoński, M. A Critical Overview of Current Theoretical Methods of Estimating the Energy of Intramolecular Interactions. *Molecules* **2020**, *25*, 5512. [CrossRef]
44. Gatti, C.; May, E.; Destro, R.; Cargnoni, F. Fundamental Properties and Nature of CH··O Interactions in Crystals on the Basis of Experimental and Theoretical Charge Densities. The Case of 3,4-Bis(dimethylamino)-3-cyclobutene-1,2-dione (DMACB) Crystal. *J. Phys. Chem. A* **2002**, *106*, 2707–2720. [CrossRef]
45. Afonin, A.V.; Vashchenko, A.V.; Sigalov, M.V. Estimating the energy of intramolecular hydrogen bonds from ^1H NMR and QTAIM calculations. *Org. Biomol. Chem.* **2016**, *14*, 11199–11211. [CrossRef]
46. Emamian, S.; Lu, T.; Kruse, H.; Emamian, H. Exploring Nature and Predicting Strength of Hydrogen Bonds: A Correlation Analysis Between Atoms-in-Molecules Descriptors, Binding Energies, and Energy Components of Symmetry-Adapted Perturbation Theory. *J. Comput. Chem.* **2019**, *40*, 2868–2881. [CrossRef]
47. Keith, T.A. *AIMAll (Version 15.05.18)*; TK Gristmill Software: Overland Park, KS, USA, 2015.

molecules

MDPI

Article

Internal Vibrations of Pyridinium Cation in One-Dimensional Halide Perovskites and the Corresponding Halide Salts

Anna Yu. Samsonova [1,*], Alena Yu. Mikheleva [1], Kirill M. Bulanin [1], Nikita I. Selivanov [1], Anton S. Mazur [2], Peter M. Tolstoy [3], Constantinos C. Stoumpos [1,4,*] and Yury V. Kapitonov [1]

[1] Photonics of Crystals Laboratory, Saint Petersburg State University, Ulyanovskaya d.1, 198504 St. Petersburg, Russia
[2] Magnetic Resonance Research Center, Saint Petersburg State University, Universitetskiy pr. 26, 198504 St. Petersburg, Russia; a.mazur@spbu.ru
[3] Institute of Chemistry, Saint Petersburg State University, Universitetskiy pr. 26, 198504 St. Petersburg, Russia
[4] Department of Materials Science and Technology, University of Crete, Voutes, GR-70013 Heraklion, Greece
* Correspondence: a.y.samsonova@spbu.ru (A.Y.S.); cstoumpos@materials.uoc.gr (C.C.S.)

Abstract: We investigate vibrations of the pyridinium cation $PyH^+ = C_5H_5NH^+$ in one-dimensional lead halide perovskites $PyPbX_3$ and pyridinium halide salts PyHX ($X^- = I^-$, Br^-), combining infrared absorption and Raman scattering methods at room temperature. Internal vibrations of the cation were assigned based on density functional theory modeling. Some of the vibrational bands are sensitive to perovskite or the salt environment in the solid state, while halide substitution has only a minor effect on them. These findings have been confirmed by ^1H, ^{13}C and ^{207}Pb solid-state nuclear magnetic resonance (NMR) experiments. Narrower vibrational bands in perovskites indicate less disorder in these materials. The splitting of NH-group vibrational bands in perovskites can be rationalized the presence of nonequivalent crystal sites for cations or by more exotic phenomena such as quantum tunneling transition between two molecular orientations. We have shown how organic cations in hybrid organic–inorganic crystals could be used as spectators of the crystalline environment that affects their internal vibrations.

Keywords: halide perovskites; Raman scattering; IR absorption; solid-state NMR; molecular vibrations

1. Introduction

The last decade has witnessed an increase in the interest in hybrid organic–inorganic lead-based halide perovskites due to their optoelectronic and photovoltaic applications [1–4]. This family of materials is valuable due to the wide variety of crystal structures, the most studied and used of which are three-dimensional (3D) lead halide perovskites with the chemical formula $APbX_3$ ($X^- = I^-$, Br^-, Cl^-). The 3D crystal framework is formed by the corner-shared lead-halide octahedra and small organic cations A^+ between them ($A^+ = CH_3NH_3^+ = MA^+$, $HC(NH_2)^+ = FA^+$). Substitution of the organic cation A^+ by a larger cation leads to the formation of low-dimensional perovskite-like structures, consisting of two-dimensional (2D) sheets [5,6] or one-dimensional (1D) chains [7,8] of lead-halide octahedra and the organic cations filling the space between them. Further, for brevity, such structures are designated as low-dimensional perovskites.

The movement of organic cations in hybrid lead halide perovskites and their low-dimensional analogues is the key to understanding the features of the crystal structure and its phase transitions. In case of 3D hybrid lead halide perovskites, order–disorder changes in the methylammonium (MA^+) and formamidinium (FA^+) cations were shown to be responsible for the distortion of the lead halide inorganic framework leading to the dynamic disorder at high temperatures [9–13]. MA^+ and FA^+ cations behave similarly in the high-temperature and intermediate-temperature phases, rotating isotropically and reorienting between their preferred orientations, respectively [9,14–19]. In the low-temperature phase,

MA^+ cations still have an ordered three-fold rotational axis [17,20], while the FA^+ cations are locally disordered, exhibiting a glass-like state in the arrangement of the cations [14]. In 2D perovskites, large organic cations have greater freedom of motion compared to 3D perovskites and tend to interact with each other by weak intermolecular forces and with the inorganic framework through strong electrostatic interactions [5,6,21–23]. In addition, the motion of organic cations in these materials is responsible for order–disorder structural phase transitions with their symmetry breaking due to the concerted alignment of organic cations across a specific dipole moment vector direction [24–26]. For 1D hybrid halide perovskites, a major group of which are perovskites with face-shared octahedra assembled into the inorganic chains, organic cations are found to demonstrate behavior similar to that in their 2D counterparts [27–30].

To study the organic subsystem of hybrid halide perovskites, neutron [9] and X-ray diffraction (XRD) [9,10,24–32], solid-state nuclear magnetic resonance (NMR), quasielastic neutron scattering (QENS) [14–17], Raman [10–13,21–24,33] scattering methods and density functional theory (DFT) modeling [22,31,34] are widely used. The vibrational properties of inorganic perovskite frameworks have been studied in the literature [5,10,21,23]. Insufficient attention has been paid to the use of such relatively experimentally convenient methods as infrared (IR) and Raman spectroscopy for the study vibrations of the organic subsystem in perovskites. Organic cations are spectators of the crystalline environment [22,35]. The polarized Raman spectroscopy of organic vibrations could be used to determine the crystal orientation [33] and the cation movement [20].

Of particular interest is the movement of the simple aromatic pyridinium cation $PyH^+ = C_5H_5NH^+$, which stabilizes both perovskite-like compounds and PyHX ($X^- = I^-, Br^-$) salts [36–40]. The hybrid halide perovskites $PyPbX_3$ ($X^- = I^-, Br^-$) have the 1D structure of face-shared octahedra chains [31,32,41,42].

In this work, we studied the vibrational properties of the pyridinium cation in hybrid halide perovskites $PyPbX_3$ ($X^- = I^-, Br^-$). IR and Raman spectra of these perovskites are compared with ones for PyHX ($X^- = I^-, Br^-$) salts. Based on theoretical modeling, we identified the observed vibrations. Frequencies were identified that were most sensitive to changing the crystalline environment of the PyH^+ cation from salt to perovskite, while replacing halogen usually had less effect. The latter statement is also confirmed by the solid-state NMR examination of organic cations. We also address the possible origins of the splitting of several bands involving cation NH-vibrations in perovskites.

2. Results and Discussion

2.1. Synthesis and Crystal Structure

Pyridinium lead trihalide $PyPbX_3$ ($X^- = I^-, Br^-$) single crystals (see photos of the samples on Figure S1) were grown by the slow counterdiffusion of ions from individual solutions of lead(II) halides and pyridine in hydrohalic acid in the silica gel filled U-tube. Pyridine hydrohalides PyHX ($X^- = I^-, Br^-$) were synthesized by adding concentrated hydrogen halogenides drop-wise with constant stirring to the solutions of ethanol and pyridine. More details on the synthesis can be found in the Methods section (Section 3).

The typical crystal structure of PyHX salts ($X^- = I^-, Br^-$) is shown in Figure 1a [43]. Upon heating from cryogenic to room temperature, these materials undergo an order–disorder phase transition from the monoclinic to the rhombohedral phase, in which the previously frozen PyH^+ cations begin to rotate around the pseudo-six-fold axis ("C_6") (Figure 1c) [37–39]. The order–disorder phase transitions and molecular motions of the PyH^+ cation in various compounds have been studied [36–40,44–47]. Pyridinium internal vibrational modes are shown to be sensitive to such phase transitions [48–51].

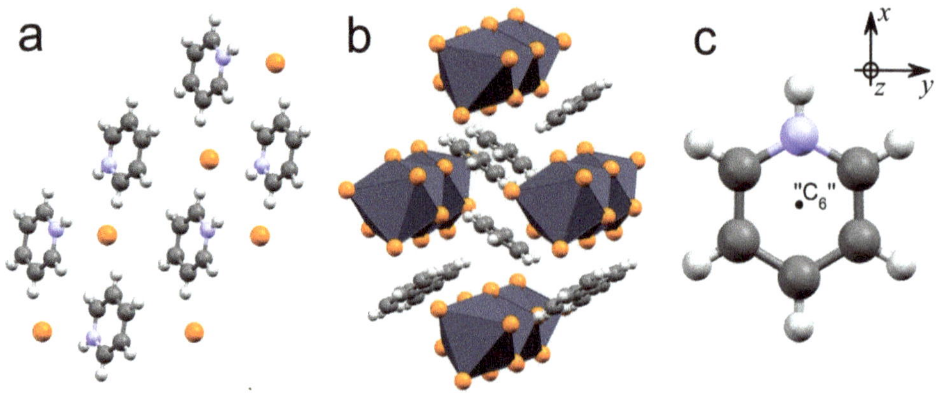

Figure 1. Typical crystal structure of PyHX [43,52] (**a**) and $PyPbX_3$ [31,32,41,42] (X^- = I^-, Br^-) (**b**). (**c**) PyH^+ cation. Gray—carbon atoms; blue—nitrogen atoms; white—hydrogen atoms; orange—halide atoms; dark grey—PbX_6 octahedra.

The crystal structure of $PyPbX_3$ (X^- = I^-, Br^-) perovskites consists of 1D chains of face-shared lead halide octahedra and tightly packed PyH^+ cations between them (Figure 1b) [31,32,41,42]. At room temperature, the compounds crystallize in the orthorhombic space group *Pnma* with each unit cell containing two chains running down the crystallographic *b*-axis and isolated from one another by individual PyH^+ cations [32]. In the case of the simple salts, pyridinium cations can be found in a totally disordered state (PyHI, 293 K) [43] or a totally ordered state (PyHBr, 100 K) [52] depending on whether the intermolecular forces are strong enough to dominate over the thermal motion.

In both cases, the pyridinium cations remain essentially isolated from their surroundings. Thus, to analyze their internal vibrations, we will consider the vibrations of free cations as a first approximation and then estimate the influence of the crystalline environment on them. Another possible approach is to calculate the vibrations of cations in clusters, which are fragments of the crystal lattice [34].

2.2. Symmetry Considerations

To identify vibrations of pyridinium cations, we first consider the benzene (C_6H_6) molecule. We would like to note that the assignment of vibrational modes in the spectra of these molecules is a subject of discussion in the literature. To denote the vibrational modes of the PyH^+ cation, we follow the Wilson notation [53], which was introduced to denote vibrational modes of the benzene C_6H_6 molecule and has been commonly used for the vibrational modes assignment of C_6H_6 derivatives [49,51,54].

The benzene C_6H_6 molecule belongs to the D_{6h} point group and contains a main C_6 axis which contains S_6 and S_3 axes. Three C_2 and three $C_{2'}$ axes are perpendicular to the C_6 axis and are passing through the middle of the bonds between carbon atoms and carbon atoms themselves, respectively. There are one σ_h, three σ_v and three σ_d planes, an inversion center, as well as twelve irreducible representations (see the character table in Table S1).

The PyH^+ cation (Figure 1c) can be considered as a C_6H_6 molecule in which one of the carbon atoms is replaced by a nitrogen atom. It leads to the loss of benzene symmetry operations except for the reflection in the one vertical plane σ_v, rotation around the $C_{2'}$ axis lying in the σ_v plane and reflection in the horizontal plane σ_h (Figure S2). These symmetry operations form the C_{2v} point group (see the character table in Table S2) and should be renamed as $C_{2'} \rightarrow C_2$, $\sigma_h \rightarrow \sigma_d$ and $\sigma_v \rightarrow \sigma_v$. The twelve irreducible representations of the D_{6h} point group are reduced to the four irreducible representations of the C_{2v} point group (Table S3), which are A_1, A_2, B_1, B_2. Here, we align the C_2 axis with the *x*-axis in the Cartesian coordinate system. Indexes 1 and 2 denote the symmetry and asymmetry with respect to the reflection in the σ_v plane. Indexes A and B denote the symmetry and asymmetry with respect to the rotation around the C_2 axis.

The C_6H_6 molecule and PyH^+ cation both have $N = 12$ atoms and $3N = 36$ degrees of freedom. In the case of an isolated cation or molecule, only the change in the position of the atoms relative to each other matters. Therefore, 36 degrees of freedom can be reduced by 6, of which 3 are translational movements and 3 are rotational movements of the molecule or cation as a whole. This results in $3N - 6 = 30$ internal vibrational modes of an isolated C_6H_6 molecule or PyH^+ cation. These modes are generated by one of the irreducible representations of the corresponding point symmetry group. The C_6H_6 molecule has doubly degenerate normal modes. For the PyH^+ cation, this degeneracy is removed and resulting modes are denoted by indexes *a* or *b*. This notation was introduced by Wilson [53] and is widely used [54,55].

2.3. DFT Modeling

IR and Raman wavenumbers and intensities of internal vibrational modes of the single C_6H_6 molecule and PyH^+ cation were calculated using DFT modeling (see Table 1). We use C_6H_6 modes notations recently refined by Gardner et al. [55] based on Wilson's original work [53]. Correlating the vibrations of pyridinium with the established Wilson nomenclature for benzene requires care. We have assigned Wilson notation to pyridinium vibrational modes based on the correspondence between irreducible representations of the C_6H_6 molecule and PyH^+ cation (Table S3), calculated IR and Raman intensities for both molecules (Table 1) and the motions of atoms (Figure S3). Closely lying vibrations with the same symmetries could interact and mix upon transition from benzene to pyridinium. We denote such modes with a dash. In the strict sense, these modes could not be considered as pure Wilson modes, but they are combinations of them. It also could be seen from the comparison of atomic motions in Figure S3 and in Wilson modes [53,55].

Table 1. Calculated internal vibrational modes of C_6H_6 molecule and PyH^+ cation, and experimental bands for pyridinium salts and perovskites.

Benzene C_6H_6 (Calculated)					Pyridinium PyH^+ (Calculated)					Experiment, Freq., cm^{-1} (IR,R)			
	Sym.	Freq., cm^{-1}	$I_{th}IR$, arb.u.	$I_{th}R$, arb.u.		Sym.	Freq., cm^{-1}	$I_{th}IR$, arb.u.	$I_{th}R$, arb.u.	PyHI	PyHBr	$PyPbI_3$	$PyPbBr_3$
ν_{16}	E_{2u}	410.06	0.0	0.0	ν_{16b}	B_1	389.19	0.851	0.269	390 (-,w)	399 (-,w)	380 (-,s)	379 (-,w)
					ν_{16a}	A_2	401.53	0.0	0.001	(-,0)	(-,0)	(-,0)	(-,0)
ν_6	E_{2g}	622.21	0.0	4.718	ν_{6a}	A_1	622.88	0.014	2.883	607 (-,s)	607 (-,s)	607 (-,s)	608 (-,s)
					ν_{6b}	B_2	646.15	0.398	5.125	633 (-,s)	635 (-,s)	635 (-,s)	634 (-,s)
ν_{11}	A_{2u}	686.61	725.940	0.0	$\nu_{11'}$	B_1	674.76	92.175	0.019	681 (-,w)	685 (-,w)	657 (-,w)	662 (-,w)
ν_4	B_{2g}	718.77	0.0	0.001	$\nu_{4'}$	B_1	743.62	88.368 *	0.022	753 (-,w)	757 (-,w)	726 (-,w)	727 (-,w)
ν_{10}	E_{1g}	862.47	0.0	0.977	$\nu_{10b'}$	B_1	855.74	6.213 *	0.119	887 (-,s)	891 (-,w)	(-,0)	835 (-,w)
					ν_{10a}	A_2	882.32	0.0	0.005	(-,0)	(-,0)	(-,0)	(-,0)
ν_{17}	E_{2u}	987.00	0.0	0.0	$\nu_{17b'}$	B_1	989.76	1.795	0.049	993 (s,0)	993 (s,0)	968 (s,0)	971 (s,0)
					ν_{17a}	A_2	1008.38	0.0	0.031	(0,0)	(0,0)	(0,0)	(0,0)

Table 1. *Cont.*

Benzene C_6H_6 (Calculated)					Pyridinium PyH^+ (Calculated)					Experiment, Freq., cm^{-1} (IR,R)			
	Sym.	Freq., cm^{-1}	$I_{th}IR$, arb.u.	$I_{th}R$, arb.u.		Sym.	Freq., cm^{-1}	$I_{th}IR$, arb.u.	$I_{th}R$, arb.u.	PyHI	PyHBr	$PyPbI_3$	$PyPbBr_3$
ν_{1}	A_{1g}	1011.61	0.0	98.763	ν_{1}	A_1	1018.61	3.442 *	52.390	1007 (s,s)	1008 (s,s)	1008 (s,s)	1008 (s,s)
ν_{5}	B_{2g}	1016.96	0.0	0.0	$\nu_{5'}$	B_1	1046.99	0.573	0.0	(0,0)	(0,0)	(0,0)	1016 (m,0)
ν_{12}	B_{1u}	1022.46	0.0	0.0	ν_{12}	A_1	1048.31	1.280	10.132 *	1027 (s,s)	1028 (s,s)	1025 (s,s)	1026 (s,s)
ν_{19}	E_{1u}	1059.14	24.137	0.0	ν_{19a}	A_1	1078.43	3.178	1.006	1052 (s,s)	1054 (s,s)	1058 (s,m)	1058 (s,m)
					ν_{19b}	B_2	1081.89	5.159	0.317			1041 (s,m)	1046 (s,m)
ν_{14}	B_{2u}	1174.67	0.0	0.0	ν_{14}	B_2	1198.20	3.550	0.913	1150 (s,s)	1151 (s,s)	1156 (s,s)	1158 (s,s)
ν_{8}	E_{2g}	1197.33	0.0	5.225	ν_{8a}	A_1	1225.74	2.207	5.432	1186 (s,s)	1188 (s,s)	1188 (s,s)	1189 (s,s)
													1193 (s,s)
					ν_{8b}	B_2	1290.23	4.616	3.836	1236 (s,s)	1238 (s,s)	1240 (s,s)	1252 (w,s)
										1245 (s,s)	1247 (s,s)	1248 (s,s)	1258 (w,s)
ν_{15}	B_{2u}	1337.39	0.0	0.0	ν_{15}	B_2	1359.06	10.665 *	0.483	1327 (s,w)	1330 (s,w)	1323 (s,w)	1324 (s,w)
ν_{3}	A_{2g}	1380.96	0.0	0.0	ν_{3}	B_2	1414.57	3.511	0.166	1364 (s,w)	1367 (s,w)	1371 (s,0)	1375 (s,0)
ν_{18}	E_{1u}	1510.48	18.87	0.0	ν_{18a}	A_1	1516.10	21.546	0.322	1478 (s,w)	1480 (s,w)	1478 (s,0)	1479 (s,w)
													1482 (s,w)
					ν_{18b}	B_2	1573.78	39.154	0.135	1526 (s,w)	1527 (s,0)	1527 (s,0)	1529 (s,s)
												1532 (s,0)	1536 (s,s)
ν_{9}	E_{2g}	1634.35	0.0	12.694	ν_{9b}	B_2	1647.59	38.561 *	9.756	1602 (s,s)	1600 (s,s)	1600 (s,s)	1603 (s,s)
					ν_{9a}	A_1	1664.58	34.814 *	15.265	1629 (s,s)	1632 (s,s)	1632 (s,w)	1633 (s,s)
												1635 (s,w)	1637 (s,s)
ν_{13}	B_{1u}	3156.58	0.0	0.0	$\nu_{13'}$	A_1	3201.51	0.254	60.939 *	?	?	?	?
ν_{7}	E_{2g}	3166.22	0.0	133.318	$\nu_{7b'}$	B_2	3217.55	0.758	79.950	?	?	?	?
					$\nu_{7a'}$	A_1	3550.84	163.241 *	77.094	2934 ** (s,?)	2818 ** (s,?)	3102 ** (s,?)	3064 ** (s,?)
ν_{20}	E_{1u}	3181.82	47.147	0.0	$\nu_{20a'}$	A_1	3219.39	11.864	33.911 *	?	?	?	?
					$\nu_{20b'}$	B_2	3229.71	21.724	10.687 *	?	?	?	?
ν_{2}	A_{1g}	3192.01	0.0	421.879	$\nu_{2'}$	A_1	3231.40	0.273	240.672	3075 (?,s)	3084 (?,s)	3087 (?,s)	3096 (?,s)

The table is sorted by calculated frequencies of benzene. Symbols: ′—mixed modes; *—modes with substantial growth in activity in PyH^+ in comparison with C_6H_6; **—modes position calculated as center-of-mass of the broad band; s—strong band; w—weak band; 0—vibration is not observed in the experiment; —no data for this region; ?—unable to determine frequency.

For some PyH^+ cation vibrational modes, a substantial growth in activity is observed compared to those for the C_6H_6 molecule. Such modes are marked with an asterisk in Table 1. However, in several cases, the PyH^+ cation mode inherits the activity of a vibrational mode of the C_6H_6 molecule, which was used as an additional criterion for mode designation. Next, the calculated pyridinium modes were correlated with the experimentally obtained ones.

2.4. Vibrational Spectra Summary

The experimental IR and Raman spectra were measured for PyHX salts and $PyPbX_3$ perovskites (X^- = I^-, Br^-) at room temperature (Figures 2–4). The description of IR and Raman experiments could be found in the Methods section (Section 3). In this work, we focus on internal vibrations of the PyH^+ cation lying above 350 cm^{-1}. We will discuss three spectral regions: the low-frequency region 80–930 cm^{-1} (Figure 3, only Raman), medium-frequency region 930–1700 cm^{-1} with fingerprint vibrations (Figure 2, both IR and Raman) and high-frequency region of hydrogen modes 1700–3600 cm^{-1} (Figure 4, both IR and Raman). Correlating calculated cation vibrations with observed bands is a rather difficult task, but simultaneous analysis of IR and Raman spectra in four different materials makes this correlation more reliable. Below, we will consider in more detail the various spectral regions and give reasons for the chosen correlation presented in Table 1. Here, we adopted the notation of ν_8, ν_9, ν_{14}, ν_{15}, ν_{18} and ν_{19} modes of benzene from Gardner et al.'s work [55]. Notation in other works may differ [51,54].

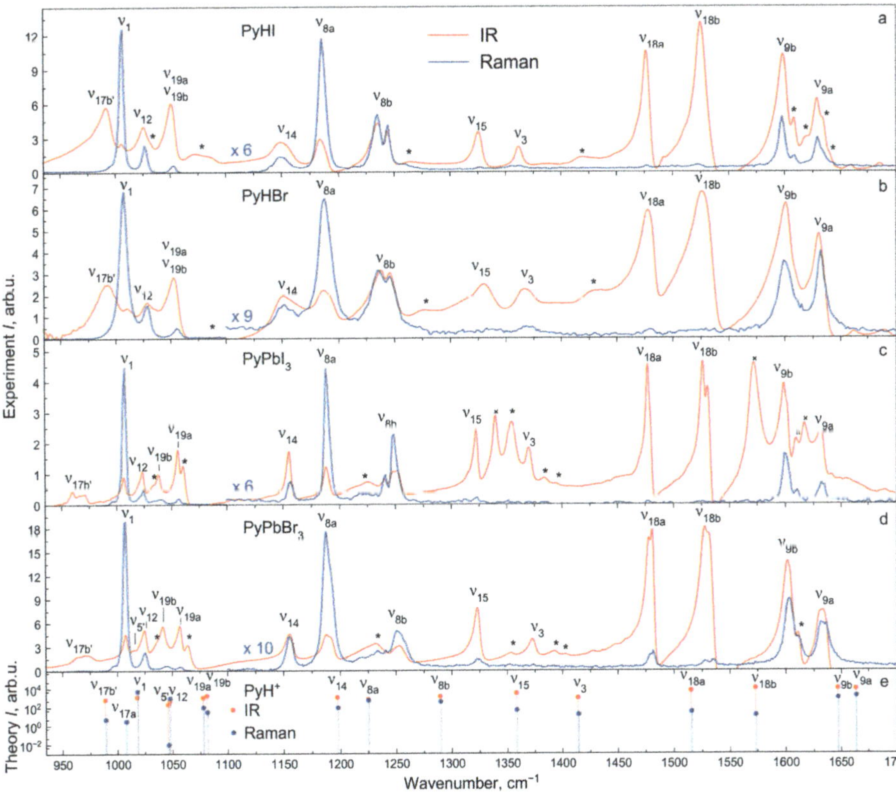

Figure 2. IR and Raman spectra of PyHX salts (**a**,**b**) and $PyPbX_3$ perovskites (X^- = I^-, Br^-) (**c**,**d**) at T = 300 K; *—combination modes; ×—KI contamination bands (observed also in KI powder). (**e**) Calculated IR (red) and Raman (blue) intensities of PyH^+ cation modes.

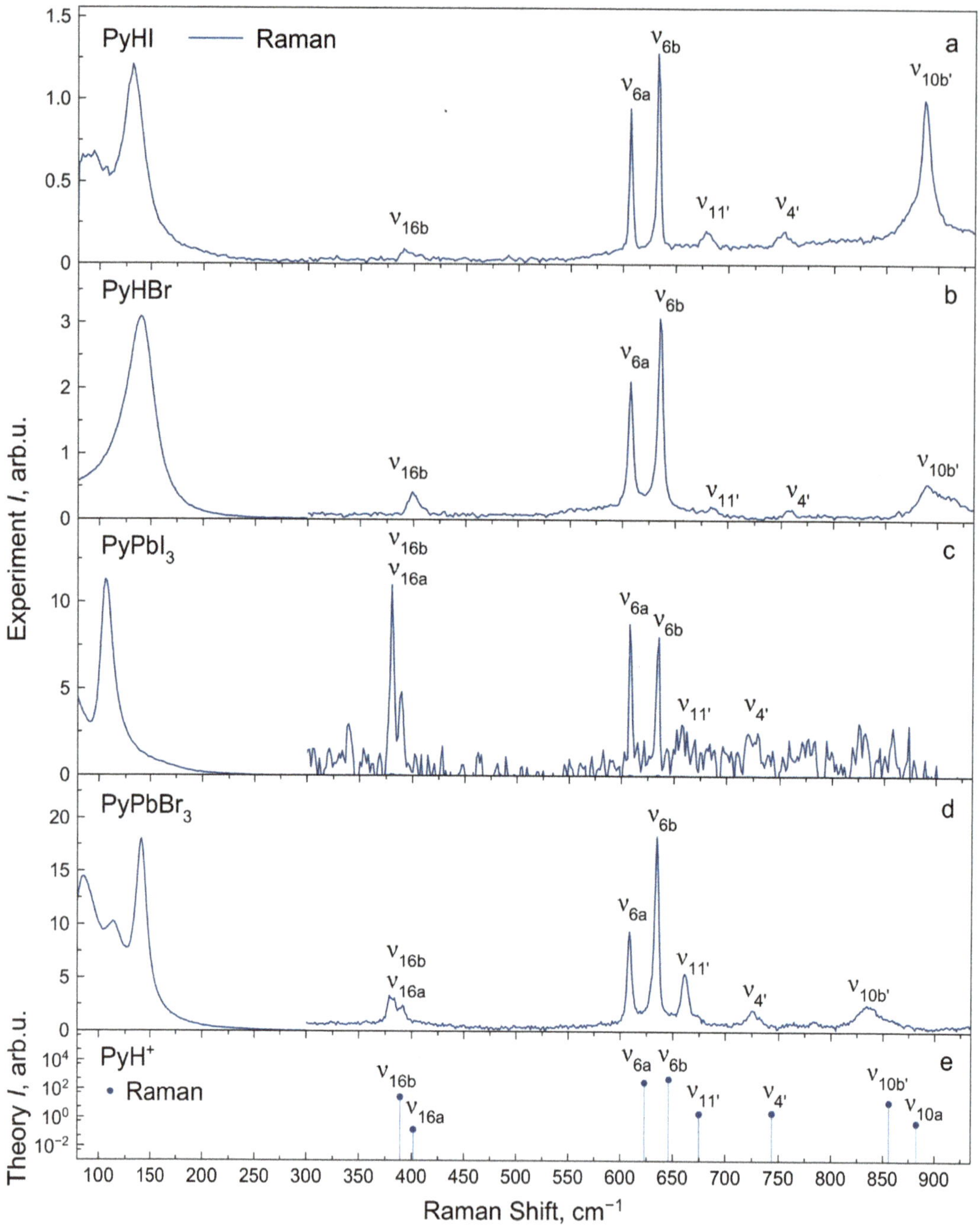

Figure 3. Raman spectra of PyHX salts (**a**,**b**) and $PyPbX_3$ perovskites ($X^- = I^-, Br^-$) (**c**,**d**) at T = 300 K in the low-frequency region. (**e**) Calculated Raman intensities of PyH^+ cation modes.

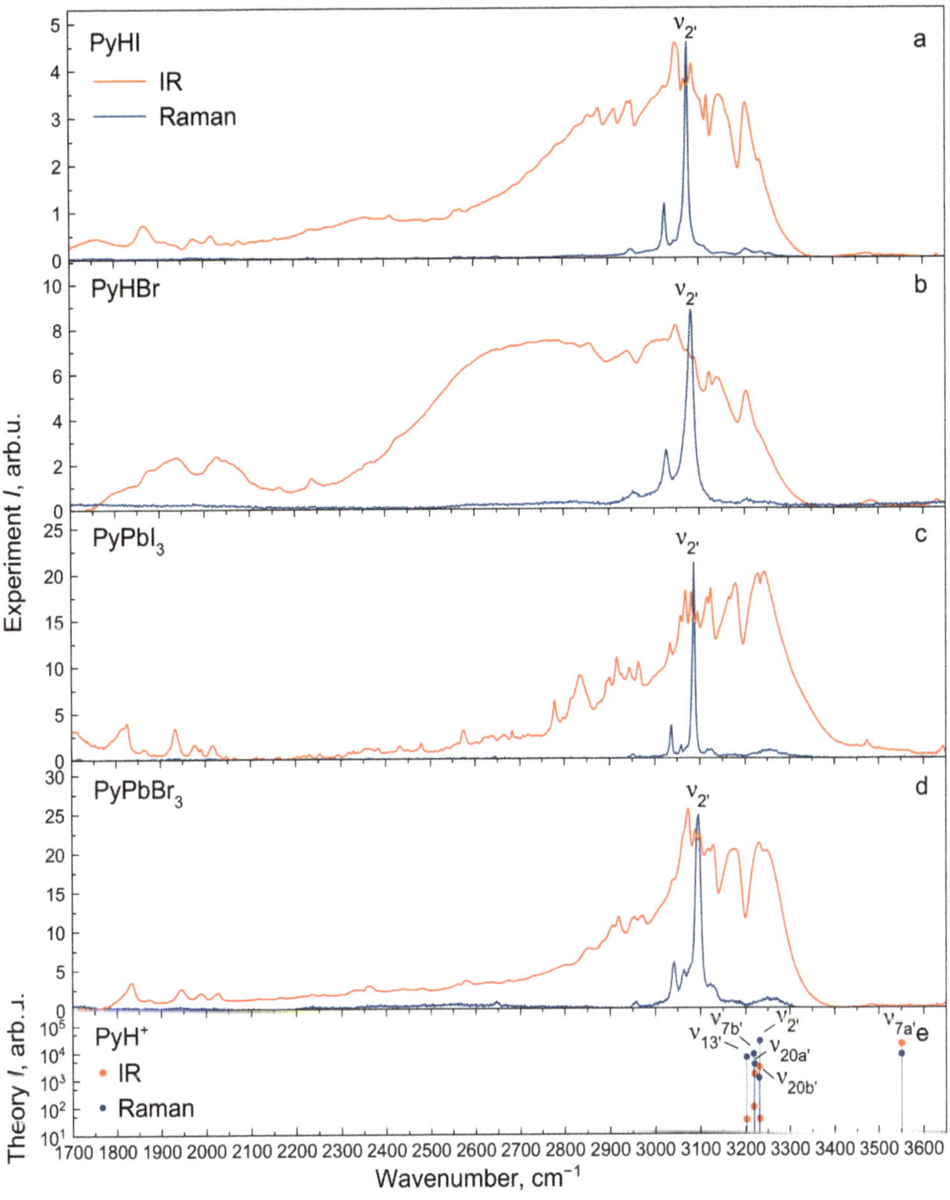

Figure 4. IR and Raman spectra of PyHX salts (**a**,**b**) and $PyPbX_3$ perovskites ($X^- = I^-, Br^-$) (**c**,**d**) at T = 300 K in the high-frequency region. (**e**) Calculated IR (red) and Raman (blue) intensities of PyH^+ cation modes.

Three vibration modes (ν_{16a}, ν_{10a}, and ν_{17a}) have the A_2 symmetry. The vibrations of this symmetry are forbidden in IR spectra. Calculations also showed that the Raman intensity of these modes is negligible. Therefore, these bands were not identified in the spectra. The intensities of unidentified vibrations in the experimental spectra are marked with the symbol "0" in Table 1.

2.5. Vibrational Spectra in Medium-Frequency Region

Let us consider the medium-frequency region 930–1700 cm^{-1} (Figure 2). The fully symmetric A_1 vibrational modes ν_1, ν_{12}, ν_{8a}, ν_{18a} and ν_{9a} are clearly recognized in all IR and Raman spectra with approximately the same frequencies in different compounds, which indicate that they are insensitive to the crystalline environment.

The doubly degenerate benzene mode ν_{19} splits into ν_{19a} and ν_{19b} modes of different symmetry in the pyridinium cation. The calculated splitting of these modes is 3.5 cm^{-1}. This is quite consistent with the broad unresolved band observed in salts around 1050 cm^{-1}. However, in perovskites, the crystal field causes a more significant splitting of these modes for around 7 cm^{-1} so that they could be potentially resolved in the IR spectra.

The vibrations ν_{8b}, ν_{18b}, ν_{9b} (mainly consisting of stretching ν_{CN} and in-plane bending δ_{NH} modes) and ν_{19b}, ν_{14}, ν_{15}, ν_3 (mainly consisting of stretching ν_{CC} and in-plane bending δ_{CH} modes) have the B_2 symmetry. Except for the ν_{19b} vibration, all of them have the same frequency in the vibrational spectra of both perovskites and salts.

In the medium-frequency region, several overtones and combination modes of intense low-frequency vibrations are also observed. In order to identify these modes, the wavenumbers and intensities of normal modes and their combinations were refined by modeling in the anharmonic approximation (Table S4). The combination modes found in the spectra are summarized in Table S5 and marked by an asterisk in Figure 2.

For most of the modes with the A_1 and B_2 symmetries in this frequency range, no significant differences in position are observed for perovskites and salts. The $\nu_{5'}$ and $\nu_{17b'}$ vibrations (out-of-plane bending γ_{NH} and γ_{CH} modes, respectively) have B_1 symmetry and low intensities in Raman spectra. The $\nu_{5'}$ vibrational band fully overlaps with other bands in the 1000–1050 cm^{-1} region of the IR spectra of salts. Narrower vibrational bands in the IR spectra of perovskites make it possible to assign the band at 1016 cm^{-1} to the $\nu_{5'}$ mode. In contrast, the frequency position of the $\nu_{17b'}$ vibration is clearly defined in the IR spectra. It significantly lowers its frequency for perovskites compared to salts (from 993 to 969 cm^{-1}) and changes its shape. For PyHX ($X^- = I^-$, Br^-) salts, it has been noticed that there is a connection between the behavior of the vibrational band at 993 cm^{-1} and stretching ν_{NH} mode with the strength of the hydrogen bond [54]. The $\nu_{17b'}$ vibration frequency is lower in perovskites as a manifestation of hydrogen bond weakening in comparison to salts.

2.6. Vibrational Spectra in Low-Frequency Region

For the low-frequency region of 80–950 cm^{-1}, only the Raman spectra were recorded (Figure 3). Experimental IR intensities of the vibrational bands in this region are marked with the symbol "-" in Table 1. The $\nu_{10b'}$, $\nu_{4'}$ and $\nu_{11'}$ vibrations with B_1 symmetry in this region involve the γ_{NH} mode and lower the frequency in perovskites compared to salts. Accurate determination of the positions of these modes is important because they have high IR activity, which will lead to the appearance of overtones and combination modes in the spectrum. Measuring these modes with basic IR spectroscopy instruments is not possible. However, they are observed as weak lines in Raman spectra. Measured frequencies were used to determine spectral positions of overtones and combination frequencies in the spectrum.

The ν_{6b} (in-plane bending δ_{CC} mode) and ν_{6a} (out-of-plane bending γ_{CC} and γ_{CN} modes) can be clearly assigned in Raman spectra, and its wavenumbers are almost the same for salts and perovskites. The wavenumbers of ν_{16b} vibrations (out-of-plane bending γ_{CC} and γ_{CN} modes) can also be estimated for all compounds in Raman spectra, but the manifestation of ν_{16a} vibrations (in-plane bending δ_{CC} and δ_{CN} modes) in spectra of perovskites is uncertain.

The low-frequency intense band at 135 cm^{-1} in Raman spectra of halide salts is known as the rotational mode and does not depend on halide anion or hydrogen bond strength [56]. In perovskites, vibrations below 200 cm^{-1} correspond to the motion of the lead halide network. Its frequencies show dependency on halide atom substitution and are lower in $PyPbI_3$ compared to $PyPbBr_3$.

2.7. Vibrational Spectra in High-Frequency Region

The vibrational structure of the IR spectrum of halide salts and perovskites in the region of 1700–3400 cm^{-1} (Figure 4) is complex. Therefore, there is no accurate assumption of the stretching vibration frequency positions, which is marked with the symbol "?" in Table 1. The exceptions are the strongest vibrational band at 3075–3096 cm^{-1} in the Raman spectra ($\nu_{2'}$ mode) and the broadest vibrational band at 1700–3300 cm^{-1} in the IR spectra ($\nu_{7a'}$ mode).

The complexity of the IR spectrum of halide salts is explained by Fermi resonances of the stretching ν_{NH} mode ($\nu_{7a'}$) with overtones and combinations of PyH^+ internal modes [56–58]. An increase in the frequency and a decrease in the intensity of the $\nu_{7a'}$ vibrational structure in the IR spectra indicates a significant hydrogen bond weakening in perovskites compared to salts. This trend and less intense and narrower Fermi resonances at the same frequencies also reveal a slight hydrogen bond weakening in $PyPbI_3$ perovskite compared to $PyPbBr_3$. The low-frequency component of the $\nu_{7a'}$ vibrational structure at 1700–2100 cm^{-1} (so-called "C-band") [58] for $PyPbBr_3$ perovskite is slightly shifted to the higher frequencies compared to $PyPbI_3$, which is a manifestation of the higher vibrational coupling of the ν_{NH} mode with other modes in this spectral region. The bands at 1800–1880 cm^{-1}, 1920–1990 cm^{-1} and 2010–2030 cm^{-1} are formed by the stretching ν_{NH} mode coupling with $2\nu_{10b'}$, $2\nu_{17b'}$ and $2\nu_1$ modes, respectively. The higher frequency bands at 2880–3400 cm^{-1} are formed by the ν_{NH} mode coupling with $2\nu_{18a}$, $2\nu_{18b}$, $2\nu_{9b}$, $2\nu_{9a}$ modes and combinations of the last ones with lower modes.

For all IR and Raman spectra, narrower vibrational bands in the spectra of perovskites are observed compared to salts, which is due to the less disorder in perovskites. It can also be emphasized that most internal vibrational modes are not sensitive to the crystal structure (salt or perovskite) and halide anion (iodide or bromide). The exceptions are vibrational bands containing out-of-plane bending γ_{NH} ($\nu_{11'}$, $\nu_{4'}$, $\nu_{10b'}$, $\nu_{17b'}$, $\nu_{5'}$) and stretching ν_{NH} modes ($\nu_{7a'}$), which are shown to be strongly affected by the environment [48–51,54,55]. Thus, the observation of these modes can provide information about the crystal structure of the material.

2.8. Vibration Splitting Effect

The intriguing feature of $PyPbX_3$ ($X^- = I^-, Br^-$) vibrational spectra is the splitting of the ν_{8a}, ν_{8b}, ν_{18a}, ν_{18b}, ν_{9b}, ν_{9a} vibrational bands containing the in-plane δ_{NH} mode (Figure 5). Such an effect can originate from one of several sources: (i) intermolecular coupling of vibrations of molecules in equivalent positions in the unit cell at $Z > 1$ (Davydov splitting); (ii) the existence of two crystallographically non-identical cation positions in the unit cell (crystal cite effect); and (iii) the switching of cations between non-equivalent positions, leading to the splitting of involved vibrational modes.

The large interaction distance in the perovskite crystal structure speaks against the Davydov splitting effect. Splitting of the abovementioned bands was observed in the vibrational spectra of pyridinium salts of tungstophosphoric acid [59,60]. A possible explanation for this phenomenon is the mechanism of quantum tunneling of the cation between two positions, leading to the splitting of vibrations containing the NH-group bending mode. Possible mechanisms to explain the splitting observed in perovskites are the switching of pyridinium cation between non-equivalent positions and the presence of two nonequivalent crystal sites of pyridinium cations. Determining the specific nature of these splittings is possible by combining XRD and IR studies at lower temperatures. In hybrid organic–inorganic perovskites, at temperatures below room temperature, order–disorder phase transitions could be observed, caused by the "freezing out" of the movement of organic cations in the lattice. The corresponding change in the symmetry could be found from XRD data. A related issue is the presence of a pseudo two-fold axis passing through the center of C-N bond in the pyridinium cation in perovskite crystals at room temperature, which is a manifestation of the frustrated rotation of the cation between two symmetry equivalent positions.

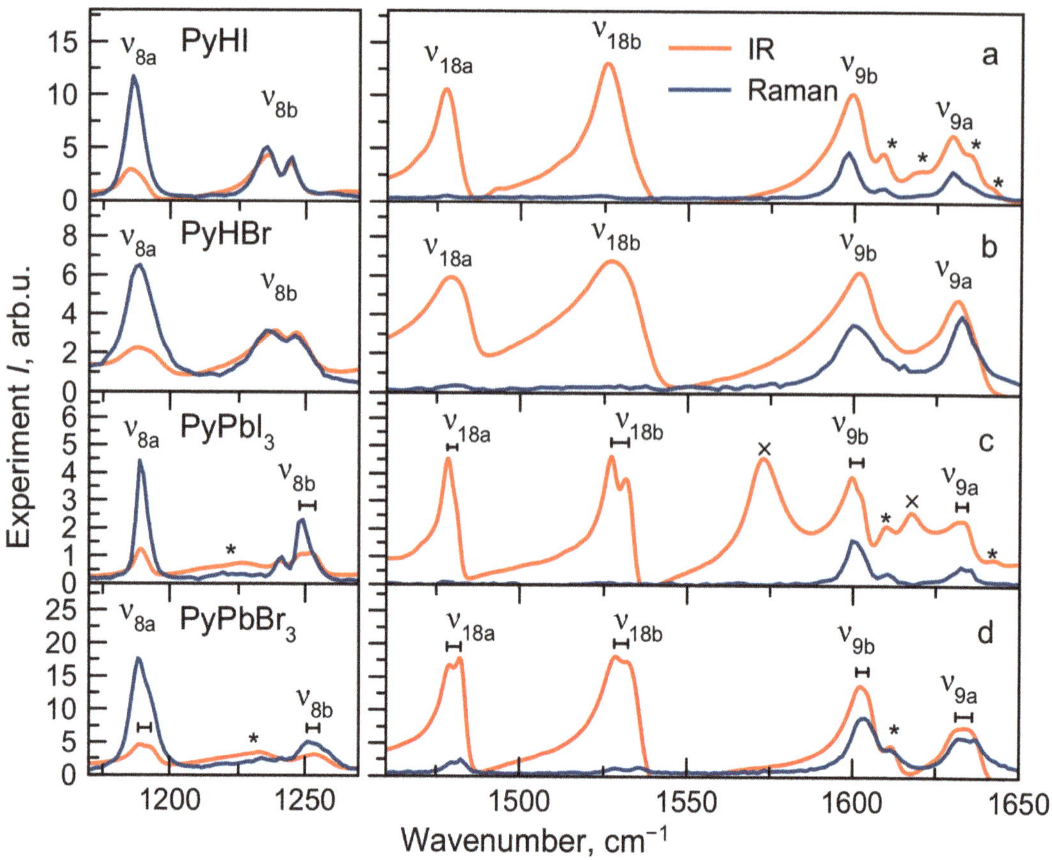

Figure 5. IR and Raman spectra of PyHX salts (**a**,**b**) and $PyPbX_3$ perovskites (X^- = I^-, Br^-) (**c**,**d**) at T = 300 K in the region of split bands. *—combination modes; ×—KI contamination bands (observed also in KI powder).

2.9. Solid-State NMR Study

The ^{1}H, ^{13}C and ^{207}Pb NMR spectra of $PyPbX_3$ (X^- = I^-, Br^-) with the assignment of the signals are shown in Figure 6. There seems to be no significant difference in the state of the pyridinium cation in bromine- and iodine-based perovskite samples, except for a somewhat larger line width in the case of bromine-based perovskite, which is probably due to a larger number of structural defects in the crystalline sample. Nevertheless, in both cases, the spectra are reasonably well resolved, and only one set of relatively narrow ^{1}H and ^{13}C NMR signals is observed. The chemical shifts of ^{207}Pb NMR signals are noticeably different from those of 3D perovskites. We speculate that the values around –200 ÷ –300 ppm for $PyPbBr_3$ and around 900–1000 ppm for $PyPbI_3$ could be indicative for the formation of 1D chains of lead-halide octahedra.

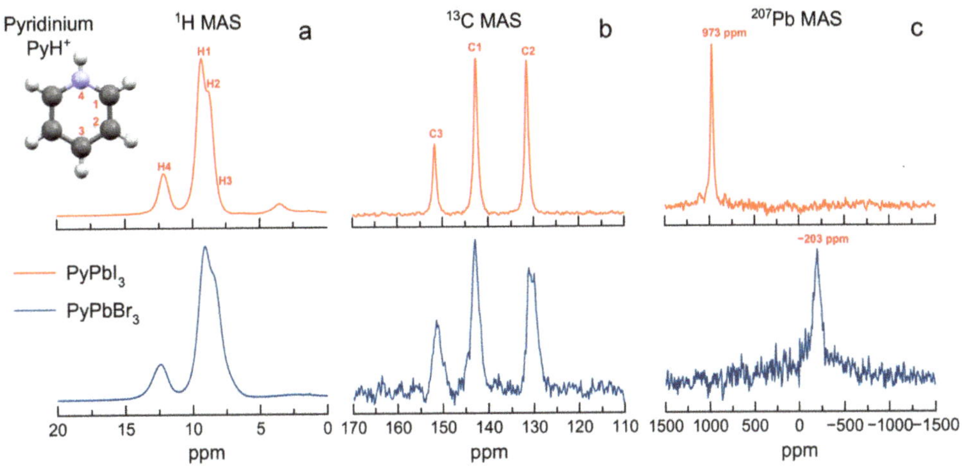

Figure 6. The ^{1}H (one-pulse) (**a**), ^{13}C (CP) (**b**) and ^{207}Pb (Hahn-echo) (**c**) MAS (12.5 kHz) NMR spectra of $PyPbI_3$ (red) and $PyPbBr_3$ (blue) recorded at T = 298 K.

3. Methods

Synthesis. Pyridinium lead trihalide $PyPbX_3$ (X^- = I^-, Br^-) single crystals were grown by the slow counterdiffusion of ions from individual solutions of lead(II) halides and pyridine in hydrohalic acid with 1 M concentrations in the silica gel filled U-tube. This counterdiffusion-in-gel crystallization (CGC) method could be used for synthesis of high quality hybrid organic–inorganic low-dimensional perovskites and 3D perovskites [61]. XRD data are presented in the Supplementary Materials (Figure S4). Pyridine hydrohalides PyHX (X^- = I^-, Br^-) were synthesized by adding concentrated hydrogen halogenides drop-wise with constant stirring to the solutions of 5 mL of ethanol and 1 mL of pyridine. The resulting solutions were kept at a temperature of 3–5 °C for 12 h, then filtrated and rinsed with ethanol and dried. More details on the PyHX synthesis could be found in Selivanov et al. [31].

DFT Calculations. IR and Raman spectra for the PyH^+ cation and C_6H_6 molecule were simulated in the harmonic approximation using the Gaussian 16 software package. The wavenumbers and IR intensities of normal and combination modes were refined in the anharmonic approximation for PyH^+ cation. For all calculations, the B3LYP functional and 6-311+G(d,p) basis set with one set of polarizing functions for heavy atoms (d-type) and hydrogen (p-type) were used.

Raman Measurements. The experimental unpolarized Raman spectra of PyHX and $PyPbX_3$ (X^- = I^-, Br^-) were acquired at T = 300 K using the Horiba Jobin-Yvon LabRam HR800 confocal Raman spectrometer (Horiba Jobin-Yvon, Oberursel, Germany) in backscattering geometry (see Figure S5 for the scheme of the Raman measurements). The equipment used a diffraction grating with a cell of 1800 L/mm and the aperture was 150 × 150 µm. Laser radiation was focused through a 100× objective lens. A solid-state laser with a wavelength of 532 nm was utilized as the radiation source. The actual laser power applied to the sample was around 6 mW, with 10 s of accumulation time and 6 repetitions.

IR Measurements. PyHX and $PyPbX_3$ crystals were subjected to grinding in a mortar and mixed with KX powders (where X^- = I^-, Br^-) in a weight ratio of 1:10 for I and 1:3 for Br, respectively. IR absorption spectra were recorded in the 850–4000 cm^{-1} spectral range with a spectral resolution of 2 cm^{-1} applying the Happ–Gensel apodization using a research-grade FT-IR spectrometer Thermo Nicolet iS50 (Thermo Scientific, Waltham, MA, USA). Spectrometer was equipped with a DTGS detector and a KBr beamsplitter.

Solid-state NMR Measurements. The ^{1}H, ^{13}C and ^{207}Pb NMR measurements were performed using Bruker Avance III 400WB NMR spectrometers (Bruker, Ettlingen, Germany) (working frequency 400.23 MHz for ^{1}H, 100.65 MHz for ^{13}C and 83.73 MHz for ^{207}Pb). The spectra were recorded under magic angle spinning conditions (MAS; spinning rate 12 kHz) at room temperature using 4.0 mm rotor. ^{13}C MAS NMR spectra were measured using cross-polarization (CP) technique (2 ms contact time); ^{207}Pb NMR spectra were measured using Hahn echo technique. The relaxation delays were set to 120 s for ^{1}H, 2–5 s for ^{13}C and 1 s for ^{207}Pb. Liquid TMS at 0 ppm was used as an external reference for ^{1}H and ^{13}C spectra. $Pb(NO_3)_2$ at −3482 ppm (under 4 kHz spinning with 25 °C room temperature) was used as an external reference for ^{207}Pb spectra.

4. Conclusions

We conducted a study of the internal vibrations of the pyridinium cation PyH^+ in four materials: PyHX salts and $PyPbX_3$ ($X^- = I^-, Br^-$) 1D-perovskites. The bands experimentally observed in the Raman scattering and IR absorption spectra were identified by analyzing the symmetry of vibrations and comparing them with the results of DFT calculations for a free benzene molecule and a pyridinium cation. Most internal vibrations of PyH^+ are not significantly influenced by the crystalline environment. However, vibrational modes $\nu_{11'}$, $\nu_{4'}$, $\nu_{10b'}$, $\nu_{17b'}$, $\nu_{5'}$ and $\nu_{7a'}$ undergo strong shifts upon the transition from salts to perovskites. What these modes have in common is the presence of NH-group movements in them, which, apparently, are influenced by the environment, in particular, by the changes in the strength of the hydrogen bond. In perovskites, splitting of non-degenerate modes ν_{8a}, ν_{8b}, ν_{18a}, ν_{18b}, ν_{9b}, ν_{9a} is observed. This phenomenon can be explained by different mechanisms, which could be distinguished by a joint XRD and IR study at lower temperatures. This work illustrates the different crystalline environment influence on organic cations behavior such as vibrational band shifts, changes in their intensities and width and splitting of non-degenerate levels. In this way, organic cations act as spectators, probes of the crystalline environment in hybrid organic–inorganic crystals, including halide perovskites, and provide valuable information on the dynamics of the system. Thus, spectroscopy of internal vibrations of organic cations can be a convenient and informative tool for studying these new materials towards new photonics applications.

Supplementary Materials: The following supporting information can be downloaded at: https://www.mdpi.com/article/10.3390/molecules29010078/s1, Figure S1: Photos of $PyPbI_3$ (a) and $PyPbBr_3$ (b) single crystals; Table S1: Character table of D_{6h} point symmetry group; Figure S2: Transformation of C_6H_6 to PyH^+ by symmetry elements; Table S2: Character table of C_{2v} point symmetry group; Table S3: Correspondence between the irreducible representations of the D_{6h} and C_{2v} point symmetry groups; Figure S3: Calculated atomic displacements in internal normal modes of the PyH^+ cation. The notation in Wilson nomenclature and symmetry of the modes are labeled. Red crosses and dots indicate out-of-plane atomic displacements; Table S4: Calculated anharmonic fundamental modes (from 1 to 30), overtones (Over) and combination modes (Comb) of the PyH^+ cation; Table S5: Combination modes recognized in the experimental vibrational spectra; Figure S4: Powder XRD patterns simulated from the single crystal data (a,c) and measured for grounded single crystals (b,d) for $PyPbI_3$ (a,b) and $PyPbBr_3$ (c,d); Figure S5: The scheme of the Raman measurement experiment.

Author Contributions: Conceptualization, Y.V.K.; Investigation, A.Y.S., A.Y.M., K.M.B., N.I.S., A.S.M., P.M.T. and Y.V.K.; Writing—original draft, A.Y.S.; Writing—review & editing, A.Y.S., C.C.S. and Y.V.K.; Visualization, A.Y.S.; Supervision, Y.V.K. All authors have read and agreed to the published version of the manuscript.

Funding: This research has been supported by the Ministry of Science and Higher Education of the Russian Federation (Megagrant no. 075-15-2022-1112).

Institutional Review Board Statement: Not applicable.

Informed Consent Statement: Not applicable.

Data Availability Statement: Data are contained within the article and Supplementary Materials.

Acknowledgments: This work was carried out on the equipment of the SPbU Research Centers "Nanophotonics", "Center for Optical and Laser materials research", "X-ray Diffraction Studies",

"Geomodel", "Magnetic Resonance Research Centre", "Centre for Diagnostics of Functional Materials for Medicine, Pharmacology and Nanoelectronics", "Thermogravimetric and Calorimetric Research Centre", "Computer Center of SPbU" (http://cc.spbu.ru, accessed on 19 December 2023) and research facilities of the laboratory "Photoactive Nanocomposite Materials" (SPbU ID: 91696387).

Conflicts of Interest: The authors declare no conflict of interest.

References

1. Yang, C.; Yin, J.; Li, H.; Almasabi, K.; Gutiérrez-Arzaluz, L.; Gereige, I.; Brédas, J.-L.; Bakr, O.M.; Mohammed, O.F. Engineering Surface Orientations for Efficient and Stable Hybrid Perovskite Single-Crystal Solar Cells. *ACS Energy Lett.* **2022**, *7*, 1544–1552. [CrossRef]
2. Pang, P.; Jin, G.; Liang, C.; Wang, B.; Xiang, W.; Zhang, D.; Xu, J.; Hong, W.; Xiao, Z.; Wang, L.; et al. Rearranging Low-Dimensional Phase Distribution of Quasi-2D Perovskites for Efficient Sky-Blue Perovskite Light-Emitting Diodes. *ACS Nano* **2020**, *14*, 11420–11430. [CrossRef] [PubMed]
3. Qin, C.; Sandanayaka, A.S.D.; Zhao, C.; Matsushima, T.; Zhang, D.; Fujihara, T.; Adachi, C. Stable Room-Temperature Continuous-Wave Lasing in Quasi-2D Perovskite Films. *Nature* **2020**, *585*, 53–57. [CrossRef] [PubMed]
4. Furlan, F.; Nodari, D.; Palladino, E.; Angela, E.; Mohan, L.; Briscoe, J.; Fuchter, M.J.; Macdonald, T.J.; Grancini, G.; McLachlan, M.A.; et al. Tuning Halide Composition Allows Low Dark Current Perovskite Photodetectors With High Specific Detectivity. *Adv. Opt. Mater.* **2022**, *10*, 2201816. [CrossRef]
5. Selivanov, N.I.; Samsonova, A.Y.; Kevorkyants, R.; Krauklis, I.V.; Chizhov, Y.V.; Stroganov, B.V.; Triantafyllou-Rundell, M.E.; Bahnemann, D.W.; Stoumpos, C.C.; Emeline, A.V.; et al. Hybrid Organic–Inorganic Halide Post-Perovskite 3-Cyanopyridinium Lead Tribromide for Optoelectronic Applications. *Adv. Funct. Mater.* **2021**, *31*, 2102338. [CrossRef]
6. Jung, M.-H. Diammonium spacer-induced stable zigzag type 2D Dion–Jacobson lead/tin-based perovskite solar cells. *Dalton Trans.* **2023**, *52*, 17258–17272. [CrossRef]
7. Biswas, A.; Bakthavatsalam, R.; Shaikh, S.R.; Shinde, A.; Lohar, A.; Jena, S.; Gonnade, R.G.; Kundu, J. Efficient Broad-Band Emission from Contorted Purely Corner-Shared One Dimensional (1D) Organic Lead Halide Perovskite. *Chem. Mater.* **2019**, *31*, 2253–2257. [CrossRef]
8. Qi, Z.; Gao, H.; Yang, X.; Chen, Y.; Zhang, F.-Q.; Qu, M.; Li, S.-L.; Zhang, X.-M. A One-Dimensional Broadband Emissive Hybrid Lead Iodide with Face-Sharing PbI_6 Octahedral Chains. *Inorg. Chem.* **2021**, *60*, 15136–15140. [CrossRef]
9. Whitfield, P.; Herron, N.; Guise, W.; Page, K.; Cheng, Y.Q.; Milas, I.; Crawford, M.K. Structures, Phase Transitions and Tricritical Behavior of the Hybrid Perovskite Methyl Ammonium Lead Iodide. *Sci. Rep.* **2016**, *6*, 35685. [CrossRef]
10. Driscoll, E.H.; Orera, A.; Anderson, P.A.; Sanjuánb, M.L.; Slater, P.R. Raman Spectroscopy Insights into the α- and δ-Phases of Formamidinium Lead Iodide ($FAPbI_3$). *Dalton Trans.* **2021**, *50*, 3315. [CrossRef]
11. Mączka, M.; Ptak, M. Temperature-Dependent Raman Studies of $FAPbBr_3$ and $MAPbBr_3$ Perovskites: Effect of Phase Transitions on Molecular Dynamics and Lattice Distortion. *Solids* **2022**, *3*, 111–121. [CrossRef]
12. Ghosh, S.; Rana, D.; Pradhan, B.; Donfack, P.; Hofkens, J.; Materny, A. Raman Spectroscopy of Formamidinium-Based Lead Mixed-Halide Perovskite Bulk Crystals. *ChemPhysChem* **2023**, *24*, e202300303. [CrossRef]
13. Naqvi, F.H.; Junaid, S.B.; Ko, J.-H. Influence of Halides on Elastic and Vibrational Properties of Mixed-Halide Perovskite Systems Studied by Brillouin and Raman Scattering. *Materials* **2023**, *16*, 3986. [CrossRef] [PubMed]
14. Lavén, R.; Koza, M.M.; Malavasi, L.; Perrichon, A.; Appel, M.; Karlsson, M. Rotational Dynamics of Organic Cations in Formamidinium Lead Iodide Perovskites. *J. Phys. Chem. Lett.* **2023**, *14*, 2784–2791. [CrossRef]
15. Brown, K.L.; Parker, S.F.; García, I.R.; Mukhopadhyay, S.; Sakai, V.G.; Stock, C. Molecular Orientational Melting within a Lead-Halide Octahedron Framework: The Order-Disorder Transition in $CH_3NH_3PbBr_3$. *Phys. Rev. B* **2017**, *96*, 174111. [CrossRef]
16. Schuck, G.; Lehmann, F.; Ollivier, J.; Mutka, H.; Schorr, S. Influence of Chloride Substitution on the Rotational Dynamics of Methylammonium in $MAPbI_{3-x}Cl_x$ Perovskites. *J. Phys. Chem. C* **2019**, *123*, 11436–11446. [CrossRef]
17. Chen, T.; Foley, B.J.; Ipek, B.; Tyagi, M.; Copley, J.R.D.; Brown, C.M.; Choi, J.J.; Lee, S.-H. Rotational Dynamics of Organic Cations in the $CH_3NH_3PbI_3$ Perovskite. *Phys. Chem. Chem. Phys.* **2015**, *17*, 31278–31286. [CrossRef]
18. Even, J.; Carignano, M.; Katan, C. Molecular disorder and translation/rotation coupling in the plastic crystal phase of hybrid perovskites. *Nanoscale* **2016**, *8*, 6222–6236. [CrossRef] [PubMed]
19. Cui, Y.; Liu, Y.-Y.; Deng, J.-P.; Zhang, X.-Z.; Yang, R.-B.; Li, Z.-Q.; Wang, Z.-W. Microscopic theory of Raman scattering for the rotational organic cation in metal halide perovskites. *Phys. Rev. B* **2023**, *107*, 094306. [CrossRef]
20. Kim, Y.; Bae, S.; Park, J.; Nguyen, T.T.T.; Jung, H.R.; Jo, W.; Kim, Y.-H.; Raebiger, H.; Yoon, S. Parallel Alignment of Methylammonium Cations in an Orthorhombic $CH_3NH_3PbCl_3$ Single Crystal Observed by Polarized Micro-Raman Scattering Spectroscopy. *Chem. Mater.* **2022**, *34*, 2972–2980. [CrossRef]
21. Menahem, M.; Dai, Z.; Aharon, S.; Sharma, R.; Asher, M.; Diskin-Posner, Y.; Korobko, R.; Rappe, A.M.; Yaffe, O. Strongly Anharmonic Octahedral Tilting in Two-Dimensional Hybrid Halide Perovskites. *ACS Nano* **2021**, *15*, 10153–10162. [CrossRef] [PubMed]

22. Lavan, S.N.; Sanni, A.M.; Rury, A.S.; Liu, Z.-F. Characterization of the Ammonium Bending Vibrations in Two-Dimensional Hybrid Lead-Halide Perovskites from Raman Spectroscopy and First-Principles Calculations. *J. Phys. Chem. C* **2021**, *125*, 223–236. [CrossRef]
23. Dhanabalan, B.; Leng, Y.-C.; Biffi, G.; Lin, M.-L.; Tan, P.-H.; Infante, I.; Manna, L.; Arciniegas, M.P.; Krahne, R. Directional Anisotropy of the Vibrational Modes in 2D-Layered Perovskites. *ACS Nano* **2020**, *14*, 4689–4697. [CrossRef] [PubMed]
24. Fedoruk, K.; Drozdowski, D.; Maczka, M.; Zareba, J.K.; Stefańska, D.; Gagor, A.; Sieradzki, A. [Methylhydrazinium]$_2$PbCl$_4$, a Two-Dimensional Perovskite with Polar and Modulated Phases. *Inorg. Chem.* **2022**, *61*, 15520–15531. [CrossRef] [PubMed]
25. Li, H.-J.; Liu, Y.-L.; Chen, X.-G.; Gao, J.-X.; Wang, Z.-X.; Liao, W.-Q. High-Temperature Dielectric Switching and Photoluminescence in a Corrugated Lead Bromide Layer Hybrid Perovskite Semiconductor. *Inorg. Chem.* **2019**, *58*, 10357–10363. [CrossRef] [PubMed]
26. Tao, K.; Han, S.; Ji, C.; Liu, X.; Wu, Z.; Zhang, J.; Luo, J.; Sun, Z. Structural Phase Transition and Switchable Dielectric Properties of a Unique Two-Dimensional Organic–Inorganic Hybrid Perovskite Compound [$C_6H_{11}NH_2CH_3$]$_4$$Pb_3I_{10}$. *Cryst. Growth Des.* **2018**, *18*, 7316–7322. [CrossRef]
27. Zeb, A.; Sun, Z.; Khan, A.; Zhang, S.; Khan, T.; Asghara, M.A.; Luo, J. [$C_6H_{14}N$]PbI_3: A One-Dimensional Perovskite-Like Order–Disorder Phase Transition Material with Semiconducting and Switchable Dielectric Attributes. *Inorg. Chem. Front.* **2018**, *5*, 897–902. [CrossRef]
28. Peng, H.; Liu, Y.-H.; Huang, X.-Q.; Liu, Q.; Yu, Z.-H.; Wang, Z.-X.; Liao, W.-Q. Homochiral One-Dimensional ABX_3 Lead Halide Perovskites with High-T_c Quadratic Nonlinear Optical and Dielectric Switchings. *Mater. Chem. Front.* **2021**, *5*, 4756–4763. [CrossRef]
29. Fedoruk, K.; Zelewski, S.J.; Zaręba, J.K.; Ptak, M.; Mączka, M.; Sieradzki, A. Getting the Details Right: Optical, Dielectric, and Vibrational Outcomes of Structural Phase Transition in One-Dimensional Pyrrolidinium Lead Iodide and the Role of Defects. *J. Mater. Chem. C* **2022**, *10*, 10519–10529. [CrossRef]
30. Zienkiewicz, J.A.; Kałduńska, K.; Fedoruk, K.; Barros dos Santos, A.J.; Stefanski, M.; Paraguassu, W.; Muzioł, T.M.; Ptak, M. Luminescence and Dielectric Switchable Properties of a 1D (1,1,1-Trimethylhydrazinium)PbI_3 Hybrid Perovskitoid. *Inorg. Chem.* **2022**, *61*, 20886–20895. [CrossRef]
31. Selivanov, N.I.; Rozhkova, Y.A.; Kevorkyants, R.; Emeline, A.V.; Bahnemann, D.W. The Effect of Organic Cations on the Electronic, Optical and Luminescence Properties of 1D Piperidinium, Pyridinium, and 3-Hydroxypyridinium Lead Trihalides. *Dalton Trans.* **2020**, *49*, 4390–4403. [CrossRef] [PubMed]
32. Geselle, M.; Fuess, H. Crystal structure of pyridinium triiodoplumbate(II), $C_5H_6NPbI_3$. *Z. Kristallogr. New Cryst. Struct.* **1997**, *212*, 235. [CrossRef]
33. Samsonova, A.Y.; Krauklis, I.V.; Chizhov, Y.V.; Selivanov, N.I.; Emeline, A.V.; Kapitonov, Y.V. Polarized Raman Study of Internal Vibrations of the Organic Cation in 3-Cyanopyridinium Lead Tribromide Post-perovskite. *J. Phys. Chem. Lett.* **2023**, *14*, 3445–3451. [CrossRef] [PubMed]
34. Krauklis, I.V.; Samsonova, A.Y.; Selivanov, N.I.; Kapitonov, Y.V.; Chizhov, Y.V. Cluster approach for the density functional theory study of organic cation vibrations in hybrid halide post-perovskite 3-cyanopyridinium lead tribromide. *J. Raman Spectrosc.* **2023**, *1*. [CrossRef]
35. Moral, R.F.; Germino, J.C.; Bonato, L.G.; Almeida, D.B.; Therézio, E.M.; Atvars, T.D.Z.; Stranks, S.D.; Nome, R.A.; Nogueira, A.F. Influence of the Vibrational Modes from the Organic Moieties in 2D Lead Halides on Excitonic Recombination and Phase Transition. *Adv. Opt. Mater.* **2020**, *8*, 2001431. [CrossRef]
36. Wąsicki, J.; Nawrocik, W.; Pająk, Z.; Natkaniec, I.; Belushkin, A.V. Neutron Scattering Study of Some Pyridinium Salts. *Phys. Stat. Sol. A* **1989**, *114*, 497–504. [CrossRef]
37. Ripmeester, J.A. ^{1}H and ^{2}H NMR Study of Pyridinium Iodide. Disorder and Molecular Motion between Inequivalent Sites. *J. Chem. Phys.* **1986**, *85*, 747–750. [CrossRef]
38. Mukhopadhyay, R.; Mitra, S.; Tsukushi, I.; Ikeda, S. Order–Disorder Transition in Pyridinium Iodide: Quasi-Elastic Neutron Scattering Study. *Chem. Phys. Lett.* **2001**, *341*, 45–50. [CrossRef]
39. Szafrański, M.; Szafraniak, I. Phase Transitions and Pressure Effects in Simple Pyridinium Salts. *J. Phys. Condens. Matter* **2003**, *15*, 5933. [CrossRef]
40. Pajzderska, A.; Gonzalez, M.A.; Wąsicki, J. Dynamics in Molecular and Molecular-Ionic Crystals: A Combined Experimental and Molecular Simulation Study of Reorientational Motions in Benzene, Pyridinium Iodide, and Pyridinium Nitrate. *J. Chem. Phys.* **2013**, *138*, 024508. [CrossRef]
41. Selivanov, N.I.; Murashkina, A.A.; Kevorkyants, R.; Emeline, A.V.; Bahnemann, D.W. Pyridinium Lead Tribromide and Pyridinium Lead Triiodide: Quasi-One-Dimensional Perovskites with an Optically Active Aromatic π-System. *Dalton Trans.* **2018**, *47*, 16313–16319. [CrossRef]
42. Yan, Z.S.; Long, J.Y.; Gong, Y. Three in situ—Synthesized novel inorganic-organic hybrid materials based on metal (M = Bi, Pb) iodide and organoamine using one-pot reactions: Structures, band gaps and optoelectronic properties. *New J. Chem.* **2018**, *42*, 699–707. [CrossRef]
43. Hartl, V.H. Die Kristallstruktur von Pyridiniumjodid. *Acta Cryst.* **1975**, *B31*, 1781. [CrossRef]
44. Pająk, Z.; Szafrańska, B.; Czarnecki, P.; Mayer, J.; Kozak, A. Polymorphic Phase Transitions and Molecular Motion in Pyridinium Chlorochromate. *Chem. Phys. Lett.* **1997**, *274*, 106–111.

45. Asaji, T.; Fujimori, H.; Ishida, H.; Eda, K.; Hashimoto, M.; Oguni, M. Calorimetric and Single Crystal X-Ray Study of the Phase Transition of $(PyH)_2PdCl_4$. *J. Phys. Chem. Solids* **2005**, *66*, 869–875. [CrossRef]
46. Wąsicki, J.; Pajzderska, A.; Fojud, Z. Temperature Dependence of Spontaneous Polarization in Order−Disorder Pyridinium Periodate Extracted from ^2H NMR Data. *J. Phys. Chem. C* **2008**, *112*, 7503–7508.
47. Pająk, Z.; Wolak, J.; Połomska, M. NIR Raman Study of Ferroelectric Phase Transitions in Pyridinium Tetrafluoroborate. *Ferroelectrics* **2004**, *303*, 207–210.
48. Wong, P.T.T.; Ripmeester, J.A. The Phase Transition and Reorientational Fluctuations in Pyridinium Iodide Crystal Investigated by Raman Scattering. *Chem. Phys. Lett.* **1980**, *72*, 122–126. [CrossRef]
49. Jóźków, J.; Jakubas, R.; Baran, J. Infrared Investigation of the Order–Disorder Phase Transitions in $(C_5H_5NH)_6Bi_4Cl_{18}$. *J. Mol. Struct.* **2000**, *555*, 273–279. [CrossRef]
50. Tarasiewicz, J.; Jakubas, R.; Bator, G.; Zaleski, J.; Baran, J.; Medycki, W. Structural Characterization, Thermal, Dielectric, Vibrational Properties and Molecular Dynamics of $(C_5H_5NH)_3BiCl_6$. *J. Mol. Struct.* **2009**, *932*, 6–15. [CrossRef]
51. Jesariew, D.; Ilczyszyn, M.M. Crystal Structure, Phase Transition, and Disorder in Pyridinium Methanesulfonate. *J. Phys. Chem. Solids* **2017**, *104*, 304–314. [CrossRef]
52. Groom, C.R.; Bruno, I.J.; Lightfoot, M.P.; Ward, S.C. The Cambridge Structural Database. *Acta Cryst.* **2016**, *B72*, 171–179. [CrossRef] [PubMed]
53. Wilson, E.B., Jr. The Normal Modes and Frequencies of Vibration of the Regular Plane Hexagon Model of the Benzene Molecule. *Phys. Rev.* **1934**, *45*, 706. [CrossRef]
54. Cook, D. Vibrational Spectra of Pyridinium Salts. *Can. J. Chem.* **1961**, *39*, 2009–2024. [CrossRef]
55. Gardner, A.M.; Wright, T.J. Consistent Assignment of the Vibrations of Monosubstituted Benzenes. *J. Chem. Phys.* **2011**, *135*, 114305. [CrossRef] [PubMed]
56. Foglizzo, R.; Novak, A. Low Frequency Infrared and Raman Spectra of Hydrogen Bonded Pyridinium Halides. *J. Chem. Phys.* **1969**, *50*, 5366. [CrossRef]
57. Glazunov, V.P.; Odinokov, S.E. Infrared Spectra of Pyridinium Salts in Solution-I. The Region of Middle Frequencies. *Spectrochim. Acta* **1982**, *38A*, 399–408. [CrossRef]
58. Glazunov, V.P.; Odinokov, S.E. Infrared Spectra of Pyridinium Salts in Solution-II. Fermi Resonance and Structure of ν_{NH} Bands. *Spectrochim. Acta* **1982**, *38A*, 409–415. [CrossRef]
59. Pichon, C.; Mialane, P.; Marrot, J.; Binet, C.; Vimont, A.; Travert, A.; Lavalley, J.-C. Combined XRD and Infrared Studies of Pyridinium Species in $(PyH)_3[PW_{12}O_{40}]$ Single Crystals. *Phys. Chem. Chem. Phys.* **2011**, *13*, 322–327. [CrossRef]
60. Binet, C.; Travert, A.; Daturi, M.; Lavalley, J.-C.; Vimont, A. Unusual IR Ring Mode Splittings for Pyridinium Species in $H_3PW_{12}O_{40}$ Heteropolyacid: Involvement of the δNH Internal Mode. *RSC Adv.* **2014**, *4*, 19159. [CrossRef]
61. Selivanov, N.I.; Murzin, A.O.; Yudin, V.I.; Kapitonov, Y.V.; Emeline, A.V. Counterdiffusion-In-Gel Growth of High Optical and Crystal Quality $MAPbX_3$ ($MA = CH_3NH_3^+$, $X = I^-$, Br^-) Lead-Halide Perovskite Single Crystals. *CrystEngComm* **2022**, *24*, 2976–2981. [CrossRef]

molecules

MDPI

Review

Structures and Dynamics of Complex Guest Molecules in Confinement, Revealed by Solid-State NMR, Molecular Dynamics, and Calorimetry

Nadia B. Haro Mares [1], Sonja C. Döller [1], Till Wissel [1], Markus Hoffmann [2,*], Michael Vogel [3,*] and Gerd Buntkowsky [1,*]

1 Eduard-Zintl-Institut für Anorganische und Physikalische Chemie, Technische Universität Darmstadt, Peter-Grünberg-Str. 8, D-64287 Darmstadt, Germany; haromares@chemie.tu-darmstadt.de (N.B.H.M.); sonja.doeller@tu-darmstadt.de (S.C.D.); wissel@chemie.tu-darmstadt.de (T.W.)

2 Department of Chemistry and Biochemistry, State University of New York at Brockport, Brockport, NY 14420, USA

3 Institute for Condensed Matter Physics, Technische Universität Darmstadt, Hochschulstr. 6, D-64289 Darmstadt, Germany

* Correspondence: mhoffman@brockport.edu (M.H.); michael.vogel@physik.tu-darmstadt.de (M.V.); gerd.buntkowsky@chemie.tu-darmstadt.de (G.B.)

Abstract: This review gives an overview of current trends in the investigation of confined molecules such as water, small and higher alcohols, carbonic acids, ethylene glycol, and non-ionic surfactants, such as polyethylene glycol or Triton-X, as guest molecules in neat and functionalized mesoporous silica materials employing solid-state NMR spectroscopy, supported by calorimetry and molecular dynamics simulations. The combination of steric interactions, hydrogen bonds, and hydrophobic and hydrophilic interactions results in a fascinating phase behavior in the confinement. Combining solid-state NMR and relaxometry, DNP hyperpolarization, molecular dynamics simulations, and general physicochemical techniques, it is possible to monitor these confined molecules and gain deep insights into this phase behavior and the underlying molecular arrangements. In many cases, the competition between hydrogen bonding and electrostatic interactions between polar and non-polar moieties of the guests and the host leads to the formation of ordered structures, despite the cramped surroundings inside the pores.

Keywords: confinement; NMR; molecular dynamics; mesoporous silica

1. Introduction

Ordered periodical mesoporous silica (PMS) [1,2], like MCM-41 (Mobil Composition of Matter No. 41) [3] and SBA-15 (Santa Barbara Amorphous) [4,5] and their many derivates, exhibit characteristic narrow pore-diameter distributions and large specific surface areas. Their high chemical stability makes them easy to handle under ambient conditions. Their reactive surface silanol groups (Si-OH) provide an easy pathway to chemical functionalization and tailored surface design, e.g., by post-synthetic grafting of functional groups such as amino, amide, carboxyl, phosphate [6,7], or by co-condensation with molecules containing such groups [8,9]. They have a high application potential in many technical processes, such as heterogeneous catalysis, separation technology, encapsulation of molecules, drug delivery, or selective adsorption [10–12]. Moreover, they provide an ideal model environment investigating the physicochemical properties of fluid guest molecules confined in a porous environment, showing strong competition with solid–liquid and liquid–liquid interactions or in biomineralization [13].

Revealing these properties necessitates the combination of several analytical and computational techniques, such as X-ray (XRD) and neutron diffraction techniques for the investigation of crystallinity and long-range order [14–17], small angle scattering (SAXS and

SANS) for the characterization of the pore geometry and pore ordering, differential scanning calorimetry (DSC) [18] for the study of phase or glass transition processes in confinement, gas adsorption (BET, BJH) for the characterization of the pore ordering, specific surface areas and pore diameters [19,20], solid-state NMR (SSNMR), and NMR diffusometry [21–25], possibly supported by Dynamic Nuclear Polarization (DNP) [26–33], to boost the NMR sensitivity or chemical shift calculations to help in the interpretation [34,35], or the study of the local ordering and dynamics on the molecular level and molecular dynamics (MD) simulations for the modelling of the dynamics and structures of the confined guests in the host material.

Hydrogen-bonded liquids in nanoscale confinements are a highly topical field of research [36]. In view of their enormous relevance to science and application, particularly intensive research efforts ascertained the properties of neat water and aqueous solutions under such circumstances [37,38]. Also, the properties of confined alcohol molecules in dependence on the size and chemistry of the confining framework were often a research focus [36,39]. In such investigations, phase behaviors, structures, and dynamics of hydrogen-bonded liquids were addressed, applying a wide range of experimental and computational methods [39]. For studies of dynamical aspects, broadband dielectric spectroscopy [40], quasielastic neutron scattering [41], NMR spectroscopy and diffusometry [42], and MD simulations [43,44] were very suitable methods. Various comprehensive review articles summarized these research efforts [36–41,43,44].

Here, we review NMR approaches to the structural and dynamical properties of hydrogen-bonded liquids in porous frameworks. In doing so, we build upon several review articles covering this topic [45–49]. Using isotope selective approaches in NMR spectroscopy, both rotational motion and translational diffusion in confinement were successfully determined. To characterize the reorientation of confined molecules, ^{2}H NMR studies of deuterated compounds proved to be a powerful tool [50–52]. In particular, when combining ^{2}H spin-lattice relaxation (SLR) analysis, including ^{2}H field-cycling relaxometry [53,54], with ^{2}H stimulated-echo experiments (STE), it was possible to follow MD over a broad range of correlations times [55–57], $\tau \approx 10^{-11}$–10^0 s. Furthermore, NMR experiments in magnetic field gradients enabled measurements of self-diffusion coefficients [23]. In these approaches, it was advantageous to use a static field gradient (SFG) rather than a pulsed one [58,59]. The SFG method enabled an application of stronger gradients and, in this way, an observation of diffusion on smaller-length scales down to roughly 100 nm. By exploiting these capabilities, it was possible to ensure that diffusion inside a particular framework is probed, e.g., inside a given pore, whereas distorting effects from an escape of the confinement can be neglected, e.g., fast displacements in empty space between mesoporous silica particles.

While these NMR approaches provide a quantitative evaluation of the molecular motions present, their qualitative interpretation on the molecular level can be challenging. To overcome this challenge, MD simulations provide a powerful theoretical approach to gain such molecular level insights. In MD simulations, an ensemble of molecules is allowed to evolve in time to obtain essentially a movie that reveals the present dynamic processes, intermolecular interactions, and the resulting structural patterns. The potentials of all bonding and non-bonding interactions between all present atoms must be defined at each time increment in order to obtain a new set of velocities at which each atom moves during the next simulation step. In ab-initio MD (AIMD) simulations, the potentials are recalculated ab-initio at each simulation time step [60]. The AIMD method is computationally very demanding, limiting its use to smaller-sized systems. For that reason, more commonly used are MD simulations that employ classical potential functions during the simulation. Bonding interactions typically consist of harmonic oscillator functions to describe chemical bonds and bond-angle vibrations and sinusoidal functions to describe the energy barriers for dihedral rotations. Non-bonding interactions are typically comprised of the Coulomb potential between (partial) charges and the Lennard Jones potential for describing the London dispersion forces. Numerous sets of parameters classically describing all of these interactions have been developed over time, and are referred to as force fields. Some of the most popular force fields include AMBER [61,62], OPLS/AA [63], CHARMM [64], and GROMOS [65]. To reduce computation times, some forcefields lump groups of atoms together, as is the case for the GROMOS force field where CH_2 groups for example are described as one constituent. Even more coarse-grained force fields have been developed, as well such as the MARTINI [66] force field that was specifically optimized for simulating polymers. The quality of the classical force field is assessed by comparison between simulated and experimental data. Aside reproducing experimental data, the most common analysis tasks of the MD simulations include the evaluation of radial distribution functions, which provide direct insights into the structural organization of the studied systems, and thus the present interactions, as well as the inspection of various correlations functions, from which time constants of present dynamics can be extracted and compared with experimental values. In the case of systems that engage in hydrogen bonding, these can be directly assessed from MD simulations, which experimentally is very difficult to achieve [67]. The interested reader is referred to several references for more details about MD simulations, such as the application of periodic boundary conditions and the pressure and temperature equilibration procedures [68,69].

In a recent review [47] some of us gave an extensive overview about the state of the art of confinement studies of small molecules, such as confined water, small aromatic molecules, alcohols, or carbonic acids [50,70–77] hosted in these materials, and discussed how the confinement affects thermophysical properties such as freezing and melting points of the guest molecules. While confinement effects on small guest molecules with simple physicochemical properties in mesoporous environments are well investigated, until recently not much was known about the local structures of more complex molecules, such as surfactants, in mesoporous confinement. In continuation with the previous review, the present paper gives an overview of a series of newer studies, where more complex molecules are confined inside these materials. The larger number of functional molecular sites permits a larger number of possible interactions, which enables these molecules to form more complex or richer structures than simple small molecules like benzene or pyridine. Of particular interest here is the competition between surface–guest hydrogen bonds and intermolecular (and possibly also intramolecular) guest–guest hydrogen bonds. This review only summarizes the findings since 2020. For a very extensive overview of older work, the reader is referred to the previous review [47].

The rest of this review is organized as follows: Section 2 gives an introduction into the preparation and surface modification of the mesoporous host materials and the investigated surfactants; Section 3 discusses the physicochemical properties of these guest molecules in their bulk phases and their behavior inside the confinement; and the review concludes with a Summary and Outlook into possible future developments in the field.

2. Materials and Methods

2.1. Host-Materials

Mesoporous silica materials like MCM-41 or SBA-15 combine large and adjustable (via the preparation) pore sizes, specific volumes, and specific surface areas with high thermal stability, low specific weights, and narrow pore-diameter distributions [78–80]. Both types of materials are relatively easy to prepare and to functionalize, following, e.g., the synthesis protocol by Grünberg et al. [81,82] or Grün et al. [83] (for details, see refs. [36,80,84]). Their quality, pore dimensions, and surface parameter can be easily determined by the combination of nitrogen adsorption (BET and BJH), ^{29}Si SSNMR spectroscopy, and SAXS. Important to note is that freshly prepared samples contain a substantial amount of surface-bound water molecules [49,85,86], which in general have to be removed for confinement [22] studies employing special drying protocols for the preparation of "water-free" silica samples [87].

2.2. Probe Molecules

The probe molecules considered in this review are water, octanol, ethylene glycol, and the surfactants E_5, polyethylene glycol, $C_{10}E_6$, and Triton-X (see Figure 1). Each of these chemical structures contains hydroxy as well as ether moieties, which both can engage in hydrogen-bonding interactions. While octanol, $C_{10}E_6$, and Triton-X have only a single hydroxyl group, which can interact with the silica surfaces, water, ethylene glycol (EG), and its polymers (PEGs) can interact via two terminal hydroxyl groups. Moreover, in the case of E_5, $C_{10}E_6$, and PEG, there is a length-dependent number of ether-oxygens, which can serve as a hydrogen-bond acceptor in competition to the hydroxyl groups.

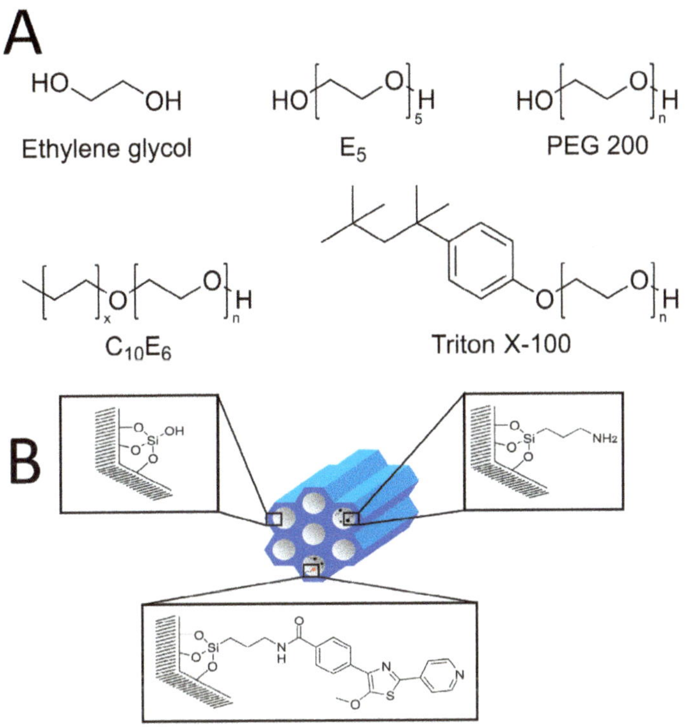

Figure 1. (**A**) Structures of 1-Octanol, EG, and the surfactants studied in this work. Except for E_5, the surfactants are polydisperse mixtures (exact compositions are given in ref. [88]). (**B**) Sketch of neat and functionalized mesoporous silica material.

2.2.1. 1-Octanol

1-octanol, an unbranched saturated fatty alcohol with the molecular formula $CH_3(CH_2)_7OH$, is commonly employed in the synthesis of esters. Owing to the hydrophilic hydroxy-group and the lipophilic alkyl chain, it is an ideal small model surfactant. It is often employed for evaluating the lipophilicity of pharmaceutical products. A quantitative measure for this is the water octanol partition coefficient or *p*-value K_{ow} [89]. It can be employed, e.g., for estimations of the partitioning of dissolved drug molecules between the cytosol and lipid membranes of living systems in pharmacology, or the behavior of water/oil mixtures in geology or environmental science [90]. Water–octanol mixtures are ideal model systems of the phase behavior of immiscible liquids in bulk and confined phases by combinations of solid-state NMR techniques such as 1D ^{1}H-MAS NMR, ^{29}Si-CP-MAS (Cross-Polarization Magic Angle Spinning) NMR, and ^{1}H/^{29}Si-HETCOR-(FSLG)-NMR (Heteronuclear Correlation by Frequency Switched Lee–Goldburg Decoupling) combined with MD simulations are commonly employed in such studies. The combination of these techniques reveals important information such as the distributions of the two liquids inside a confinement, as was shown recently by Kumari et al. in a series of papers [91–93].

2.2.2. Ethylene Glycol, Pentaethylene Glycol, and Polyethylene Glycol (PEG)

Ethylene glycol (EG) is the smallest vicinal diol and the simplest example of a polyhydric alcohol. EG finds wide application in the production of polyester fibers and for antifreeze formulations. Owing to the presence of the two hydroxyl groups, it can be employed as a simple model for confined liquids that can interact via several hydrogen bonds with host surface groups. Typical examples of these studies include NMR studies of small confined molecules inside zeolites [22,94–96] or mesoporous silica materials [45,97–99] and their functionalized derivates [100,101]. The low molecular weight representatives of polyethylene glycols (PEG, $H\text{-}[O\text{-}CH_2\text{-}CH_2]_n\text{-}OH$) possess environmentally benign properties including no toxicity, low vapor pressure, reducing exposure through inhalation and biodegradability. PEG is widely and relatively inexpensively available with an industrial annual production of about 500,000 tons per year [102]. Commercial PEG is manufactured as polydisperse mixtures, where the average molar weight is part of the product name. For example, PEG200 has an average molar weight of approximately 200 g mol^{-1}. PEG is a very good solvent for a wide variety of chemicals including some mineral salts [103], which facilitates its use, e.g., in transition metal catalyzed reactions, including heterogeneously catalyzed reactions where the transition metal catalyst is immobilized on a solid material of large surface area [104]. Accordingly, PEG has been increasingly used as a very attractive alternative solvent for Green Chemistry [105]. Its properties as an alternative solvent for chemical synthesis were reviewed in several recent articles [106–108]. To further aid these efforts in heterogenous catalysis using PEG as a solvent, they need to be studied under confinement to understand how their physical and chemical properties change under these conditions as a function of the degree of polymerization. Such studies should combine experimental studies employing NMR and thermodynamic measurements with theoretical methodologies such as MD simulation in order to be able to correctly interpret experimental results at the molecular level. In this review, confinement studies of the monomer EG, two different EG polymers, namely E_5, a monodisperse polymer with chain length of five and a polydisperse polymer with a distribution of chain lengths (PEG200), are reported and compared to the commercial surfactants $C_{10}E_6$ and Triton X-100 (see Figure 1).

2.3. *Simulation Methods*

As there are a number of excellent reviews on MD simulations [109–111], here, only the salient features relevant for this review are shortly summarized. The simulations were carried out using the GROMACS platform [112,113]. For liquids, the typical simulation protocol consisted of the steps of randomly inserting usually 1000 molecules into a virtual box that is chosen in size to be too large, removing close contacts between atoms that may have arisen during the random molecule insertion, letting the density equilibrate at the desired constant simulation temperature and pressure, and finally, after establishing the average density, simulating long enough at constant temperature and volume corresponding to the average density so that the system reaches the diffusive regime, which then allows extraction of the self-diffusion coefficient. For PEG200, simulation times of 300 ns were typical at a temperature of 328 K saving at least 10,000 frames for analysis. The simulations results summarized in this review were mostly obtained with the OPLS/AA force field. As is common, periodic boundary conditions were employed along with a Verlet cutoff scheme [114], and treatment of long-rang electrostatic interactions were employed with a smooth Particle-Mesh Ewald (PME) grid-wise cubic interpolation [115]. Temperature control was established with the Bussi–Donadio–Parinello velocity-rescaling thermostat [116], while pressure was controlled with the Parrinello-Rahman barostat [117,118]. Analysis of the obtained simulations was carried out mainly using modules available with the

GROMACS platform augmented by some self-developed script files that can be found along with a very detailed description of the simulation details in Hoffmann et al. [119].

2.4. Differential Scanning Calorymetry (DSC)

The DSC analysis were performed using a Heat Flux DSC. In this type of DSC, the sample and the reference are heated through the same heating source. Nitrogen was used as the purge and protective gas during the experiments.

The samples mentioned in this review were prepared and packed under inert conditions and tested within a temperature range of 100 to 300 °C under various heating/cooling rates, depending on the sample. For more detailed information about a specific sample, the reader is referred to the original papers [48,120].

3. Exemplary Studies

When studying the temperature-dependent dynamics of confined molecules that readily crystallize, such as water, it is important to consider that confinement usually affects the freezing and melting behaviors. This necessitates the determination whether specific findings relate to the fully liquid state above the melting temperature (T_m) or the partially frozen state below this temperature. In partially frozen states, crystalline regions near the pore center coexist at equilibrium with a liquid layer at the pore wall [45,52,121–123]. As specific examples of NMR reorientation and diffusion studies on fully liquid or partially frozen states in nanoscale confinements, we will discuss the dynamical properties of water (H_2O and D_2O), ethylene glycol, and LiCl aqueous solutions in native and functionalized mesoporous silica. This approach will provide detailed insights into the dependence of the rotational and diffusive motions of hydrogen-bonded liquids on the size, i.e., the diameter *d*, of the pores and the chemistry of their walls.

3.1. Water and Ion Dynamics

In partially frozen states, the molecules of the liquid layer interact with the porous framework in their immediate neighborhood very closely, enabling detailed insights into the influence of the confinement chemistry on liquid dynamics. Exploiting this possibility, ^{2}H NMR was utilized to investigate D_2O reorientation near different biomimetic interfaces, specifically near silica walls functionalized with various amino acids [58,124]. It was found that the rotational correlation times τ obtained from ^{2}H SLR analysis were longer in the functionalized than the pristine silica pores, but with values strongly dependent on the type of the amino acid; see Figure 2. The longest correlation times were observed for lysine (LYS), followed by alanine (ALA) and, finally, glutamic acid (GLU) functionalization. Based on these results, it was concluded that the flexibility of the surface groups is not the decisive parameter for the mobility of neighboring water molecules [58]. Instead, it was proposed that water reorientation is slower near amino acids with basic residues than near those with acidic ones. In addition, it is evident from Figure 2 that, independent of the functionalization, water dynamics exhibited a high temperature dependence in the narrow interfacial layers between the pore walls and the ice cores, described by an activation of about 1 eV, corresponding to nearly 100 kJ/mol.

Other studies employed very narrow confinements with diameters of $d \approx 2$ nm to fully suppress crystallization of water [53,56,57]. In these cases, a combination of ^{2}H SLR and STE studies provided access to the slowdown of water reorientation when approaching a glass transition. An important question of such research was to what degree the dynamics of confined water resembles that of bulk water in the supercooled regime, which is difficult to access due to rapid crystallization. In particular, it was vigorously debated whether or not dynamical crossovers observed for confined water [40,125,126] may be taken as evidence for the existence of a second critical point associated with a liquid–liquid phase transition of bulk water, which was proposed to be at the origin of water's anomalies [127].

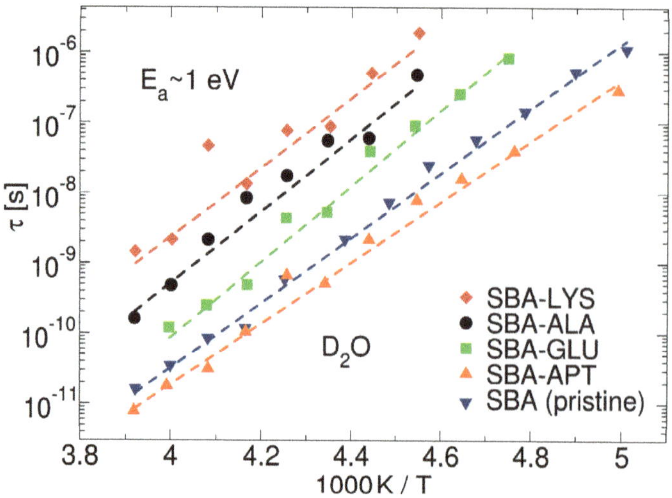

Figure 2. Rotational correlation times τ of D_2O in pristine (d = 5.4 nm) [57] and functionalized [124] mesoporous silica. For the functionalization, SBA-15 silica was functionalized via co-condensation with 3-(aminopropyl)triethoxysilane (APTES), yielding SBA-APT (d = 6.8 nm). Afterwards, the amino acids lysine (LYS, d = 5.8 nm), alanine (ALA, d = 5.9 nm), or glutamic acid (GLU, d = 5.6 nm) were coupled to this SBA-APT batch. The surface densities of the linked amino acids were 0.4–0.5 nm^{-2}. The dashed lines are Arrhenius fits with activation energies of ca. 1 eV.

Electrolyte solutions in nanoscale confinements are of enormous relevance across various fields. Their applications in heterogeneous catalysis and energy conversion are fundamental to modern society, and ion channels are crucial for the biological functions of living cells. However, many properties of electrolyte solutions in interfaces remain insufficiently understood to this day. The isotope selectivity of NMR was exploited to separately analyze water and ion dynamics for LiCl solutions in the bulk [128] and various confinements [129,130]. Figure 3 shows diffusion coefficients of LiCl–7H_2O solution obtained from ^{1}H and ^{7}Li SFG studies, which reflect the mobility of the water molecules and lithium ions, respectively. This composition, which is close to the eutectic one, was chosen to suppress crystallization and enable investigations in a broad temperature range. For the bulk solution, it was found that the ^{1}H diffusivity was slightly larger than the ^{7}Li diffusivity, and both exhibited a prominent non-Arrhenius temperature dependence typical of many glass-forming liquids. When confining the LiCl–7H_2O solution to a pristine silica pore with a diameter of d = 3.0 nm, the difference of the ^{1}H and ^{7}Li diffusivities increased to nearly an order of magnitude, and the temperature dependence again changed to an Arrhenius behavior with an activation energy of E_a = 0.26 eV [129]. It was argued that these prominent confinement effects resulted from Stern layer formation at the usually negatively charged silica walls. In a silica material with dye molecules grafted to the inner surfaces, even smaller ^{7}Li diffusion coefficients were reported [130]. To rationalize this finding, it was conjectured that these bulky functional groups protrude into the interior of the pores and, in this way, form obstacles for the long-range transport of the highly hydrated lithium ions.

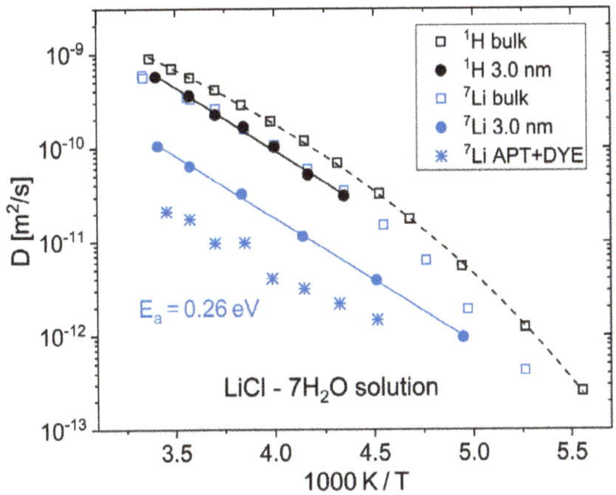

Figure 3. ^{1}H and ^{7}Li diffusion coefficients of a LiCl-7·H_2O solution in the bulk [128], in pristine silica pores with a diameter of d = 3.0 nm [129], and in functionalized silica pores (APT + DYE, d = 5.8 nm) [130]. For the functionalization, SBA-15 was functionalized with APTES via co-condensation, and then further modified by adding 4-(5-methoxormatioridin-4-yl)thiazol-4-yl)benzoic acid dye molecules, yielding a grafting density of ~1 dye molecule per nm^2. The dashed line is a VFT interpolation of the ^{1}H diffusion coefficients of the bulk solution. The solid lines are Arrhenius fits of the ^{1}H and ^{7}Li diffusion coefficients in the pristine pores, yielding the same activation energy of E_a = 0.26 eV.

Despite their high practical relevance, a comprehensive understanding of the partially frozen states of confined liquids is still lacking. For example, it is unclear whether these two-phase states are ergodic, i.e., whether the crystalline and liquid phases exchange molecules over time. ^{2}H NMR spectroscopy yielded important information about the ice–water equilibrium in silica nanopores below the melting temperature T_m [131]. The method exploited the fact that the 1D ^{2}H NMR line shape enabled a discrimination between molecules in the different phases; see Figure 4. Specifically, molecules of the less mobile ice phase contributed a broad Pake pattern ($\nu \neq 0$, in general), while those of the more mobile water phase added a narrow Lorentzian line ($\nu \approx 0$). Under such circumstances, 2D ^{2}H NMR spectra provided access to an exchange between both fractions. Specifically, molecules that belonged to water prior to the mixing time (t_m) exhibited $\nu_1 \approx 0$ and became ice during this period of the 2D experiment so that they showed $\nu_2 \neq 0$, produced a Pake spectrum along the frequency axis $\nu_1 = 0$. Vice versa, molecules that were a part of the ice phase before the mixing time t_m and of the water phase afterwards contributed such line shape along $\nu_2 = 0$. Together, a cross-like 2D spectral intensity along the frequency axes indicated ice–water exchange during the mixing time t_m. In a 2D ^{2}H NMR spectrum of D_2O in SBA-15 pores measured at 220 K for a mixing time of t_m = 5 ms, a cross-like intensity along the frequency axes was clearly observed; [131] see Figure 4. Unlike in a ^{1}H NMR approach [132], where spin diffusion was faster and at the underlying process of an exchange between the magnetizations of ice and water phases, the ^{2}H NMR findings could be traced back to an exchange of molecules between both phases. Explicitly, a detailed analysis of the mixing-time dependence of the cross-like and other 2D spectral intensities revealed that the residence time of a molecule in either phase was characterized by an exchange time of 5.7 ms at 220 K. Thus, the ice–water equilibrium was highly dynamic, or, in other words, ergodicity restoration occurred relatively fast for the two-phase state of water in nanoscale confinement. In this context, it should be mentioned that the crystal structure of confined ice was discussed for years, and the existence of stack-disordered ice comprising interlaced layers of cubic ice (I_c) and hexagonal (I_h) ice was proposed in recent studies [133,134].

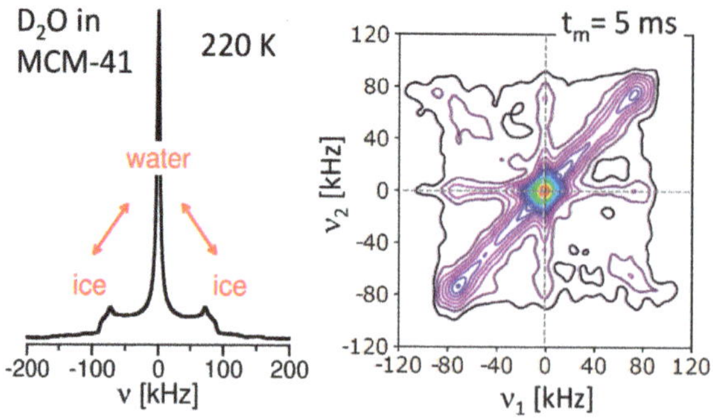

Figure 4. 1D and 2D ^{2}H NMR spectrum of D_2O in SBA-15 silica pores with a diameter of d = 5.4 nm at 220 K [131]. A mixing time of t_m = 5 ms was used to record the 2D spectrum. The dashed lines indicate the frequency axes $\nu_1 = 0$ and $\nu_2 = 0$ of the 2D spectrum.

3.2. Octanol

In the following, it is described how solid-state NMR techniques combined with MD simulations and thermodynamic techniques can be employed to investigate the structural arrangement and dynamics of confined molecules, employing monohydric alcohol 1-octanol [92,135] as a model compound for surfactants. The techniques employed for these investigations were originally developed for the study of confined isobutyric acid [136,137].

The structural arrangement of the octanol molecules [92] can be investigated with the application of ^{1}H/^{13}C CP-MAS FSLG HETCOR experiments (see Figure 5, left panel). These HETCOR experiments are sensitive to the magnetic dipolar interaction, thus providing information about the distance between the carbon nuclei of the octanol and the carbon and silica protons of the octanol and the silica host, respectively. By variation of the contact time in these experiments, it is possible to sense how close various moieties of the confined solvent molecules are to the pore surface. The analysis of the resulting spectra allows to deduce neighborship relations, which can be interpreted in terms of orientations and arrangements of confined molecules relative to the pore surfaces of the silica host (Figure 5, right panel).

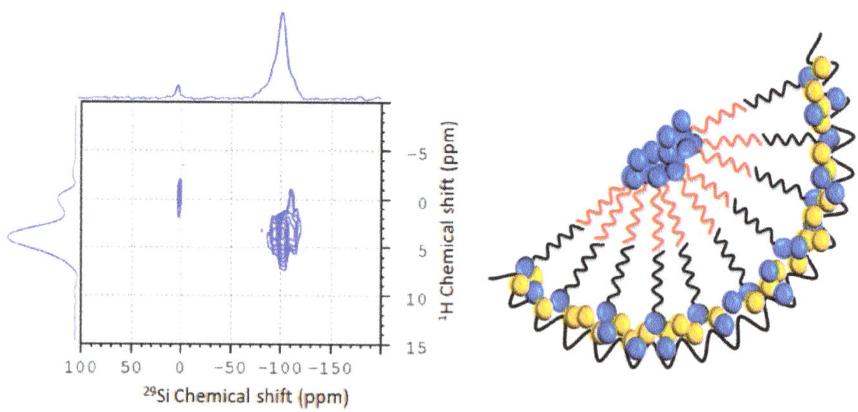

Figure 5. Left: Low temperature ^{1}H/^{29}Si CP-MAS FSLG HETCOR obtained with 9 ms contact time of an 80:20 mol% 1-octanol:water mixture confined in SBA-15. Referencing of the ^{1}H-dimension was performed by employing the technique described in ref. [93]. **Right**: Graphical visualization of a feasible bilayer formation of 1-octanol (blue) inside the pore. Water molecules (blue) are concentrated near the pore wall, as well as in the pore center (adapted from Kumari et al. [92]).

The dynamics of 1-octanol-d_{17} in its bulk phase and confined in mesoporous silica SBA-15 were also investigated by a combination of DSC experiments and ^{2}H-variable temperature solid-state NMR (solid-echo and MAS NMR experiments) in the region of the solid–liquid phase transition [135]. Compared to previous studies of smaller molecules such as benzene [50,138], naphtalene [139], or bipyridine [140], where relatively broad activation energy distributions were observed, the DSC results could be modeled by the Kissinger model, employing a single activation energy of (313.6 ± 2.1) kJ mol^{-1} for the bulk and a single activation energy of (172 ± 17) kJ mol^{-1} for the confined octanol [141]. The smaller activation energy of the confined octanol is reflected in its lower melting point, which is approximately 38 K below the bulk value (e.g., 219.7 K versus 257.3 K at a heating rate of 5 K min^{-1}). The larger uncertainty in the value of the confined molecules is already an indication of larger structural disorder and the coexistence of different species with different melting points [142]. In order to gain further insights into the effects of the confinement on the octanol, the ^{2}H-solid state NMR spectra of bulk and confined octanol are compared in Figure 6A,B. Generally, the melting points observed in the NMR experiments agree with those determined using the DSC measurements. Their line-shape analysis (Figure 6C) reveals a superposition of different spectral components. The static spectra below the melting point display a superposition of two Pake patterns with different width and intensity. The broader Pake pattern shows $C_Q \approx 170$ kHz just until the melting point, a value typical for an immobile deuteron of a -CD bond [143]. The quadrupolar coupling constant of $C_Q \approx 55$ kHz of the narrower Pake pattern is characteristic for a CD_3-group moving around its C_3-axis in a three-fold jump [144]. Accordingly, the two Pake patterns are assigned to the methyl and methylene deuterons of the alkyl chain. An additional narrow Lorentzian signal appears close to the melting point. This type of signal is characteristic for the onset of melting and the presence of mobile molecules. Below temperatures of 195 K for bulk octanol-d_{17} and 170 K for octanol-d_{17} in SBA-15, an additional broad and unstructured component is present in the static spectra. The latter is attributed to deuterons whose motions are falling into the intermediate exchange regime and have relatively short effective T_2 values [145,146]. This interpretation is corroborated with the ^{2}H MAS NMR spectra, where only the large and the small Pake pattern are visible, plus the narrow Lorentzian signal close to the melting point of the respective compound, but not the broad unstructured component, due to its short T_2. The distribution of activation energies for the melting process is calculated from the mole-fractions of the spectral components employing the Roessler model [147] (see ref. [135] for details).

The resulting distributions of activation energies of melting corroborated the results of the Kissinger analysis. For the bulk octanol-d_{17}, a narrow distribution of a well-ordered crystalline solid was observed [135], and for the confined octanol a broad distribution denoting a melting process involving species in a distribution of different environments and activation energies, and possibly a distribution of rigidly and less-rigidly ordered molecules [71,142], was observed. Moreover, the melting curves for the confined octanol-d_{17} exhibit clear deviations between the experimental and calculated curves towards lower temperature, which are indicative for a non-Gaussian distribution of activation energies. Additionally, this non-Gaussian distribution is illustrated by the numerical derivative of the experimental data (Figure 6, right), especially for the confined octanol-d_{17} under MAS conditions. For this sample, a shoulder on the low-temperature flank of the main curve is visible, indicating lower melting points for the molecules involved in the pre-melting process compared to the full melting.

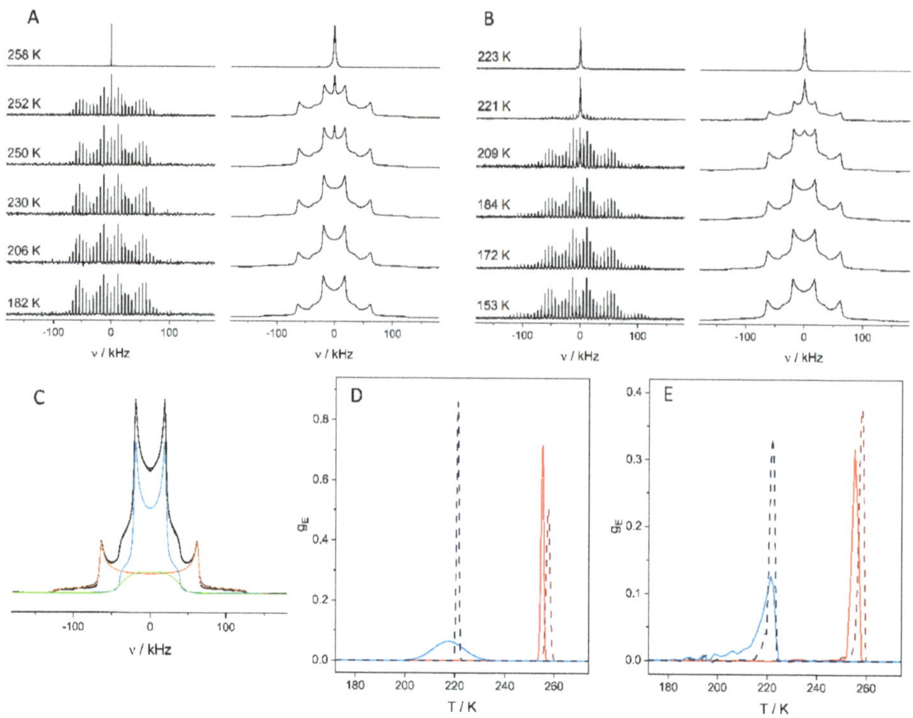

Figure 6. Comparison of ^{2}H MAS NMR spectra and solid echo ^{2}H spectra of (**A**) bulk octanol-d_{17} and (**B**) octanol-d_{17} confined in mesoporous SBA-15 as a function of temperature. Exemplary line-shape analysis of a static spectrum and calculated distributions of activation energies are shown in the bottom panels: (**C**) line-shape analysis of the solid echo ^{2}H NMR spectrum of bulk octanol-d_{17} at 150 K (black) revealing (blue) narrow Pake pattern of methyl deuterons; (orange) broad Pake pattern of methylene deuterons and unstructured component (green) of deuterons with short relaxation times. (**D**,**E**): Distribution of activation energies for the melting process, calculated with the Roessler model (blue: octanol-d_{17} in SBA-15, red: bulk octanol-d_{17}, solid lines: MAS conditions, dashed lines: static conditions). (Adapted from Döller et al. [135]).

In studies of guests that are not fully isotope labelled, it is often necessary to employ DNP enhancement for the characterization of the confined guest molecules. The efficacy of these DNP enhancements in hyperpolarized NMR crucially depends on a good compatibility between the polarizing agent (PA), which is typically a dissolved organic mono- or bi-radical, and the employed frozen solvents, often called the DNP matrix. A homogeneous distribution of the radicals, both in space and orientation, is necessary to achieve good enhancements. In order to investigate this, Döller et al. [148] studied the behavior of four different commercially available DNP polarizing agents confined in the non-ionic model surfactant 1-octanol as analyte and established a novel relative quantification method for the comparison of the proportion of the direct and indirect polarization transfer pathway efficacies, which is able to take concentration effects into account. This study revealed that the hydrophilicity of the PA is the key factor in the way polarization is transferred from the polarizing agent to the analyte.

3.3. Ethylene Glycol

As the starting point of the investigation of larger confined guest molecules who are capable to perform several hydrogen bonding interactions, a study involving partially deuterated ethylene glycol monomer (EG-d_4) in its bulk phase, confined in SBA-15, as well as APTES-modified SBA-15 was conducted. A combination of thermodynamic measurements, solid-state NMR, and MD simulations were employed [120]. The phase behavior (i.e., melting, crystallization, glass formation, etc.) of EG-d_4 in these three systems was studied using DSC (Figure 7). Through line-shape analysis of the ^{2}H ssNMR spectra recorded at different temperatures, two signal patterns were identified for each of the three investigated systems: a Lorentzian pattern, indicative of a liquid-like state, and a Pake pattern characteristic of a solid-like state. Using a two-phase model, the distribution of activation energies for the dynamic processes in each system was calculated. The spectra reveal an interesting behavior of the confined EG. On the one hand, similar to the previously studied confined 1-octanol [135], the ^{2}H NMR spectra indicated the formation of a crystalline solid inside the pores at reduced phase transition temperatures compared to unconfined EG. On the other hand, DSC scans of the same samples, shown exemplarily in Figure 7, indicated the formation of an amorphous glass under rapid cooling. Interestingly, during the heating phase of the DSC temperature cycle, the formed amorphous glass relaxes to form a crystalline solid that later melts at further elevated temperatures. Moreover, the behavior of the EG depends strongly on the surface modification of the SBA-15 host. On non-functionalized surfaces, strong hydrogen-bonding interactions between EG and surface silanol groups were revealed by causing a slowing down of EG dynamics [149].

In contrast, in the case of APTES-functionalized surfaces, where the polar surface-silanol groups are, to a large extent, removed by the binding of the APTES, the surface is far less polar and has a much lower capacity to form hydrogen bonds with EG-molecules [150]. This results in weaker interactions between pore-surface and EG molecules, placing more importance to the EG–EG interactions and a higher tendency to form crystalline EG phases. As a result, a substantially larger portion of EG in the pore remains solid after the first melting event. These effects are schematically shown in the lower panel of Figure 7 (see ref. Haro et al. [120] for more details).

The rotational motion and translational diffusion of EG in mesoporous silica is studied in detail in refs. [120,149]. To determine the pore-size dependence of these dynamics unaffected by crystallization, a ^{1}H and ^{2}H NMR study focused on the fully liquid state above T_m [149]. Correlation times (τ) from ^{2}H SLR measurements indicated that the molecular reorientation mildly slows down as the pore size decreases to d = 2.4 nm; see Figure 8. A slightly more pronounced pore-size dependence was observed for the diffusion coefficients (D) from ^{1}H SFG experiments. However, both reorientation and diffusion exhibited similar temperature dependencies. In both cases, the non-Arrhenius temperature behavior of the bulk liquid turned into Arrhenius behavior in sufficiently severe confinements. In more detail, the Stokes–Einstein–Debye (SED) relation,

$$D\tau = \frac{2}{9}R_H^2$$

was obeyed not only in the bulk liquid, where the experimental values of τ and D indicated a hydrodynamic radius of R_H = 1.15 Å, but also in the silica pores. However, the different pore-size dependencies of the rotational and translation motions manifested themselves in reduced hydrodynamic radii, e.g., R_H = 0.8 Å for the narrowest confinement with d = 2.4 nm. Correlation times τ of EG in lysozyme and elastin matrices obtained from ^{2}H NMR SLR and STE studies resembled those in silica pores [151]. In particular, an Arrhenius temperature dependence with an activation energy of $E_a \approx 0.6$ eV was also observed. Thus, the dynamics of EG does not depend on the exact chemistry of a confining framework, at least as long as the latter allows for a formation of hydrogen bonds.

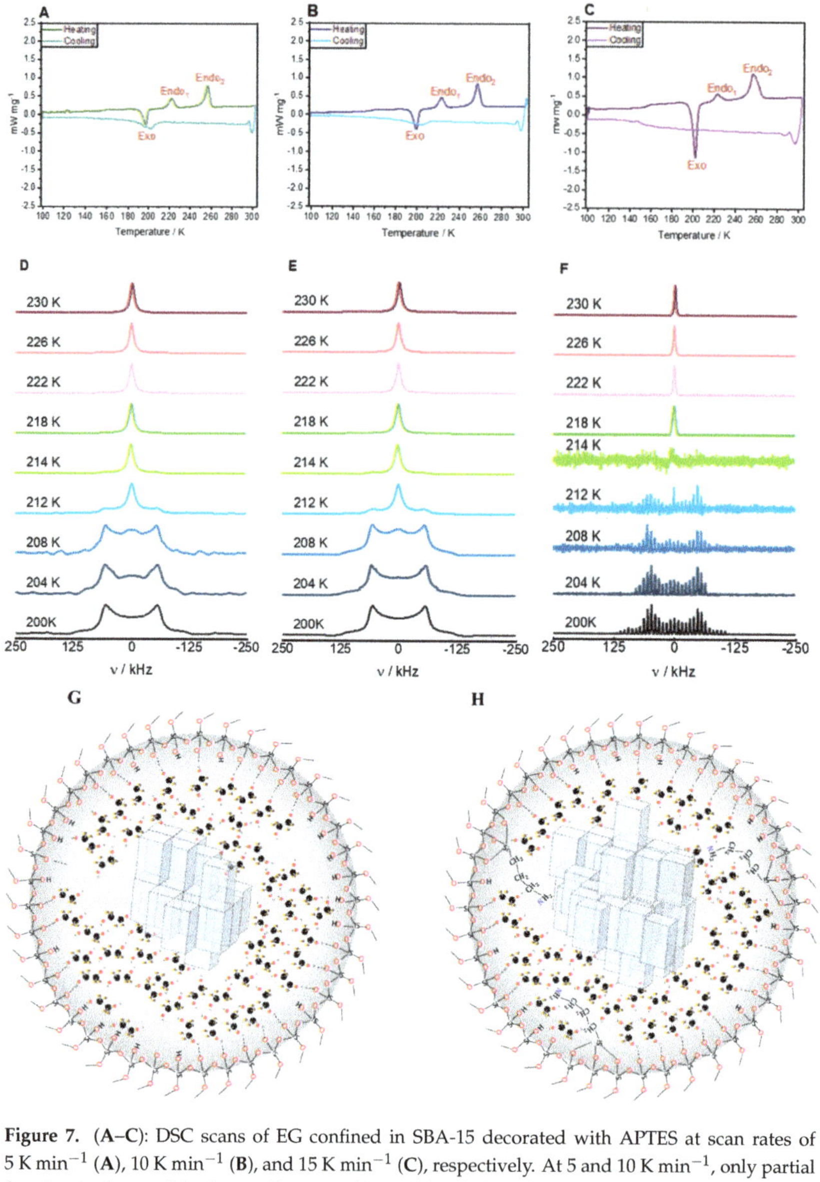

Figure 7. **(A–C)**: DSC scans of EG confined in SBA-15 decorated with APTES at scan rates of 5 K min^{-1} **(A)**, 10 K min^{-1} **(B)**, and 15 K min^{-1} **(C)**, respectively. At 5 and 10 K min^{-1}, only partial freezing is observed during cooling, as evidenced by the broad negative peak near 205 K. As the sample relaxes during heating, the release of heat near 200 K indicates the formation of a crystalline solid that then melts again in two steps near 200 K and 260 K, where the latter is presumed to indicate the presence of EG not confined in the pores. At 15 K min^{-1} (and higher rates, not shown) the small step between 150–160 K during cooling indicates formation of a glass. **(D–F)**: ^{2}H ssNMR spectra obtained in the temperature range between 200 and 230 K for EG-d4 in SBA-15 (sample 2). **(D)** ^{2}H static NMR experimental data, **(E)** fitted ^{2}H static NMR spectra, and **(F)** ^{2}H MAS experimental data. **(G,H)**: Schematic illustration of the arrangement of EG-d$_4$ molecules inside the pores of **(G)** pristine SBA-15 and **(H)** APTES-functionalized SBA-15. This picture includes the formation of crystal-like structures of EG-d$_4$ in the middle of the pores, as well as amorphous structures of EG-d$_4$ close to the pore wall.

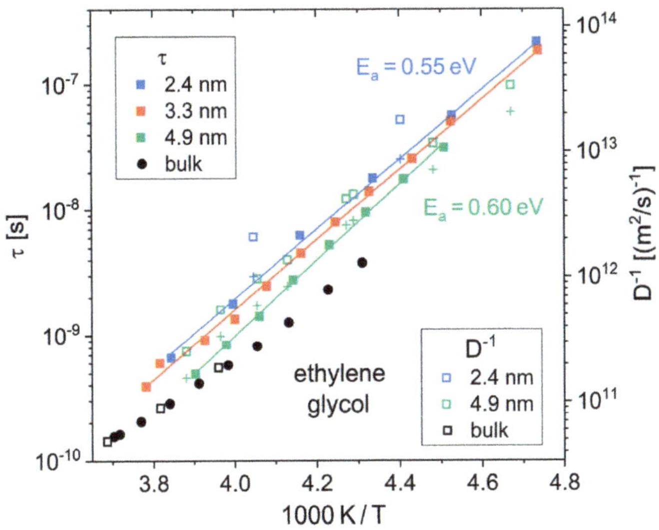

Figure 8. Temperature-dependent correlation times τ (solid symbols) and diffusion coefficients D (open symbols) of EG in the bulk liquid and in silica pores with the indicated diameters d [149]. The axis scaling superimposes τ (from ^{2}H SLR on EG-d$_4$) and D^{-1} (from ^{1}H SFG on EG-h$_6$) data given the SED relation is valid, and the hydrodynamic radius amounts to R_H = 1.15 Å. The solid lines are Arrhenius fits of the correlation times, yielding activation energies of E_a = 0.55 eV for d = 2.4 nm and d = 3.3 nm and E_a = 0.60 eV for d = 4.9 nm. The crosses are correlation times τ calculated from the diffusion coefficients D using a hydrodynamic radius of R_H = 0.8 Å and R_H = 0.9 Å for the pore diameters of d = 2.4 nm and d = 4.9 nm, respectively.

3.4. Polyethylene Glycol and Related Surfactants

3.4.1. Experimental Studies

In the next step, the EG polymers E5 and PEG200 were studied in confinement and compared to the commercially available surfactants $C_{10}F_6$ and Triton X-100 [152]. Two different mesoporous silica materials (SBA-15 and MCM 41) were impregnated with these surfactants. DSC was employed to confirm the confinement of the surfactants in the pores of their host materials. DNP enhanced solid state ^{13}C MAS-NMR spectra were recorded for these materials, showing that both the direct as well as the indirect polarization transfer pathways are active for the carbons of the polyethylene glycol moieties of the surfactants (see Figure 9). The presence of the indirect polarization pathway implies the presence of molecular motion with correlation times faster than the inverse Larmor frequency of the observed signals. The intensities of the signals were determined, and an approach based on relative intensities was employed to ensure comparability throughout the samples. From these data, the interactions of the surfactants with the pore walls could be determined. Additionally, a model describing the surfactants' arrangement in the pores was developed. It was concluded that all carbons of the hydrophilic surfactants, E_5 and PEG200, interact with the silica walls in a similar fashion, leading to similar polarization transfer pathway patterns for all observed signals. For the amphiphilic surfactants $C_{10}E_6$ and Triton X-100, the terminal hydroxyl group mediates the majority of the interactions with the pore walls and the polarizing agent. From these data, models of the distribution of the surfactants and PA in confinement can be built (see Figure 9).

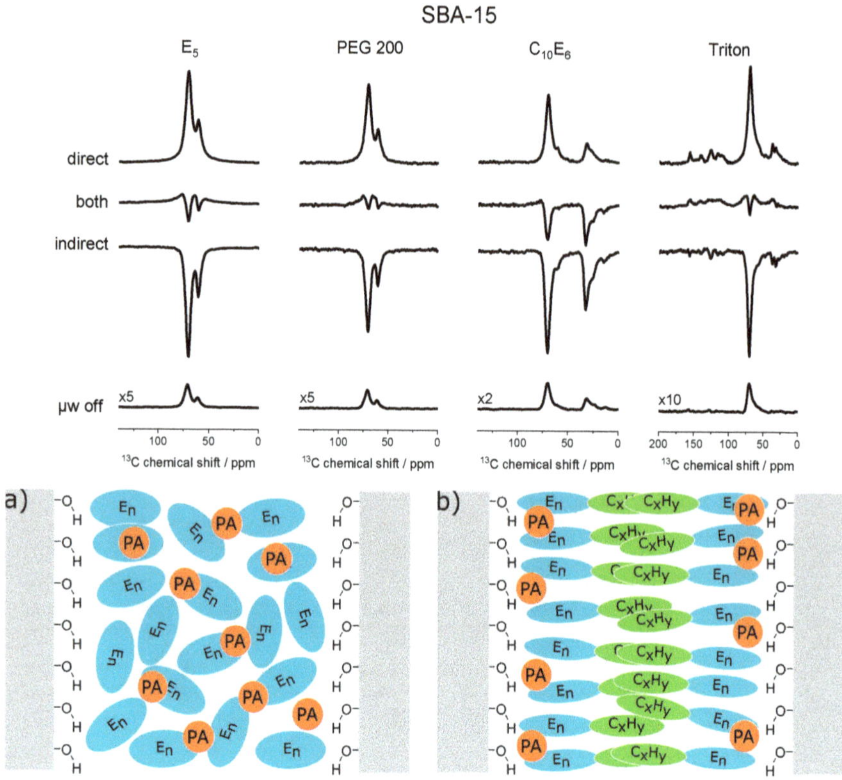

Figure 9. Upper panel: DNP-enhanced ^{13}C MAS ssNMR spectra revealing the competition between direct and indirect polarization of the confined surfactants. **Lower panel**: Schematic illustration of the distribution of the PA and the surfactant in the pores: (**a**) the hydrophilic surfactants E_5 and PEG200, and (**b**) the amphiphilic surfactants $C_{10}E_6$ and Triton. Adapted from ref. [152].

Finally, in ref. [153] it is shown how linewidth and DNP intensities are able to provide very detailed structural information about confined surfactants, employing $C_{10}E_6$ confined in SBA-15 modified with aminopropyltriethoxysilane (APTES) as an example. The structural interpretations were largely deduced from the trends in linewidths and intensities of the observed dynamic DNP-enhanced solid state ^{13}C spectral features originating from the direct and indirect polarization transfer. These findings illustrate that DNP may not only be used to boost signal intensity, but also as a tool to obtain chemical information related to structure and intermolecular interactions. Specifically, for the case of $C_{10}E_6$, as shown in Figure 10, it was found that at low surface coverage of SBA-15 with APTES ("SBA-5%" where 5: 95 equivalents of APTES: tetraethyl orthosilicate (TEOS) were used during synthesis) the polar portion of the $C_{10}E_6$ molecules interact competitively with the polar polarizing agent as well as the abundantly present hydroxy surface functional groups. These competitive interactions result in a more disordered assembly of $C_{10}E_6$ molecules around the polarizing agent, as shown in Figure 10a, compared to the situation shown in Figure 10b where the APTES surface coverage is increased by about 4-fold ("SBA-20%", i.e., 20:80 equivalents of APTES:TEOS), and thus is overall much less polar. In another set of experiments, we tried to repeat a similar study with 1-octanol as the confined molecule. Unfortunately, the DNP enhancements were too small for this purpose. However, we discovered that the ratio of direct and indirect spectral intensities along with their build-up times of bulk unconfined 1-octanol, can be used as a polarity measure of the polarization agent [148].

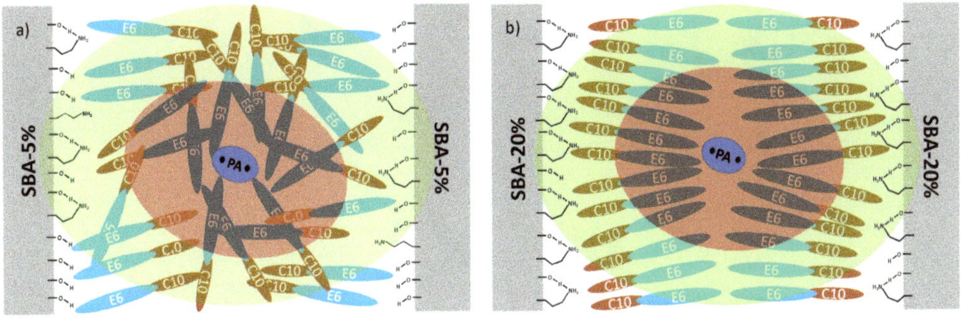

Figure 10. Illustration of (**a**) SBA-5%, (**b**) SBA-20% impregnated with a hydrophilic polarizing agent (PA) in $C_{10}E_6$. The larger APTES coverage in SBA-20% renders the pore surface to be less polar than in SBA-5%, which supports formation of a structured bilayer arrangement of $C_{10}E_6$ within the pore. The gray blocks represent the silica pore wall, and the red oval area represents the region around the polarizing agent where nuclei cannot be detected using NMR. The green oval represents the region to which nuclei may receive polarization directly from the polarizing agent. Adapted from Hoffmann et al. [119].

3.4.2. Molecular Dynamics Simulations of PEG

The force fields used in MD simulations to describe the bonded and nonbonded interactions between all involved atoms of PEG need to be validated against experimental results. This initial task was recently completed for PEG200, which, to the best of our knowledge, was the first report on MD simulations of mixtures of ethylene glycol oligomers [154]. Widely available forcefields were tested for their accuracy. While some force fields reproduced reasonably well experimental densities and self-diffusion coefficients for diethylene glycol, the agreement of the self-diffusion coefficients and viscosities progressively worsened with increasing oligomer length.

Adjustments to the All-atom optimized potentials for liquid simulation (OPLS-AA) force field regarding the hydroxy end-group charges and the dihedral potential barrier for rotation around the terminal C-C axis were found to improve agreement with experimental data. These adjustments had to be optimized for each oligomer. It was observed that the changes to the OPLS-AA force field reduced overall hydrogen bonding interactions and shifted them from inter- to intra-molecular hydrogen bonding. A remarkable propensity for intramolecular hydrogen bonding was observed for tri- and tetra-ethylene glycol. Typical structural configurations are shown in Figure 11, and illustrate how these two molecules maximize the proximity of the hydroxy hydrogen atoms to intramolecular oxygen atoms. The increased tendency of intramolecular interactions leads to smaller end-to-end distances and radii of gyration, explaining the observed decrease in viscosity and concurrent increase in self-diffusion. Overall, neither the modified nor the unmodified forcefield showed any indication of preferential association of oligomers to form clusters. Instead, PEG200 may be regarded as a random mixture of its ethylene glycol oligomer components. Experimental values of density, viscosity, and self-diffusion coefficient of PEG200, as well as its ethylene glycol oligomer components up to nonaethylene glycol, to which the simulation results were compared to, have only been recently become available, covering a range of temperatures from room temperature to 358 K [155,156]. The trends of these data are important for promoting PEG200 as a solvent, and warrant a brief summary here.

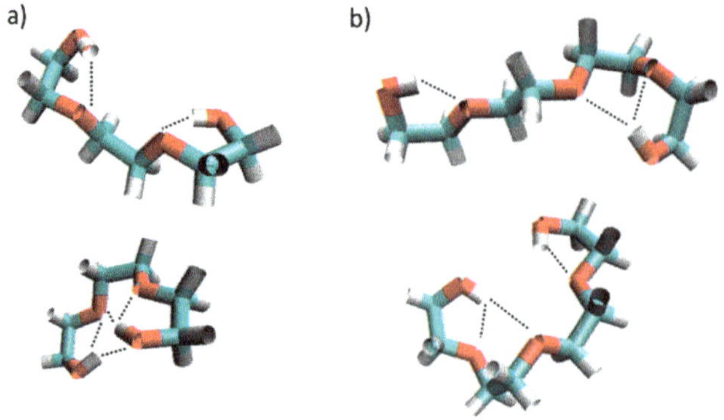

Figure 11. MD simulation sample snapshots of triethylene glycol (**a**) and tetraethylene glycol; (**b**) oligomers illustrating typical structural configurations. The dotted line highlights close proximities of hydroxyl hydrogen atoms to oxygen atoms.

Since water can readily be absorbed from the atmosphere, it is the main impurity of PEG and its oligomer components. The effect of water impurities on the physical properties was tested, and was found to hardly affect these three physical properties up to mole fractions of about 0.15. The temperature dependence showed linearity for density and followed an Arrhenius-type behavior for viscosity and self-diffusion,

$$\ln(X(T)) = \ln A \pm \frac{E_a}{RT} \quad (1)$$

where $X(T)$ represents the temperature-dependent property, and the sign before the second term of the right-hand equation is positive for viscosity and negative for the self-diffusion coefficients, T is the temperature, R the gas constant, E_a is the activation energy, and A is the pre-exponential factor. Interestingly, the E_a values were within experimental uncertainty the same for viscosity and self-diffusion, indicating that the same activation barriers exist for translational motion and momentum transfer. Moreover, with respect to neat oligomer components, the straightforward relationships of the physical properties, with respect to the number of ethylene oxide repeat units, n, shown in Equations (2)–(4), could be established allowing the prediction of these properties for higher oligomers.

$$\text{Molar Volume}, \overline{V} : \overline{V}(T)/\text{cm}^3{\cdot}\text{mol} = (0.0303\,n + 0.0079)T + 30.11\,n + 14.38 \quad (2)$$

$$\text{Viscosity}, \eta : \ln(\eta(T)/\text{mP}{\cdot}\text{s}) = -8.619 + \frac{456.1n + 28851}{RT} \quad (3)$$

$$\text{Self-diffusion coefficient}, D : \ln\left(D(T)/m^2{\cdot}s^{-1}\right) = 11.268 + \frac{456.1n + 28851}{RT} \quad (4)$$

It was also tested if the composition for PEG200 varies significantly from vendor to vendor, which was not the case. Even if it did, it would be inconsequential because of the remarkable finding that the properties of PEG do not depend on its composition as long as the average molar weight is not changed. This conclusion was based on the observation that a binary tri- and hexaethylene glycol mixture with an average molar weight of 200 g mol^{-1} possesses the same properties as PEG200, as exemplarily shown in Figure 12 for the viscosity and the self-diffusion coefficient. Finally, the binary system of tri- and hexaethylene glycol showed ideal mixing behavior. Thus, the three investigated physical properties were found to be reasonably predictable for PEG200 by mole-fraction weighted averaging of the properties of their individual ethylene glycol oligomer components, as can be seen in the left panel of Figure 12.

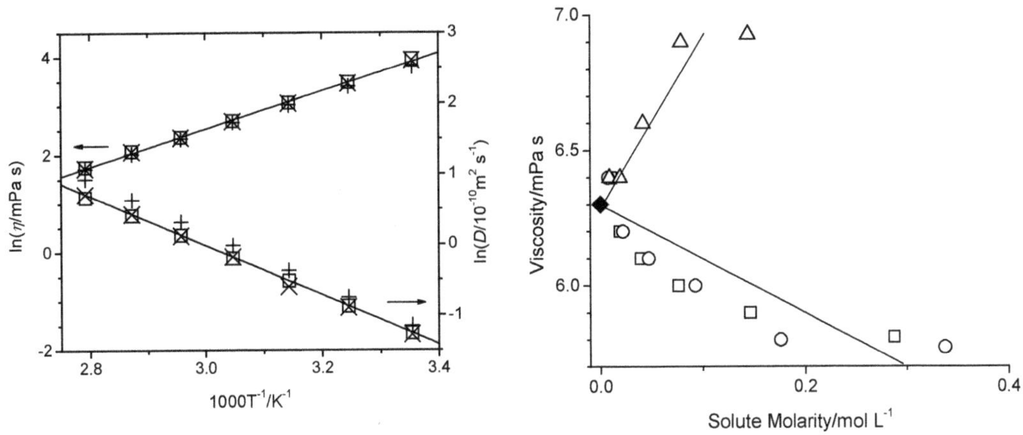

Figure 12. Left: Natural logarithms of viscosity (left y-scale) and self-diffusion coefficients (right y-scale) as a function of inverse temperature: experimentally for PEG200 (squares) with least square fit (solid line), experimentally for binary mixture of tri- and hexaethylene glycol with average molar weight of 200 g·mol^{-1} (crosses), and calculated for PEG200 from mole-fraction weighted averages of the individual component properties obtained from Equations (3) and (4) (plusses). **Right**: Bulk viscosity as a function of solute molarity in PEG200 at 358 K for added TEMPO (from BP Bio as squares and AA Blocks as circles) and 5-TBIPA (triangles). The viscosity of neat PEG200 (solid diamond) is taken from Hoffmann et al. [156], and the solid lines serve as guides for the eye.

Overall, these findings illustrate that applications using PEG200 as solvent should be robust with respect to tolerance to batch-to-batch variations from the same or different vendor sources and levels of water impurity. In preparation for future MD simulations of solutions with PEG as the solvent, experimental data have also been measured for two different solutes of interest in PEG200, namely 2,2,6,6-tetramethylpiperidinyloxyl (TEMPO) and 5-tert-butylisophthalic acid (5-TBIPA) [157]. 5-TBIPA was chosen as a model reagent for the synthesis of MOFs [158]. TEMPO is a widely used stable free radical with applications in chemical transformations [159,160], particularly as a redox catalyst [161]. It was primarily chosen in that study because many polarization agents in DNP-enhanced NMR studies are based on TEMPO as the free radical moiety [162–168]. As exemplified for the viscosity (right panel of Figure 12), the addition of 5-TBIPA increases solution density and viscosity. Combined with the finding that 5-TBIPA consistently self-diffuses at about half the rate as PEG200 at all investigated experimental conditions [157], this suggests strong attractive solute–solvent interactions, likely through hydrogen bonding interactions. In contrast, the addition of TEMPO causes lower solution densities and viscosities, suggesting that the solute–solvent interactions of TEMPO lead to an overall weakening of the intermolecular interactions present compared to neat PEG200. Furthermore, while the viscosity increases with 5-TBIPA addition, it was found that the PEG self-diffusion remains essentially unaffected, contradicting the Stokes–Einstein relation

$$D = \frac{k_B T}{\xi \pi \eta r}, \tag{5}$$

where viscosity and self-diffusion are inversely proportional to each other at constant temperature (k_B: Boltzmann constant, T: temperature, r: hydrodynamic solute radius, ξ: a constant typically between values of 4–6).

Deviations from the Stokes–Einstein relation were also reported for the PEG-related surfactant $C_{10}E_6$ dissolved in cyclohexane [169,170]. In contrast to PEG, the additional alkyl chain leads to the formation of reverse micellar aggregates. The size of the aggregates increases with increasing surfactant concentration. Addition of water further increases the size of the aggregates, as water preferentially resides inside the reverse micelle due to its interactions with the hydrophilic heads of the surfactant compared to the hydrophobic cyclohexane continuous phase. Thus, the water is in a prison consisting, in this case, of reverse micelles in a nonpolar fluid medium. Using the independently measured surfactant self-diffusion coefficients and the solution viscosity, the calculated radii using the Stokes–Einstein relation undergo a maximum as a function of surfactant concentration instead of continuously increasing. This can be explained by a change in mass transport where at low solute concentrations all surfactant solute species (monomers, dimers, small aggregates) contribute to the average self-diffusion coefficient, while at a high-solute concentration, the aggregates have become too large to significantly contribute to the mass transport that is thus now dominated by the monomer species (along with dimers/lower oligomers). Pictorially, at a high-solute concentration, the surfactant monomers are hopping from aggregate to aggregate, which in comparison are stationary. This also implies that there remain significant numbers of surfactant monomers present even at high surfactant concentrations, which shows that reverse micellar formation proceeds through a series of chemical equilibria steps, monomer to dimer, dimer to trimer, etc. Such gradual formation of the reverse micelle is in sharp contrast to the typical micelle formation in aqueous solution where a large number of surfactant solutes form in one step one very large micellar aggregate, and the chemical equilibrium is shifted so far to the micelle formation that the concentration of unaggregated surfactant monomers is negligibly small. The same phenomenon of an apparent maximum of aggregate size with increasing solute concentration has also been observed for some ionic liquids dissolved in solvents of low polarity [71,72,171,172], as shown in Figure 13 exemplary for 1-ethyl-3-methylimidazolium bis(trifluoromethylsulfonyl)amide in dichloromethane, illustrating that this phenomenon is more broadly applicable. Interestingly, more reasonable aggregate radii could be obtained when employing the ratio of surfactant solute and cyclohexane solvent self-diffusion coefficients, i.e., without the viscosities entering the evaluations. This shows how viscosity and self-diffusion are decoupled in these systems thus leading to an apparent breakdown of the Stokes–Einstein relation.

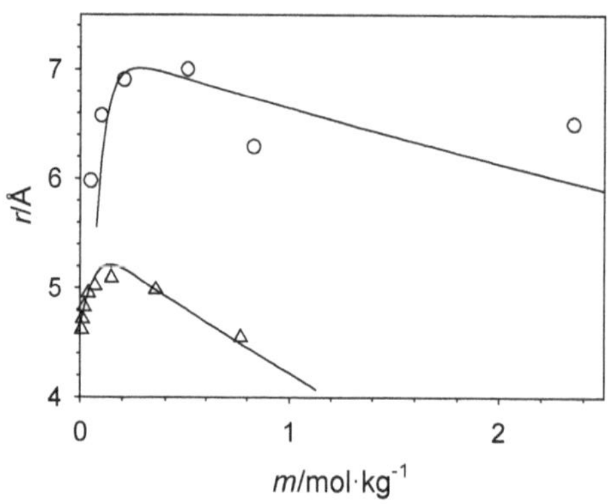

Figure 13. Apparent average solute radii as a function of solute concentration for $C_{10}E_6$ in cyclohexane (circles) at 283.15 K and for 1-ethyl-3-methylimidazolium bis(trifluoromethylsulfonyl)amide in dichloromethane at 288.15 K (triangles). The lines are guides for the eye.

4. Summary and Outlook

This paper reviews recent advances in characterizing, in particular, surfactants confined in microporous and mesoporous materials, employing solid-state NMR techniques, calorimetry, and MD simulations. It is shown that the combination of these techniques is capable of revealing detailed insights into the structural arrangement and dynamics of the confined guest molecules, which arise from the interplay between guest–guest and guest–host interactions. It is shown that the simultaneous presence of multiple hydrogen-bonding sites and polar/non-polar moieties in the guests leads to the formation of ordered structures, despite the cramped surroundings inside the pores. A number of examples from the groups of the authors were given to highlight this fascinating emergence of order in the narrow pores. Finally, the review concludes with some thoughts on the future direction of the field. It is anticipated that, due to advancements in MD simulation techniques including the recently reviewed advancements of extracting NMR relevant parameters directly from aiMD simulations [173], it will be more and more feasible to predict the structural arrangements of the confined guest molecules a priori, and later probe these predictions by the combination of calorimetry, solid-state NMR, and DNP.

Funding: Parts of the review are based upon work supported by the National Science Foundation under Grant No. [1953428] and the Deutsche Forschungsgemeinschaft (DFG) under grant Bu 911/24-3. The latter included a Mercator fellowship for MMH to support research stays at the Technical University Darmstadt.

Conflicts of Interest: The authors declare no conflicts of interest.

References

1. Le Page, M.; Beau, R.; Duchene, J. Porous Silica Particles Containing a Crystallized Phase and Method. U.S. Patent 3,493,341, 3 February 1970.
2. Chiola, V.; Ritsko, J.E.; Vanderpool, C.D. Process for Producing Low-Bulk DENSITY Silica. U.S. Patent 3,556,725, 19 January 1971.
3. Beck, J.S.; Vartuli, J.C.; Roth, W.J.; Leonowicz, M.E.; Kresge, C.T.; Schmitt, K.D.; Chu, C.T.; Olson, D.H.; Sheppard, E.W. A new family of mesoporous molecular sieves prepared with liquid crystal templates. *J. Am. Chem. Soc.* **1992**, *114*, 10834–10843. [CrossRef]
4. Zhao, D.Y.; Huo, Q.S.; Feng, J.L.; Chmelka, B.F.; Stucky, G.D. Nonionic triblock and star diblock copolymer and oligomeric surfactant syntheses of highly ordered, hydrothermally stable, mesoporous silica structures. *J. Am. Chem. Soc.* **1998**, *120*, 6024–6036. [CrossRef]
5. Zhao, D.Y.; Feng, J.L.; Huo, Q.S.; Melosh, N.; Fredrickson, G.H.; Chmelka, B.F.; Stucky, G.D. Triblock copolymer syntheses of mesoporous silica with periodic 50 to 300 angstrom pores. *Science* **1998**, *279*, 548–552. [CrossRef] [PubMed]
6. Vinu, A.; Miyahara, M.; Ariga, K. Assemblies of biomaterials in mesoporous media. *J. Nanosci. Nanotechnol.* **2006**, *6*, 1510–1532. [CrossRef] [PubMed]
7. Vinu, A.; Hossain, K.Z.; Ariga, K. Recent advances in functionalization of mesoporous silica. *J. Nanosci. Nanotechnol.* **2005**, *5*, 347–371. [CrossRef] [PubMed]
8. Yokoi, T.; Yoshitake, H.; Tatsumi, T. Synthesis of amino-functionalized MCM-41 via direct co-condensation and post-synthesis grafting methods using mono-, di- and tri-amino-organoalkoxysilanes. *J. Mater. Chem.* **2004**, *14*, 951–957. [CrossRef]
9. Wang, X.; Lin, K.S.K.; Chan, J.C.C.; Cheng, S. Direct Synthesis and Catalytic Applications of Ordered Large Pore Aminopropyl-Functionalized SBA-15 Mesoporous Materials. *J. Phys. Chem. B* **2005**, *109*, 1763–1769. [CrossRef] [PubMed]
10. Saravanan, M.; Sudalai, S.; Dharaneesh, A.B.; Prahaaladhan, V.; Srinivasan, G.; Arumugam, A. An extensive review on mesoporous silica from inexpensive resources: Properties, synthesis, and application toward modern technologies. *J. Sol-Gel Sci. Technol.* **2023**, *105*, 1–29. [CrossRef]
11. Matlahov, I.; Geiger, Y.; Goobes, G. Trapping RNase A on MCM41 pores: Effects on structure stability, product inhibition and overall enzymatic activity. *Phys. Chem. Chem. Phys.* **2014**, *16*, 9031–9038. [CrossRef] [PubMed]
12. Shimon, D.; Chen, C.-H.; Lee, J.J.; Didas, S.A.; Sievers, C.; Jones, C.W.; Hayes, S.E. 15N Solid State NMR Spectroscopic Study of Surface Amine Groups for Carbon Capture: 3-Aminopropylsilyl Grafted to SBA-15 Mesoporous Silica. *Environ. Sci. Technol.* **2018**, *52*, 1488–1495. [CrossRef]
13. Ravera, E.; Martelli, T.; Geiger, Y.; Fragai, M.; Goobes, G.; Luchinat, C. Biosilica and bioinspired silica studied by solid-state NMR. *Coordin. Chem. Rev.* **2016**, *327–328*, 110–122. [CrossRef]
14. Treacy, M.M.J.; Higgins, J.B.; von Ballmoos, R. *Collection of Simulated XRD Powder Patterns for Zeolites*; Elsevier: London, UK, 1996.
15. Marler, B.; Oberhagemann, U.; Vortmann, S.; Gies, H. Influence of the sorbate type on the XRD peak intensities of loaded MCM-41. *Microporous Mater.* **1996**, *6*, 375–383. [CrossRef]
16. Yao, M.H.; Baird, R.J.; Kunz, F.W.; Hoost, T.E. An XRD and TEM investigation of the structure of alumina-supported ceria–zirconia. *J. Catal.* **1997**, *166*, 67–74. [CrossRef]
17. Hoffmann, I.; Malayil Kalathil, F.; Lopian, T.; Touraud, D.; Czakkel, O.; Plazanet, M.; Alba-Simionesco, C. Unexpected molecular dynamics of ethanol in hydrogen-bonded binary mixtures, ethanol-octanol and ethanol-water. *EPJ Web Conf.* **2022**, *272*, 1003. [CrossRef]
18. Höhne, G.; Hemminger, W.F.; Flammersheim, H.-J. *Differential Scanning Calorimetry*; Springer: Berlin/Heidelberg, Germany, 2003; ISBN 354000467X.

19. Brunauer, S.; Emmett, P.H.; Teller, E. Adsorption of Gases in Multimolecular Layers. *J. Am. Chem. Soc.* **1938**, *60*, 309–319. [CrossRef]
20. Barrett, E.P.; Joyner, L.G.; Halenda, P.P. The Determination of Pore Volume and Area Distributions in Porous Substances. I. Computations from Nitrogen Isotherms. *J. Am. Chem. Soc.* **1951**, *73*, 373–380. [CrossRef]
21. Shenderovich, I.G.; Mauder, D.; Akcakayiran, D.; Buntkowsky, G.; Limbach, H.-H.; Findenegg, G.H. NMR provides checklist of generic properties for atomic-scale models of periodic mesoporous silicas. *J. Phys. Chem. B* **2007**, *111*, 12088–12096. [CrossRef]
22. Freude, D.; Kärger, J. NMR Techniques. In *Handbook of Porous Solids*; Schüth, F., Sing, K.S.W., Weitkamp, J., Eds.; John Wiley & Sons, Inc.: Hoboken, NJ, USA, 2002. [CrossRef]
23. Kärger, J. Transport Phenomena in Nanoporous Materials. *ChemPhysChem* **2015**, *16*, 24–51. [CrossRef]
24. Shenderovich, I.G. For Whom a Puddle Is the Sea? Adsorption of Organic Guests on Hydrated MCM-41 Silica. *Langmuir* **2020**, *36*, 11383–11392. [CrossRef]
25. Lesnichin, S.B.; Kamdem, N.; Mauder, D.; Denisov, G.S.; Shenderovich, I.G. Studies of adsorption of 2,2′-bipyridyl on the surface of highly regulated silica matrix of the MCM-41 type by means of N-15 NMR spectroscopy. *Russ. J. Gen. Chem.* **2010**, *80*, 2027–2031. [CrossRef]
26. Overhauser, A.W. Polarization of Nuclei in Metals. *Phys. Rev.* **1953**, *92*, 411–415. [CrossRef]
27. Wind, R.A.; Duijvestijn, M.J.; Vanderlugt, C.; Manenschijn, A.; Vriend, J. Applications of dynamic nuclear-polarization in C-13 NMR in solids. *Prog. NMR Spec.* **1985**, *17*, 33–67. [CrossRef]
28. Maly, T.; Debelouchina, G.T.; Bajaj, V.S.; Hu, K.N.; Joo, C.G.; Mak-Jurkauskas, M.L.; Sirigiri, J.R.; van der Wel, P.C.A.; Herzfeld, J.; Temkin, R.J.; et al. Dynamic nuclear polarization at high magnetic fields. *J. Chem. Phys.* **2008**, *128*, 52211. [CrossRef] [PubMed]
29. Thankamony, A.S.L.; Wittmann, J.J.; Kaushik, M.; Corzilius, B. Dynamic nuclear polarization for sensitivity enhancement in modern solid-state NMR. *Prog. NMR Spec.* **2017**, *102–103*, 120–195. [CrossRef]
30. Lesage, A.; Lelli, M.; Gajan, D.; Caporini, M.A.; Vitzthum, V.; Mieville, P.; Alauzun, J.; Roussey, A.; Thieuleux, C.; Mehdi, A.; et al. Surface Enhanced NMR Spectroscopy by Dynamic Nuclear Polarization. *J. Am. Chem. Soc.* **2010**, *132*, 15459–15461. [CrossRef] [PubMed]
31. de Oliveira, M.; Herr, K.; Brodrecht, M.; Haro-Mares, N.B.; Wissel, T.; Klimavicius, V.; Breitzke, H.; Gutmann, T.; Buntkowsky, G. Solvent-free dynamic nuclear polarization enhancements in organically modified mesoporous silica. *Phys. Chem. Chem. Phys.* **2021**, *23*, 12559–12568. [CrossRef] [PubMed]
32. Kobayashi, T.; Singappuli-Arachchige, D.; Wang, Z.R.; Slowing, I.I.; Pruski, M. Spatial distribution of organic functional groups supported on mesoporous silica nanoparticles: A study by conventional and DNP-enhanced Si-29 solid-state NMR. *Phys. Chem. Chem. Phys.* **2017**, *19*, 1781–1789. [CrossRef] [PubMed]
33. Kobayashi, T.; Singappuli-Arachchige, D.; Slowing, I.I.; Pruski, M. Spatial distribution of organic functional groups supported on mesoporous silica nanoparticles (2): A study by H-1 triple-quantum fast-MAS solid-state NMR. *Phys. Chem. Chem. Phys.* **2018**, *20*, 22203–22209. [CrossRef] [PubMed]
34. Sebastiani, D. Ab-Initio Calculations of NMR Parameters in Condensed Phases. *Mod. Phys. Lett. B* **2003**, *17*, 1301–1319. [CrossRef]
35. Xu, Y.; Watermann, T.; Limbach, H.-H.; Gutmann, T.; Sebastiani, D.; Buntkowsky, G. Water and small organic molecules as probes for geometric confinement in well-ordered mesoporous carbon materials. *Phys. Chem. Chem. Phys.* **2014**, *16*, 9327–9336. [CrossRef] [PubMed]
36. Buntkowsky, G.; Vogel, M.; Winter, R. Properties of Hydrogen-Bonded Liquids at Interfaces. *Z. Phys. Chem.* **2018**, *232*, 937–972. [CrossRef]
37. Buch, V. (Ed.) *Water in Confining Geometries*; Springer: Berlin/Heidelberg, Germany, 2003; ISBN 3540004114.
38. Björneholm, O.; Hansen, M.H.; Hodgson, A.; Liu, L.-M.; Limmer, D.T.; Michaelides, A.; Pedevilla, P.; Rossmeisl, J.; Shen, H.; Tocci, G.; et al. Water at Interfaces. *Chem. Rev.* **2016**, *116*, 7698–7726. [CrossRef] [PubMed]
39. Huber, P. Soft matter in hard confinement: Phase transition thermodynamics, structure, texture, diffusion and flow in nanoporous media. *J. Phys. Condens. Matter* **2015**, *27*, 103102. [CrossRef] [PubMed]
40. Swenson, J.; Cerveny, S. Dynamics of deeply supercooled interfacial water. *J. Phys. Condens. Matter* **2015**, *27*, 33102. [CrossRef] [PubMed]
41. Amann-Winkel, K.; Bellissent-Funel, M.-C.; Bove, L.E.; Loerting, T.; Nilsson, A.; Paciaroni, A.; Schlesinger, D.; Skinner, L. X-ray and Neutron Scattering of Water. *Chem. Rev.* **2016**, *116*, 7570–7589. [CrossRef] [PubMed]
42. Kimmich, R. *NMR*; Springer: Berlin/Heidelberg, Germany, 1997; ISBN 978-3-642-64465-8.
43. Laage, D.; Elsaesser, T.; Hynes, J.T. Water Dynamics in the Hydration Shells of Biomolecules. *Chem. Rev.* **2017**, *117*, 10694–10725. [CrossRef] [PubMed]
44. Horstmann, R.; Hecht, L.; Kloth, S.; Vogel, M. Structural and Dynamical Properties of Liquids in Confinements: A Review of Molecular Dynamics Simulation Studies. *Langmuir* **2022**, *38*, 6506–6522. [CrossRef] [PubMed]
45. Vogel, M. NMR studies on simple liquids in confinement. *Eur. Phys. J.* **2010**, *189*, 47–64. [CrossRef]
46. Demuth, D.; Sattig, M.; Steinrücken, E.; Weigler, M.; Vogel, M. 2H NMR Studies on the Dynamics of Pure and Mixed Hydrogen-Bonded Liquids in Confinement. *Z. Phys. Chem.* **2018**, *232*, 1059–1087. [CrossRef]
47. Buntkowsky, G.; Vogel, M. Small Molecules, Non-Covalent Interactions, and Confinement. *Molecules* **2020**, *25*, 3311. [CrossRef] [PubMed]

48. Buntkowsky, G.; Döller, S.; Haro-Mares, N.; Gutmann, T.; Hoffmann, M. Solid-state NMR studies of non-ionic surfactants confined in mesoporous silica. *Z. Phys. Chem.* **2022**, *236*, 939–960. [CrossRef]
49. Buntkowsky, G.; Breitzke, H.; Adamczyk, A.; Roelofs, F.; Emmler, T.; Gedat, E.; Grünberg, B.; Xu, Y.; Limbach, H.H.; Shenderovich, I.; et al. Structural and Dynamical Properties of Guest Molecules confined in mesoporous Silica Materials revealed by NMR. *Phys. Chem. Chem. Phys.* **2007**, *9*, 4843–4853. [CrossRef] [PubMed]
50. Gedat, E.; Schreiber, A.; Albrecht, J.; Shenderovich, I.; Findenegg, G.; Limbach, H.-H.; Buntkowsky, G. 2H- Solid State NMR Study of Benzene-d6 confined in mesoporous Silica SBA-15. *J. Phys. Chem. B* **2002**, *106*, 1977. [CrossRef]
51. Aksnes, D.W.; Gjerdåker, L.; Kimtys, L.; Førland, K. Dynamic 1 H and 2 H NMR investigations of acetonitrile confined in porous silica. *Phys. Chem. Chem. Phys.* **2003**, *5*, 2680–2685. [CrossRef]
52. Sen, S.; Risbud, S.H.; Bartl, M.H. Thermodynamic and Kinetic Transitions of Liquids in Nanoconfinement. *Acc. Chem. Res.* **2020**, *53*, 2869–2878. [CrossRef] [PubMed]
53. Steinrücken, E.; Weigler, M.; Schiller, V.; Vogel, M. Dynamical Susceptibilities of Confined Water from Room Temperature to the Glass Transition. *J. Phys. Chem. Lett.* **2023**, *14*, 4104–4112. [CrossRef] [PubMed]
54. Becher, M.; Lichtinger, A.; Minikejew, R.; Vogel, M.; Rössler, E.A. NMR Relaxometry Accessing the Relaxation Spectrum in Molecular Glass Formers. *Int. J. Mol. Sci.* **2022**, *23*, 5118. [CrossRef]
55. Vogel, M. Origins of apparent fragile-to-strong transitions of protein hydration waters. *Phys. Rev. Lett.* **2008**, *101*, 225701. [CrossRef] [PubMed]
56. Sattig, M.; Reutter, S.; Fujara, F.; Werner, M.; Buntkowsky, G.; Vogel, M. NMR studies on the temperature-dependent dynamics of confined water. *Phys. Chem. Chem. Phys.* **2014**, *16*, 19229–19240. [CrossRef]
57. Weigler, M.; Brodrecht, M.; Buntkowsky, G.; Vogel, M. Reorientation of Deeply Cooled Water in Mesoporous Silica: NMR Studies of the Pore-Size Dependence. *J. Phys. Chem. B* **2019**, *123*, 2123–2134. [CrossRef] [PubMed]
58. Weigler, M.; Winter, E.; Kresse, B.; Brodrecht, M.; Buntkowsky, G.; Vogel, M. Static field gradient NMR studies of water diffusion in mesoporous silica. *Phys. Chem. Chem. Phys.* **2020**, *22*, 13989–13998. [CrossRef] [PubMed]
59. Winterstein, S.F.; Privalov, A.F.; Greve, C.; Siegel, R.; Pötzschner, B.; Bettermann, M.; Adolph, L.; Timm, J.; Marschall, R.; Rössler, E.A.; et al. Ultrafast Proton Conduction in an Aqueous Electrolyte Confined in Adamantane-like Micropores of a Sulfonated, Aromatic Framework. *J. Am. Chem. Soc.* **2023**, *145*, 27563–27575. [CrossRef] [PubMed]
60. Paquet, E.; Viktor, H.L. Computational Methods for Ab Initio Molecular Dynamics. *Adv. Chem.* **2018**, *2018*, 9839641. [CrossRef]
61. Jämbeck, J.P.M.; Lyubartsev, A.P. Update to the general amber force field for small solutes with an emphasis on free energies of hydration. *J. Phys. Chem. B* **2014**, *118*, 3793–3804. [CrossRef] [PubMed]
62. Wang, J.; Wolf, R.M.; Caldwell, J.W.; Kollman, P.A.; Case, D.A. Development and testing of a general amber force field. *J. Comput. Chem.* **2004**, *25*, 1157–1174. [CrossRef] [PubMed]
63. Jorgensen, W.L.; Maxwell, D.S.; Tirado-Rives, J. Development and Testing of the OPLS All-Atom Force Field on Conformational Energetics and Properties of Organic Liquids. *J. Am. Chem. Soc.* **1996**, *118*, 11225–11236. [CrossRef]
64. Vanommeslaeghe, K.; Hatcher, E.; Acharya, C.; Kundu, S.; Zhong, S.; Shim, J.; Darian, E.; Guvench, O.; Lopes, P.; Vorobyov, I.; et al. CHARMM general force field: A force field for drug-like molecules compatible with the CHARMM all-atom additive biological force fields. *J. Comput. Chem.* **2010**, *31*, 671–690. [CrossRef]
65. Horta, B.A.C.; Merz, P.T.; Fuchs, P.F.J.; Dolenc, J.; Riniker, S.; Hünenberger, P.H. A GROMOS-Compatible Force Field for Small Organic Molecules in the Condensed Phase: The 2016H66 Parameter Set. *J. Chem. Theory Comput.* **2016**, *12*, 3825–3850. [CrossRef] [PubMed]
66. Grunewald, F.; Rossi, G.; de Vries, A.H.; Marrink, S.J.; Monticelli, L. Transferable MARTINI Model of Poly(ethylene Oxide). *J. Phys. Chem. B* **2018**, *122*, 7436–7449. [CrossRef]
67. Prada-Gracia, D.; Shevchuk, R.; Rao, F. The quest for self-consistency in hydrogen bond definitions. *J. Chem. Phys.* **2013**, *139*, 84501. [CrossRef] [PubMed]
68. Lemkul, J. From Proteins to Perturbed Hamiltonians: A Suite of Tutorials for the GROMACS-2018 Molecular Simulation Package [Article v1.0]. *Living J. Comput. Mol. Sci.* **2019**, *1*, 5068. [CrossRef]
69. Braun, E.; Gilmer, J.; Mayes, H.B.; Mobley, D.L.; Monroe, J.I.; Prasad, S.; Zuckerman, D.M. Best Practices for Foundations in Molecular Simulations Article v1.0. *Living J. Comput. Mol. Sci.* **2019**, *1*, 5957. [CrossRef] [PubMed]
70. Medick, P.; Blochowicz, T.; Vogel, M.; Roessler, E. Comparing the dynamical heterogeneities in binary glass formers and in a glass former embedded in a zeolite—A (2) HNMR study. *J. Non-Cryst. Solids* **2002**, *307*, 565–572. [CrossRef]
71. Dosseh, G.; Xia, Y.; Alba-Simionesco, C. Cyclohexane and Benzene Confined in MCM-41 and SBA-15: Confinement Effects on Freezing and Melting. *J. Phys. Chem. B* **2003**, *107*, 6445. [CrossRef]
72. Lusceac, S.A.; Koplin, C.; Medick, P.; Vogel, M.; Brodie-Linder, N.; LeQuellec, C.; Alba-Simionesco, C.; Roessler, E.A. Type a versus type B glass formers: NMR relaxation in bulk and confining geometry. *J. Phys. Chem. B* **2004**, *108*, 16601–16605. [CrossRef]
73. Alba-Simionesco, C.; Coasne, B.; Dosseh, G.; Dudziak, G.; Gubbins, K.E.; Radhakrishnan, R.; Sliwinska-Bartkowiak, M. Effects of confinement on freezing and melting. *J. Phys. Condens. Matter* **2006**, *18*, R15. [CrossRef]
74. Kiwilsza, A.; Pajzderska, A.; Gonzalez, M.A.; Mielcarek, J.; Wąsicki, J. QENS and NMR Study of Water Dynamics in SBA-15 with a Low Water Content. *J. Phys. Chem. C* **2015**, *119*, 16578–16586. [CrossRef]
75. Krzyżak, A.T.; Habina, I. Low field 1H NMR characterization of mesoporous silica MCM-41 and SBA-15 filled with different amount of water. *Microporous Mesoporous Mater.* **2016**, *231*, 230–239. [CrossRef]

76. Shenderovich, I.G.; Buntkowsky, G.; Schreiber, A.; Gedat, E.; Sharif, S.; Albrecht, J.; Golubev, N.S.; Findenegg, G.H.; Limbach, H.H. Pyridine-N-15—A mobile NMR sensor for surface acidity and surface defects of mesoporous silica. *J. Phys. Chem. B* **2003**, *107*, 11924–11939. [CrossRef]
77. Vyalikh, A.; Emmler, T.; Gedat, E.; Shenderovich, I.; Findenegg, G.H.; Limbach, H.H.; Buntkowsky, G. Evidence of microphase separation in controlled pore glasses. *Solid State Nucl. Mag.* **2005**, *28*, 117–124. [CrossRef] [PubMed]
78. Brodrecht, M.; Breitzke, H.; Gutmann, T.; Buntkowsky, G. Biofunctionalization of Nano Channels by Direct In-Pore Solid-Phase Peptide Synthesis. *Chem. Eur. J.* **2018**, *24*, 17814–17822. [CrossRef] [PubMed]
79. Brodrecht, M.; Kumari, B.; Thankamony, A.S.S.L.; Breitzke, H.; Gutmann, T.; Buntkowsky, G. Structural Insights into Peptides Bound to the Surface of Silica Nanopores. *Chem. Eur. J.* **2019**, *25*, 5214–5221. [CrossRef] [PubMed]
80. Brodrecht, M.; Kumari, B.; Breitzke, H.; Gutmann, T.; Buntkowsky, G. Chemically Modified Silica Materials as Model Systems for the Characterization of Water-Surface Interactions. *Z. Phys. Chem.* **2018**, *232*, 1127–1146. [CrossRef]
81. Grünberg, A.; Yeping, X.; Breitzke, H.; Buntkowsky, G. Solid-state NMR characterization of Wilkinson's catalyst immobilized in mesoporous SBA-3 silica. *Chemistry* **2010**, *16*, 6993–6998. [CrossRef]
82. Gutmann, T.; Grunberg, A.; Rothermel, N.; Werner, M.; Srour, M.; Abdulhussain, S.; Tan, S.L.; Xu, Y.P.; Breitzke, H.; Buntkowsky, G. Solid-state NMR concepts for the investigation of supported transition metal catalysts and nanoparticles. *Solid State Nucl. Mag.* **2013**, *55–56*, 1–11. [CrossRef] [PubMed]
83. Grün, M.; Unger, K.K.; Matsumoto, A.; Tsutsumi, K. Novel pathways for the preparation of mesoporous MCM-41 materials: Control of porosity and morphology. *Microporous Mesoporous Mater.* **1999**, *27*, 207–216. [CrossRef]
84. Werner, M.; Rothermel, N.; Breitzke, H.; Gutmann, T.; Buntkowsky, G. Recent Advances in Solid State NMR of Small Molecules in Confinement. *Isr. J. Chem.* **2014**, *54*, 60–73. [CrossRef]
85. Grünberg, B.; Emmler, T.; Gedat, E.; Shenderovich, I.; Findenegg, G.H.; Limbach, H.H.; Buntkowsky, G. Hydrogen bonding of water confined in mesoporous silica MCM-41 and SBA-15 studied by H-1 solid-state NMR. *Chem. Eur. J.* **2004**, *10*, 5689–5696. [CrossRef] [PubMed]
86. Richert, R. Dynamics of Nanoconfined Supercooled Liquids. *Annu. Rev. Phys. Chem.* **2011**, *62*, 65–84. [CrossRef]
87. Brodrecht, M.; Klotz, E.; Lederle, C.; Breitzke, H.; Stühn, B.; Vogel, M.; Buntkowsky, G. A combined Solid-State NMR, Dielectric Spectroscopy and Calorimetric Study of Water in lowly hydrated MCM-41 Samples. *Z. Phys. Chem.* **2018**, *232*, 1003–1016. [CrossRef]
88. Hoffmann, M.M.; Bothe, S.; Gutmann, T.; Hartmann, F.-F.; Reggelin, M.; Buntkowsky, G. Directly vs Indirectly Enhanced 13 C in Dynamic Nuclear Polarization Magic Angle Spinning NMR Experiments of Nonionic Surfactant Systems. *J. Phys. Chem. C* **2017**, *121*, 2418–2427. [CrossRef]
89. Hermens, J.L.; de Bruijn, J.H.; Brooke, D.N. The octanol–water partition coefficient: Strengths and limitations. *Environ. Toxicol. Chem.* **2013**, *32*, 732–733. [CrossRef] [PubMed]
90. Leo, A.; Hansch, C.; Elkins, D. Partition coefficients and their uses. *Chem. Rev.* **1971**, *71*, 525–616. [CrossRef]
91. Kumari, B.; John, D.; Hoffmann, P.; Spende, A.; Toimil-Molares, M.E.; Trautmann, C.; Hess, C.; Ruff, P.; Stark, R.; Schulze, M.; et al. Surface enhanced DNP assisted solid-state NMR of functionalized SiO_2 coated Polycarbonate Membranes. *Z. Phys. Chem.* **2018**, *232*, 1173–1186. [CrossRef]
92. Kumari, B.; Brodrecht, M.; Breitzke, H.; Werner, M.; Grunberg, B.; Limbach, H.H.; Forg, S.; Sanjon, E.P.; Drossel, B.; Gutmann, T.; et al. Mixtures of Alcohols and Water confined in Mesoporous Silica: A Combined Solid-State NMR and Molecular Dynamics Simulation Study. *J. Phys. Chem. C* **2018**, *122*, 19540–19550. [CrossRef]
93. Kumari, B.; Brodrecht, M.; Gutmann, T.; Breitzke, H.; Buntkowsky, G. Efficient Referencing of FSLG CPMAS HETCOR Spectra Using 2D H-1-H-1 MAS FSLG. *Appl. Magn. Reson.* **2019**, *50*, 1399–1407. [CrossRef]
94. Kärger, J.; Pfeifer, H. NMR self-diffusion studies in zeolite science and technology. *Zeolites* **1987**, *7*, 90–107. [CrossRef]
95. Kärger, J.; Freude, D. In situ studies of catalytic reactions in zeolites by means of PFG and MAS NMR techniques. *Stud. Surf. Sci. Catal.* **1997**, *105*, 551–558.
96. Kärger, J.; Freude, D. Mass transfer in micro- and mesoporous materials. *Chem. Eng. Technol.* **2002**, *25*, 769–778. [CrossRef]
97. Kaerger, J.; Valiullin, R. Mass transfer in mesoporous materials: The benefit of microscopic diffusion measurement. *Chem. Soc. Rev.* **2013**, *42*, 4172–4197. [CrossRef] [PubMed]
98. Findenegg, G.H.; Jaehnert, S.; Akcakayiran, D.; Schreiber, A. Freezing and Melting of Water Confined in Silica Nanopores. *ChemPhysChem* **2008**, *9*, 2651–2659. [CrossRef] [PubMed]
99. Geppi, M.; Borsacchi, S.; Mollica, G.; Veracini, C.A. Applications of Solid-State NMR to the Study of Organic/Inorganic Multicomponent Materials. *Appl. Spectrosc. Rev.* **2009**, *44*, 1–89. [CrossRef]
100. Yang, Y.; Beele, B.; Bluemel, J. Easily immobilized di- and tetraphosphine linkers: Rigid scaffolds that prevent interactions of metal complexes with oxide supports. *J. Am. Chem. Soc.* **2008**, *130*, 3771–3773. [CrossRef] [PubMed]
101. Bluemel, J. Linkers and catalysts immobilized on oxide supports: New insights by solid-state NMR spectroscopy. *Coordin. Chem. Rev.* **2008**, *252*, 2410–2423. [CrossRef]
102. Vafaeezadeh, M.; Hashemi, M.M. Polyethylene glycol (PEG) as a green solvent for carbon–carbon bond formation reactions. *J. Mol. Liq.* **2015**, *207*, 73–79. [CrossRef]
103. Sayari, A.; Hamoudi, S. Periodic mesoporous silica-based organic-inorganic nanocomposite materials. *Chem. Mater.* **2001**, *13*, 3151–3168. [CrossRef]

104. Linssen, T.; Cassiers, K.; Cool, P.; Vansant, E.F. Mesoporous templated silicates: An overview of their synthesis, catalytic activation and evaluation of the stability. *Adv. Colloid Interface Sci.* **2003**, *103*, 121–147. [CrossRef]
105. Hoffmann, M.M. Polyethylene glycol as a green chemical solvent. *Curr. Opin. Colloid Interface Sci.* **2022**, *57*, 101537. [CrossRef]
106. Campos, J.F.; Berteina-Raboin, S. Greener Synthesis of Nitrogen-Containing Heterocycles in Water, PEG, and Bio-Based Solvents. *Catalysts* **2020**, *10*, 429. [CrossRef]
107. Kardooni, R.; Kiasat, A.R. Polyethylene Glycol as a Green and Biocompatible Reaction Media for the Catalyst Free Synthesis of Organic Compounds. *Curr. Org. Chem.* **2020**, *24*, 1275–1314. [CrossRef]
108. Soni, J.; Sahiba, N.; Sethiya, A.; Agarwal, S. Polyethylene glycol: A promising approach for sustainable organic synthesis. *J. Mol. Liq.* **2020**, *315*, 113766. [CrossRef]
109. Stanley, H.E.; Buldyrev, S.V.; Franzese, G.; Kumar, P.; Mallamace, F.; Mazza, M.G.; Stokely, K.; Xu, L. Liquid polymorphism: Water in nanoconfined and biological environments. *J. Phys. Condens. Matter* **2010**, *22*, 284101. [CrossRef]
110. Geske, J.; Harrach, M.; Heckmann, L.; Horstmann, R.; Klameth, F.; Müller, N.; Pafong, E.; Wohlfromm, T.; Drossel, B.; Vogel, M. Molecular Dynamics Simulations of Water, Silica, and Aqueous Mixtures in Bulk and Confinement. *Z. Phys. Chem.* **2018**, *232*, 1187–1225. [CrossRef]
111. Baschnagel, J.; Meyer, H.; Varnik, F.; Metzger, S.; Aichele, M.; Müller, M.; Binder, K. Computer Simulations of Polymers Close to Solid Interfaces: Some Selected Topics. *Interface Sci.* **2003**, *11*, 159–173. [CrossRef]
112. Berendsen, H.; van der Spoel, D.; van Drunen, R. GROMACS: A message-passing parallel molecular dynamics implementation. *Comput. Phys. Commun.* **1995**, *91*, 43–56. [CrossRef]
113. van der Spoel, D.; Lindahl, E.; Hess, B.; Groenhof, G.; Mark, A.E.; Berendsen, H.J.C. GROMACS: Fast, flexible, and free. *J. Comput. Chem.* **2005**, *26*, 1701–1718. [CrossRef]
114. Páll, S.; Hess, B. A flexible algorithm for calculating pair interactions on SIMD architectures. *Comput. Phys. Commun.* **2013**, *184*, 2641–2650. [CrossRef]
115. Essmann, U.; Perera, L.; Berkowitz, M.L.; Darden, T.; Lee, H.; Pedersen, L.G. A smooth particle mesh Ewald method. *J. Chem. Phys.* **1995**, *103*, 8577–8593. [CrossRef]
116. Bussi, G.; Donadio, D.; Parrinello, M. Canonical sampling through velocity rescaling. *J. Chem. Phys.* **2007**, *126*, 14101. [CrossRef]
117. Parrinello, M.; Rahman, A. Polymorphic transitions in single crystals: A new molecular dynamics method. *J. Appl. Phys.* **1981**, *52*, 7182–7190. [CrossRef]
118. Nosé, S.; Klein, M.L. Constant pressure molecular dynamics for molecular systems. *Mol. Phys.* **1983**, *50*, 1055–1076. [CrossRef]
119. Hoffmann, M.M.; Too, M.D.; Paddock, N.A.; Horstmann, R.; Kloth, S.; Vogel, M.; Buntkowsky, G. On the Behavior of the Ethylene Glycol Components of Polydisperse Polyethylene Glycol PEG200. *J. Phys. Chem. B* **2023**, *127*, 1178–1196. [CrossRef] [PubMed]
120. Haro Mares, N.B.; Brodrecht, M.; Wissel, T.; Döller, S.C.; Rösler, L.; Breitzke, H.; Hoffmann, M.M.; Gutmann, T.; Buntkowsky, G. Influence of APTES-Decorated Mesoporous Silica on the Dynamics of Ethylene Glycol Molecules–Insights from Variable Temperature 2 H Solid-State NMR. *J. Phys. Chem. C* **2023**, *127*, 19735–19746. [CrossRef]
121. Vyalikh, A.; Emmler, T.; Gruenberg, B.; Xu, Y.; Shenderovich, I.; Findenegg, G.H.; Limbach, H.H.; Buntkowsky, G. Hydrogen bonding of water confined in controlled-pore glass 10-75 studied by H-1-solid state NMR. *Z. Phys. Chem.* **2007**, *221*, 155–168. [CrossRef]
122. Weigler, M.; Brodrecht, M.; Breitzke, H.; Dietrich, F.; Sattig, M.; Buntkowsky, G.; Vogel, M. ^{2}H NMR studies on water dynamics in functionalized mesoporous silica. *Z. Phys. Chem.* **2018**, *232*, 1041–1058. [CrossRef]
123. Yao, Y.; Fella, V.; Huang, W.; Zhang, K.A.I.; Landfester, K.; Butt, H.J.; Vogel, M.; Floudas, G. Crystallization and Dynamics of Water Confined in Model Mesoporous Silica Particles: Two Ice Nuclei and Two Fractions of Water. *Langmuir* **2019**, *35*, 5890–5901. [CrossRef]
124. Steinrücken, E.; Wissel, T.; Brodrecht, M.; Breitzke, H.; Regentin, J.; Buntkowsky, G.; Vogel, M. 2H NMR study on temperature-dependent water dynamics in amino-acid functionalized silica nanopores. *J. Chem. Phys.* **2021**, *154*, 114702. [CrossRef] [PubMed]
125. Cerveny, S.; Mallamace, F.; Swenson, J.; Vogel, M.; Xu, L. Confined Water as Model of Supercooled Water. *Chem. Rev.* **2016**, *116*, 7608–7625. [CrossRef]
126. Sattig, M., Vogel, M. Dynamic Crossovers and Stepwise Solidification of Confined Water: A (2)H NMR Study. *J. Phys. Chem. Lett.* **2014**, *5*, 174–178. [CrossRef]
127. Poole, P.H.; Sciortino, F.; Essmann, U.; Stanley, H.E. Phase behaviour of metastable water. *Nature* **1992**, *360*, 324–328. [CrossRef]
128. Schneider, S.; Vogel, M. NMR studies on the coupling of ion and water dynamics on various time and length scales in glass-forming LiCl aqueous solutions. *J. Chem. Phys.* **2018**, *149*, 104501. [CrossRef] [PubMed]
129. Schneider, S.; Säckel, C.; Brodrecht, M.; Breitzke, H.; Buntkowsky, G.; Vogel, M. NMR studies on the influence of silica confinements on local and diffusive dynamics in LiCl aqueous solutions approaching their glass transitions. *J. Chem. Phys.* **2020**, *153*, 244501. [CrossRef]
130. Schneider, S.; Brodrecht, M.; Breitzke, H.; Wissel, T.; Buntkowsky, G.; Varol, H.S.; Brilmayer, R.; Andrieu-Brunsen, A.; Vogel, M. Local and diffusive dynamics of LiCl aqueous solutions in pristine and modified silica nanopores. *J. Chem. Phys.* **2022**, *157*, 34503. [CrossRef] [PubMed]
131. Schiller, V.; Vogel, M. Ice-Water Equilibrium in Nanoscale Confinement. *Phys. Rev. Lett.* **2024**, *132*, 16201. [CrossRef] [PubMed]
132. Valiullin, R.; Furó, I. The morphology of coexisting liquid and frozen phases in porous materials as revealed by exchange of nuclear spin magnetization followed by H1 nuclear magnetic resonance. *J. Chem. Phys.* **2002**, *117*, 2307–2316. [CrossRef]

133. Kuhs, W.F.; Sippel, C.; Falenty, A.; Hansen, T.C. Extent and relevance of stacking disorder in "ice I(c)". *Proc. Natl. Acad. Sci. USA* **2012**, *109*, 21259–21264. [CrossRef]
134. Malkin, T.L.; Murray, B.J.; Salzmann, C.G.; Molinero, V.; Pickering, S.J.; Whale, T.F. Stacking disorder in ice I. *Phys. Chem. Chem. Phys.* **2015**, *17*, 60–76. [CrossRef] [PubMed]
135. Döller, S.C.; Brodrecht, M.; Haro Mares, N.B.; Breitzke, H.; Gutmann, T.; Hoffmann, M.; Buntkowsky, G. Deuterium NMR Studies of the Solid–Liquid Phase Transition of Octanol- d17 Confined in SBA-15. *J. Phys. Chem. C* **2021**, *125*, 25155–25164. [CrossRef]
136. Vyalikh, A.; Emmler, T.; Shenderovich, I.; Zeng, Y.; Findenegg, G.H.; Buntkowsky, G. H-2-solid state NMR and DSC study of isobutyric acid in mesoporous silica materials. *Phys. Chem. Chem. Phys.* **2007**, *9*, 2249–2257. [CrossRef]
137. Harrach, M.F.; Drossel, B.; Winschel, W.; Gutmann, T.; Buntkowsky, G. Mixtures of Isobutyric Acid and Water Confined in Cylindrical Silica Nanopores Revisited: A Combined Solid-State NMR and Molecular Dynamics Simulation Study. *J. Phys. Chem. C* **2015**, *119*, 28961–28969. [CrossRef]
138. Masierak, W.; Emmler, T.; Gedat, E.; Schreiber, A.; Findenegg, G.H.; Buntkowsky, G. Microcrystallization of benzene-d6 in mesoporous silica revealed by 2H-solid state NMR. *J. Phys. Chem. B* **2004**, *108*, 18890–18896. [CrossRef]
139. Gruenberg, B.; Gruenberg, A.; Limbach, H.-H.; Buntkowsky, G. Melting of Naphthalene Confined in Mesoporous Silica MCM-41. *Appl. Magn. Reson.* **2013**, *44*, 189–201. [CrossRef]
140. Amadeu, N.d.S.; Gruenberg, B.; Frydel, J.; Werner, M.; Limbach, H.-H.; Breitzke, H.; Buntkowsky, G. Melting of Low Molecular Weight Compounds in Confinement Observed by H-2-Solid State NMR: Biphenyl, a Case Study. *Z. Phys. Chem.* **2012**, *226*, 1169–1185. [CrossRef]
141. Kissinger, H.E. Variation of Peak Temperature with Heating Rate in Differential Thermal Analysis. *J. Res. Nat. Bur. Stand.* **1956**, *57*, 217. [CrossRef]
142. Jackson, C.L.; McKenna, G.B. The melting behavior of organic materials confined in porous solids. *J. Chem. Phys.* **1990**, *93*, 9002–9011. [CrossRef]
143. Millett, F.S.; Dailey, B.P. NMR Determination of Some Deuterium Quadrupole Coupling Constants in Nematic Solutions. *J. Chem. Phys.* **1972**, *56*, 3249–3256. [CrossRef]
144. Polson, J.M.; Fyfe, J.D.D.; Jeffrey, K.R. The reorientation of t -butyl groups in butylated hydroxytoluene: A deuterium nuclear magnetic resonance spectral and relaxation time study. *J. Chem. Phys.* **1991**, *94*, 3381–3388. [CrossRef]
145. Beshah, K.; Olejniczak, E.T.; Griffin, R.G. Deuterium NMR study of methyl group dynamics in L -alanine. *J. Chem. Phys.* **1987**, *86*, 4730–4736. [CrossRef]
146. Spiess, H.; Sillescu, H. Solid echoes in the slow-motion region. *J. Magn. Reson.* **1981**, *42*, 381–389. [CrossRef]
147. Rössler, E.; Taupitz, M.; Börner, K.; Schulz, M.; Vieth, H.-M. A simple method analyzing 2H nuclear magnetic resonance line shapes to determine the activation energy distribution of mobile guest molecules in disordered systems. *J. Chem. Phys.* **1990**, *92*, 5847–5855. [CrossRef]
148. Döller, S.C.; Gutmann, T.; Hoffmann, M.; Buntkowsky, G. A case study on the influence of hydrophilicity on the signal enhancement by dynamic nuclear polarization. *Solid State Nucl. Mag.* **2022**, *122*, 101829. [CrossRef] [PubMed]
149. Reuhl, M.; Weigler, M.; Brodrecht, M.; Buntkowsky, G.; Vogel, M. Nuclear Magnetic Resonance and Broadband Dielectric Spectroscopy Studies on the Dynamics of Ethylene Glycol in Mesoporous Silica. *J. Phys. Chem. C* **2020**, *124*, 20998–21012. [CrossRef]
150. Doadrio, A.L.; Sánchez-Montero, J.M.; Doadrio, J.C.; Salinas, A.J.; Vallet-Regí, M. A molecular model to explain the controlled release from SBA-15 functionalized with APTES. *Microporous Mesoporous Mater.* **2014**, *195*, 43–49. [CrossRef]
151. Reuhl, M.; Vogel, M. Temperature-dependent dynamics at protein-solvent interfaces. *J. Chem. Phys.* **2022**, *157*, 74705. [CrossRef] [PubMed]
152. Döller, S.C.; Brodrecht, M.; Gutmann, T.; Hoffmann, M.; Buntkowsky, G. Direct and Indirect DNP NMR Uncovers the Interplay of Surfactants with Their Mesoporous Host Material. *J. Phys. Chem. C* **2023**, *127*, 12125–12134. [CrossRef]
153. Hoffmann, M.M.; Bothe, S.; Brodrecht, M.; Klimavicius, V.; Haro-Mares, N.B.; Gutmann, T.; Buntkowsky, G. Direct and Indirect Dynamic Nuclear Polarization Transfer Observed in Mesoporous Materials Impregnated with Nonionic Surfactant Solutions of Polar Polarizing Agents. *J. Phys. Chem. C* **2020**, *124*, 5145–5156. [CrossRef]
154. Selvam, P.; Bhatia, S.K.; Sonwane, C.G. Recent advances in processing and characterization of periodic mesoporous MCM-41 silicate molecular sieves. *Ind. Eng. Chem. Res.* **2001**, *40*, 3237–3261. [CrossRef]
155. Hoffmann, M.M.; Horowitz, R.H.; Gutmann, T.; Buntkowsky, G. Densities, Viscosities, and Self-Diffusion Coefficients of Ethylene Glycol Oligomers. *J. Chem. Eng. Data* **2021**, *66*, 2480–2500. [CrossRef]
156. Hoffmann, M.M.; Kealy, J.D.; Gutmann, T.; Buntkowsky, G. Densities, Viscosities, and Self-Diffusion Coefficients of Several Polyethylene Glycols. *J. Chem. Eng. Data* **2022**, *67*, 88–103. [CrossRef]
157. Hoffmann, M.M.; Randall, N.P.; Apak, M.H.; Paddock, N.A.; Gutmann, T.; Buntkowsky, G. Solute–Solvent Interactions of 2,2,6,6-Tetramethylpiperidinyloxyl and 5-Tert-Butylisophthalic Acid in Polyethylene Glycol as Observed by Measurements of Density, Viscosity, and Self-Diffusion Coefficient. *J. Solut. Chem.* **2023**, *52*, 685–707. [CrossRef]
158. Forsyth, C.; Taras, T.; Johnson, A.; Zagari, J.; Collado, C.; Hoffmann, M.M.; Reed, C.R. Microwave Assisted Surfactant-Thermal Synthesis of Metal-Organic Framework Materials. *Appl. Sci.* **2020**, *10*, 4563. [CrossRef]
159. Beejapur, H.A.; Zhang, Q.; Hu, K.; Zhu, L.; Wang, J.; Ye, Z. TEMPO in Chemical Transformations: From Homogeneous to Heterogeneous. *ACS Catal.* **2019**, *9*, 2777–2830. [CrossRef]

160. Prakash, N.; Rajeev, R.; John, A.; Vijayan, A.; George, L.; Varghese, A. 2,2,6,6-Tetramethylpiperidinyloxyl (TEMPO) Radical Mediated Electro-Oxidation Reactions: A Review. *ChemistrySelect* **2021**, *6*, 7691–7710. [CrossRef]
161. Morey, M.S.; Davidson, A.; Stucky, G.D. Silica-based, cubic mesostructures: Synthesis, characterization and relevance for catalysis. *J. Porous Mater.* **1998**, *5*, 195–204. [CrossRef]
162. Chen, H.-T.; Huh, S.; Wiench, J.W.; Pruski, M.; Lin, V.S.-Y. Dialkylaminopyridine-functionalized mesoporous silica nanosphere as an efficient and highly stable heterogeneous nucleophilic catalyst. *J. Am. Chem. Soc.* **2005**, *127*, 13305–13311. [CrossRef] [PubMed]
163. Grünberg, A.; Breitzke, H.; Buntkowsky, G. Solid state NMR of immobilized catalysts and nanocatalysts. *Spectrosc. Prop. Inorg. Organomet. Compd.* **2012**, *43*, 289–323.
164. Mellaerts, R.; Jammaer, J.A.G.; van Speybroeck, M.; Chen, H.; van Humbeeck, J.; Augustijns, P.; van den Mooter, G.; Martens, J.A. Physical state of poorly water soluble therapeutic molecules loaded into SBA-15 ordered mesoporous silica carriers: A case study with itraconazole and ibuprofen. *Langmuir* **2008**, *24*, 8651–8659. [CrossRef]
165. Ukmar, T.; Cendak, T.; Mazaj, M.; Kaucic, V.; Mali, G. Structural and Dynamical Properties of Indomethacin Molecules Embedded within the Mesopores of SBA-15: A Solid-State NMR View. *J. Phys. Chem. C* **2012**, *116*, 2662–2671. [CrossRef]
166. Vallet-Regi, M.; Ramila, A.; Del Real, R.P.; Pérez-Pariente, J. A new property of MCM-41: Drug delivery system. *Chem. Mater.* **2001**, *13*, 308–311. [CrossRef]
167. Hall, S.R.; Walsh, D.; Green, D.; Oreffo, R.; Mann, S. A novel route to highly porous bioactive silica gels. *J. Mater. Chem.* **2003**, *13*, 186–190. [CrossRef]
168. Bothe, S.; Nowag, J.; Klimavičius, V.; Hoffmann, M.; Troitskaya, T.I.; Amosov, E.V.; Tormyshev, V.M.; Kirilyuk, I.; Taratayko, A.; Kuzhelev, A.; et al. Novel Biradicals for Direct Excitation Highfield Dynamic Nuclear Polarization. *J. Phys. Chem. C* **2018**, *122*, 11422–11432. [CrossRef]
169. Hoffmann, M.M.; Bothe, S.; Gutmann, T.; Buntkowsky, G. Combining Freezing Point Depression and Self-Diffusion Data for Characterizing Aggregation. *J. Phys. Chem. B* **2018**, *122*, 4913–4921. [CrossRef] [PubMed]
170. Hoffmann, M.M.; Too, M.D.; Vogel, M.; Gutmann, T.; Buntkowsky, G. Breakdown of the Stokes-Einstein Equation for Solutions of Water in Oil Reverse Micelles. *J. Phys. Chem. B* **2020**, *124*, 9115–9125. [CrossRef] [PubMed]
171. Scharf, N.T.; Stark, A.; Hoffmann, M.M. Ion pairing and dynamics of the ionic liquid 1-hexyl-3-methylimidazolium bis(irifluoromethylsulfonyl)amide (C6mimNTf2) in the low dielectric solvent chloroform. *J. Phys. Chem. B* **2012**, *116*, 11488–11497. [CrossRef] [PubMed]
172. Cade, E.A.; Petenuci, J.; Hoffmann, M.M. Aggregation Behavior of Several Ionic Liquids in Molecular Solvents of Low Polarity--Indication of a Bimodal Distribution. *ChemPhysChem* **2016**, *17*, 520–529. [CrossRef] [PubMed]
173. Mazurek, A.H.; Szeleszczuk, Ł.; Pisklak, D.M. A Review on Combination of Ab Initio Molecular Dynamics and NMR Parameters Calculations. *Int. J. Mol. Sci.* **2021**, *22*, 4378. [CrossRef]

MDPI AG
Grosspeteranlage 5
4052 Basel
Switzerland
Tel.: +41 61 683 77 34

Molecules Editorial Office
E-mail: molecules@mdpi.com
www.mdpi.com/journal/molecules

www.ingramcontent.com/pod-product-compliance
Lightning Source LLC
Chambersburg PA
CBHW061607210926
51850CB00009B/2240

9783725853410